《生態保育系列》

中國人與動物

Animals and the Chinese

張靜茹等 著

序(一)
Preface I

子子孫孫永寶用

■胡志強

經濟發展與文明進步的目的，應該是爲了讓人們生活得更有尊嚴。光華雜誌出版「中國人與動物」的動機，是希望今天人們在物質生活已無匱乏的同時，仍能確保一個更適合人們長久居住的環境。這也是尊重和愛護我們的下一代，表示我們了解自然資源是向後代子孫暫借，而爲之保留一塊充滿自然生機的淨土。人們開發這個世界的時候，不但要尊敬自然，進而更要尊敬與自然互動而產生的文化。

在生態保育已成爲世界大勢所趨，並凝聚爲全球性的行動之時，臺灣地區經濟的蓬勃發展有目共睹。緊接著，生活環境品質、生態保育工作，是否隨著經濟進展而逐日提升，也是我們欲儕入更進步、文明國家之林的更高標竿。我們由十年前開始規劃成立墾丁第一個國家公園，拉開了生態保育工作序幕，其間面臨的困境與努力，也有必要讓世人知曉。

從光華雜誌多年來所累積的有關生態保育的報導中，我們了解到，不論物質或精神，都必須倚賴健康的自然環境來提升；了解到只有尊重自然，才能創造健全的下一代，社會也才能更和諧；了解只有認識自己的土地，才可能愛護這塊土地，並累積社會的反省能力；也只有透過人人深層的思考，才可能化爲行動，才有辦法眞正做好生態保育。

生態關懷絕不只是口號，但保育工作要能扎根，還需要人們由內心徹底了解，關心自然生命與人性之間有密不可分的關係。自然資源一旦被破壞，再難完全重生。我們衷心希望，能讓這塊土地上永恆的自然價值，被人們寶惜尊重。

胡志強

發行人

Guarding Nature's Treasures for Future Generations

By Jason C. Hu

The purpose of economic development and civilization's progress ought to be to give people's lives greater dignity. *Sinorama* is publishing *Animals and the Chinese* in the hope that today, living free from material want, we will be sure to preserve a suitable living environment for the long term. By so doing we are showing respect and love to our future generations, showing that we understand both that natural resources are being temporarily borrowed from our children and grandchildren and the importance of keeping a clean land full of natural vitality for them. When people pursue development, they not only must respect nature but go a step further to respect the culture that comes from the interaction of man and nature.

As the ecological conservation movement has established itself worldwide, the economic prosperity brought by the development of Taiwan has been undeniable. The question today is can the quality of the living environment and ecological conservation work be pursued day in and day out with the same vigor? These higher standards will be necessary if we are to enter the ranks of the more advanced and civilized countries. Since a decade ago, when planning on the first R.O.C. national park in Kenting raised the curtain on ecological conservation work here, we have needed to let the people of the world know the difficulties we have faced and the efforts we have made.

From the experience we have gained at *Sinorama* making ecological reports, we know that both the body and the spirit need a healthy natural living environment. Only by respecting nature can we bring healthy future generations into this world, and only by so doing will society be more harmonious. By understanding the land, we can learn to care for it and society will learn how to reflect upon its actions. And if everyone thinks about these issues deeply, this can be turned into action that brings effective ecological conservation.

Concern for the ecology is not only a slogan. But if conservation work is to be done thoroughly, it requires that people have an understanding deep down inside that nature is inextricably bound together with humanity. Once a natural resource is damaged, it is extremely hard to restore completely. We sincerely hope that we can let this world's eternal natural value be treasured and respected by people.

Jason

Publisher

序(二)
Preface Ⅱ

見樹也見林

■劉克襄

一九八〇年代初，聯合報刊出韓韓、馬以工女士撰寫「我們只有一個地球」系列文章後，合乎時宜地在國內掀起一股生態環境意識熱潮。十多年後的今天看來，這股熱潮無疑是一段自然保育報導的啓蒙期；但就一個銜續著自然保育報導的工作者而言，這時期似乎還代表著一個可以比較無阻礙地闡揚保育的階段。

譬如，她們舉著保育的大纛，報導竹圍紅樹林，或東北角的九孔池等問題時，從內容的字裡行間，我們即嗅聞到遠比任何時候都更理直氣壯的氛圍；而且，在思考上比較單向，並未周延地顧及到周遭其他事物的影響。

相對的，整個社會對何謂保育仍是懵懂未知的情況下，她們也獲有更大的議論空間，發揮單方面的意見與影響力。至少，整個社會像莫名其妙犯了錯誤的小孩，充滿愧疚，做出了有誠意去改善的回應。

但這個保育至上論的時代十分短暫——彷彿不曾發生，我們還來不及察覺、定位，就已過去。此後，它一直陷入開發（廣義的如狩獵、食物均包括在內）與自然保育（更多時候是環保）何者爲要，這個典型的環境困境裡。

一九八四年，我個人以關渡沼澤區做爲旅行報導的對象時，遭遇的回應便是陷入這種兩難的困境。保育不再是道德無上、絕對正義的一方。在時空的迅速變遷下，正反兩方都有了很大的變化。這個變化說不上是環保意識的倒退，反而是國人更加明白，相關的保育措施與他們之間會有什麼樣的互動，他們又應該如何因應。

於是，當關渡沼澤區被提議將闢建成自然公園的計畫曝光後，繼之而來的便是當地人反彈、廢土傾倒或改建豬舍和鴨寮等破壞環境的行動。短短半年間，整個沼澤區變質了，水鳥的數量也大量銳減。

緊接著，民意代表的反對更使自然公園的計畫懸宕至今，整個沼澤區已無法恢復過去的景觀。這種反應，我

Seeing the Trees and the Forest

By Liu Ka-shiang

In the early eighties, after Han Han and Ma Yi-kung published the series of articles entitled ''We Only Have One Earth,'' it was as if the country had gained a consciousness about the importance of the environment just when it was needed. A decade later, we can see that the period was one of enlightenment about the environment, and from the perspective of people writing these reports, it was a period of relatively few obstacles in making a case for ecological conservation.

For example, in hoisting up conservation's flag in reports on the mangrove forest in Chuwei or the pools for small abalones on the northeast coast, the tone of moral certainty in the writing was far more apparent than at any other time. And the writers were less inclined to qualify themselves by looking at the related impacts of this ecological conservation.

With society still ignorant about what is meant by ecological conservation, these writers have grabbed at the chance to make full use of this platform for their environmental opinions. At the very least, these reports caused all of society to stop and reflect like a child bewildered at having made a mistake, full of shame and wanting to improve.

But this conservation-oriented period was extremely short lived. Almost as soon as it began, it was gone — before we had even time to become aware of it and define it. Afterwards, came the classic conflict between environmental protection and development (which broadly includes both for hunting and agriculture).

In 1984 I was making a travel report about the Kuantu Wetlands, and the reaction I witnessed typified this dilemma. Environmental protection was no longer thought of as absolutely morally correct. In this new era, the two sides were greatly transformed. These changes did not necessarily represent a step backward for environmental consciousness. Rather people in the R.O.C. had a greater understanding of measures related to conservation, how these affected their lives, how they could affect these measures and how they ought to adapt.

Therefore, when the Kuantu Wetlands became the focus of controversy as the proposed site of a nature preserve, local people countered by engaging in activities that were damaging the environment such as by dumping wastes and building pig and duck sheds. In a short half year, the entire wetlands area was changed and the number of water fowl declined sharply.

On the heels of this, the people's representatives opposing the plan for the natural park were able to put it on hold until the present, and now it will be impossible to restore the entire wetlands area to its former state. With this kind of response, we cannot simply dismiss these local opponents as flouters of the law. We must understand the power of opposition in the local community, especially in

們絕不能以當地人藐視國法、公然違紀，如此簡單的推理去認識地方上的反對力量——尤其是在面對原住民的狩獵問題上。保育不再至上，它有一種複雜性，必須放入整個社會體系的脈絡去體察。我們遭遇到的是另一個階段的來臨。

面對這種自然保育的情形，報導者在報導時也不時會遇見一個令人充滿挫折感的問題。當他們將關心的地點在媒體公佈時，常常因而加速了當地的破壞，而不是讓它在被更多人知悉下，引起更多的重視，並經由立法迅速維續下來。像關渡這樣的保育悲劇例子各地皆有。臺南四草、宜蘭無尾港與竹安溪、彰化彰濱與漢寶、新竹罟寮、大武山大小鬼湖等都是最佳的例證。

更值得注意的，很多報導者後來會產生無力感，主要便是來自於這種結果。不過，也有不少報導過的地方幸運地透過立法，有了美好的遠景。曾文溪口黑面琵鷺、華江橋的雁鴨公園就是值得借鏡的成功案例。

其實，無論成功與否，報導者也必須體認，他們在做的是「長期洗腦」的工作。這幾年來，大部份民眾透過媒體不斷披露的自然生態消息，在不知不覺中，經由這種訊息的一再告知，而約略認識了什麼是保護野生動物，愛護自然。雖然有好些地方繼續在開發的壓力下犧牲掉了，但民眾的保育意識普遍地比過去提昇了許多，的確是不爭的事實。

報導者必須有這種見樹也見林的認知，結合自己越來越紮實的自然生態知識，繼續給民眾更健康的保育觀念，這仍是一個從事自然保育報導者，目前仍要加快腳步，而且充分瞭解的現況。

作家・自然保育工作者

facing the problem of hunting by native peoples. Conservation can no longer be placed above all other considerations. With such an extremely complicated situation, we need to look at the interconnected web of all social factors. It's a new era.

In facing this situation, writers can't help at times but be filled with feelings of failure. When what they are concerned about is featured in the media, this often results in a quickening of the environmental damage rather than increased concern through awareness or preservation of the environment through the quick establishment of environmental laws. Environmental tragedies like Kuantu can be found everywhere. Szutsao in Tainan, Wuweikang and the Chuan River of Ilan, Changpin and Hanpao of Changhua, Kuliao of Hsinchu, and the big and small ghost lakes of Tawushan are all good examples.

It's important to pay attention to the feelings of powerlessness that reporters have when seeing such results. Nevertheless, there are also natural areas that have futures secured forever through law. The sanctuary for the black-faced spoonbill at the mouth of the Tsengwen River and the duck park of Huachiang Bridge are all examples of success from which much can be learned.

In fact, whether or not successful, reporters must realize that they are doing long-term "brainwashing" work. In these past few years, most people have been exposed to ecological news in the media. From being completely unaware, they have come to have sketchy understanding about conservation of wildlife and concern for nature. Although many places have fallen like dominos to the pressure of development, people are much more conscious about the environment than before. This is undeniable.

Reporters must be able to see the trees and the forest at the same time, by both strengthening their own ecological knowledge while continuing to give the public a healthier concept of ecological conservation. The situation requires environmental reporters to quicken their steps and acquire a more thorough understanding.

Liu Ka-shiang

Writer and Conservationist

序(三)
Preface Ⅲ

在建立共識的路上

Carving Out a Consensus

■張靜茹

重視自然生態，絕不是少數人或某一民族的專長。

只是，媒體如此發達、資訊如此豐富，仍然無法填補人與人之間的隔閡與偏見，而其間的了解與尊重也仍有增進的必要。

面對環境保育的問題時，不是每個人都會先試著了解他人的文化背景、歷史淵源，再下斷語。許多外國觀光客到臺灣，看過華西街的殺蛇風景，就認爲中國人都是四隻脚除了桌子不吃的化外之民；而國人往往也缺乏開闊的心胸去認知爲何過去在世界各地大量濫捕野生動物的西方，卻有越來越多的人，一聽到破壞自然環境、濫殺野生動物就顯出激憤異常的理由。遺憾的是，有些人對自己的文化也欠缺同情的理解，以爲保育是個西方思潮，何必委屈自己、一味「跟風」。十年來國內生態保育工作就在這樣的重重隔閡下，國家形象未能提升，反而海豚事件、殺老虎事件、犀牛角事件，一再重演；東西方的保育觀也未能尋找到焦點，和更跨出一步。

By Chang Chin-ju

Ecological concern is not something for only the few or only certain ethnic groups.

It's just that with the media so developed and information so abundant, there is still no way to break down the mental barriers and prejudices between people. And mutual understanding and respect is ever more necessary at a time when people are worrying about the rekindling of Western ethnic hostilities or speaking generally of diversity in culture — while unable themselves to keep an open mind.

Similarly, in facing environmental problems, not everyone tries first to understand other peoples' cultural backgrounds and histories before making judgments. When foreign tourists come to Taiwan, they want first to see the snakes being skinned at "Snake Alley," thinking that Chinese are uncivilized brutes that eat anything on four legs save tables. And Chinese lack the openmindedness to try to understand why more and more Westerners get so steamed as soon as they hear about damage to the natural environment or the indiscriminate killing of wild animals. Regretably, some people lack sympathy and understanding for their own culture. Thinking that conservation is a Western concept, they refuse to slight themselves by playing "tag along" with the West. For ten years, conservation work in this country has been obstructed by these divisions, and the R.O.C.'s national image has been held down by a succession of repeated environmental incidents including the killing of dolphins and tigers and the import of rhino horns. No understanding has been reached between Eastern and Western views of conservation, and progress has been stymied.

We hope the collection of articles in *Animals and the Chinese* will give us a chance to engage in self reflection and understand others. Besides containing reports that have appeared in *Sinorama* about ecological developments in Taiwan in recent years, there are also discussions about traditional views of the Chinese toward animals, about the relationship between Taiwan's native peoples and birds and about Western people's relationship with nature. These articles increase understanding between ethnic groups and show that in the past all cultures treated the environment in ways that put others to shame and held conceptions about nature that others find incomprehensible.

It simply doesn't matter in what country or place environmentalism started because environmental problems affect everyone. Caring for and understanding the natural environment is not something people can do only after they have full bellies, because — in addition to social and economic factors — natural disasters and development that violates natural principles are also major reasons why one-fifth of the world's population does not get enough to eat.

As for humanity's dependence on nature for life,

「中國人與動物」希望藉著一次次的報導，讓我們省察自己、也理解別人。這本書除了收集光華近年來臺灣地區生態保育發展的報導，也有部分傳統中國人的動物觀、臺灣原住民與鳥類、西方與自然關係的討論。除了希望增進不同族羣間彼此的了解，更重要的是，各個文化中，在過去都可找到令今人慚愧或尚難理解的自然法則。

事實上，不論今天的生態保育思潮起自那一個國家、那一個地區，已經不重要，因爲它將是全人類都不能不認知的問題。關心、了解自然生態絕不只是吃飽以後的事，造成今天世界仍有五分之一人口在饑餓中度日的，除了社會、經濟因素，災害的起源與人類的開發違反了當地的自然法則有重要關係。

對依賴自然維持生命的人類，生態保育就像一個人的健康、一個種族的文化、一個國家的敎育一樣，是人類之所以爲人類價値中的一部分。不一樣的是，每個地方奉行的生態原則將比文化或敎育等領域有更多共通性，更容易牽一髮動全身，不管哪一地區、民族，彼此有更多的利害關係，隨時需要互相合作。

物種滅絕對全體人類的損失、熱帶雨林減損造成的氣候改變、空氣愈不適合人類等等，已替我們證明誰也無法自外於自然保育的工作。它也許不是新的東西，與某些民族的傳統有互相呼應之處，但今天它已有了一個全新的面貌，那就是世人都須要共同朝此目標前進。

只有社會上大部分的人有這樣的認知、警覺，人類也才可能企求一個更好的生活環境。而在建立這個共識的過程中，扮演社會最敏銳角色的記者負有更大的責任。尤其在分工愈來愈細的今天，作爲大衆與專業人士、學者之間的橋樑，新聞媒體工作者的角色更顯得重要，希望這不僅是記者的認知，社會大衆、學者專家也都能認同。

the natural ecology is like a person's health, an ethnic group's culture or a nation's educational system — our valuing of it makes us human. What's different is that a nation's ecological principles more than its cultural and educational values can have an effect elsewhere. Regardless of the place or the people, things are ever more tightly interconnected. At any time there could be the need for mutual cooperation. The cutting of the tropical rain forests has affected climate, the air is less and less suitable for people to breathe, etc. These facts already prove that nobody can shut themselves off from natural conservation work. Perhaps this work isn't something new after all, but today it has an entirely new look in that the world's people must march toward this goal together.

Only when most people in society have this kind of understanding and vigilance can humanity hope for a better living environment. In the process of building this mutual understanding, those with the sensitive social role of being reporters bear a lot of responsibility. Particularly today when the division of labor is ever more specialized, the news media must serve as a bridge between the experts and the public. I hope that this is something recognized not only by reporters but by the general public and the experts as well.

Take this book, for instance. While the topics were selected by *Sinorama*, the opinions of scholars and conservationists were solicited, including such figures as Li Ling-ling and Lin Yao-sung of National Taiwan University's Department of Zoology, Wang Ying and Lu Kuang-yang of National Taiwan Normal University's Department of Biology and experts at the Institute of Zoology of the Academia Sinica, Pingtung Technical College and the Wild Bird Society. In addition to helping reporters develop their knowledge, these experts also help them take a step further in playing their role properly and well, allowing what is most needed in the country today — a concept of ecology — to become universally understood, so that natural conservation will more quickly be adopted in society and become a universally held concept among the people.

As an environmental reporter, it is most important to accurately pass along the conservation concepts that are needed to be understood today, to keep a handle on international environmental protection information and developments and to find the special character and insufficiencies of domestic environmental protection work so that there is greater understanding of the environment in society and the power to act on this understanding. Only in this way will the media be doing their duties, and it is in this direction that *Sinorama* will continue to push.

Finally we note with sadness the passing away of Peter Eberly, the English editor of *Sinorama* who was responsible for most of the translations in this book. With a deep knowledge of Chinese and Western culture and an expansive breadth of vision,

就像這一本書，每個專題雖是光華自己規劃的，但題目的製作過程，都有不吝提供意見的學者與保育人士，臺大動物系的李玲玲、林曜松，師大生物系的王穎、呂光洋等多位生態學者，與中研院動物所、屏東技術學院、中華民國野鳥學會等單位。他們不只幫助記者成長，也使媒體工作者能進一步扮演好應扮的角色，讓目前國內最需要的生態觀念能借此普及，保育也能更快深入社會、普及大衆。

身爲環保記者，最重要的是能正確無誤的傳達今天環境中需要的保育觀念，掌握國際上的環保資訊與發展，找出國內環保發展的特色與不足之處，讓社會上所有人凝聚更強的共識與行動力。如此媒體才算盡到責任，而這也是光華同仁未來會繼續努力的方向。

最後必須在此一提的是，本書英譯部分，大多出自光華資深美籍英文編輯易伯禮(Peter Eberly)的譯筆。他以對中西文化豐富的涵養和寬廣的視野，經常與作者深談，並提出新的議題與更周延的討論。不幸的是，就在本書即將校清付梓的同時，這位與光華同仁共事將近十年的夥伴，却因腦瘤手術後併發症驟逝臺大醫院。享年四十歲。

我們希望以此書紀念這位以開濶胸懷親近、引介中華文化的友人。

張靜茹

作者

he would often discuss with writers topics for new articles and other matters. Unfortunately, when this book was being proofread, Peter, our colleague at *Sinorama* for nearly a decade, died of complications after having a benign tumor removed from his brain at National Taiwan University Hospital. He was 40 years old.

We hope that this book will serve to honor this friend who approached and introduced Chinese culture with a big and open heart.

Chang Chin-ju

Author

目次

Contents

目次
Contents

中國人與動物

Our Incr-edible Animal Friends

文•張靜茹　圖•張良綱

貓熊不來已成定案。除擔心國內飼養貓熊的條件，「國際形象」更是農業委員會考慮的重要因素之一。

近年來的生態保育風氣，使各國保育人士對稀有動物的進出口極爲敏感。與立法委員提議進口貓熊的同時，有一百多個會員國的「國際瀕臨絕種動物貿易公約組織」，抗議我國商人違法進口犀角、象牙、鱷魚皮、玳瑁殼等和瀕臨絕種動物有關的產品，並派代表來我國進行「了解」。

這種事不是第一次發生。全國生態保育協會和主管保育工作的農委會，多年來爲了國人殺老虎、進口犀角不時奔忙，向指責我國的外人解釋。

就在許多國家睜大眼睛注意我們對動物的一舉一動時，師大生物系在對全國山產店進行調查後指出，國人進補的山產，數量不下從前。

不久前，美、俄兩國爭相救助兩隻被困在北極冰塊的鯨魚脫險，至今仍爲人樂道。無論救援動機爲何，生態保育已是世界潮流；並深深影響國家形象。

在各國紛紛以保護動物自重的情形下，何以國人卻給人「四隻脚的除了桌子不吃」和「只要是野生動物都可進補」的印象？中國人與動物關係到底如何？

It has already been decided not to bring the panda to Taiwan. Besides concern about domestic conditions for raising the panda, one of the important factors that the Council on Agriculture (COA) considered in the decision was "international image."

The recent trend of ecological protection has made environmentalists extremely sensitive to the trade in rare species. At the same time as a legislator raised the idea of importing the pandas, the Organization of the Convention on International Trade in Endangered Species of Wild Fauna and Flora — which includes over 100 member nations — protested illegal trading by ROC businesses in rhino horns, elephant ivory, alligator skins, tortoise shells, and other products related to endangered species. They also sent a delegation to Taiwan to "understand the situation."

This is not the first time for such an event. For many years now the Society for Wildlife and Nature, ROC, and the COA (the body charged with overseeing conservation), have been busy explaining to foreigners about Chinese who kill tigers or import rhino horns.

With many nations now focussing attention on our behavior toward animals, the biology department at National Taiwan Normal University undertook a survey of shops selling "natural products" and discovered there are just as many wild animals being sold as "dietary supplements" or tonic foods as in the past.

Not long ago, the US and USSR struggled to free two trapped whales from the polar ice cap, an act still looked upon very favorably. Whatever the motive may have been, it is clear that ecological protection is a global wave and has a great impact on a country's image. Given current conditions, with each nation moving to protect its animals, how is it that the Chinese give the impression of "eating anything with four legs except for tables" or believing that "any wild animal is a tonic food"? What after all is the relationship between Chinese and wild animals?

六、七年前，幾位回國從事保育工作的學者專家，四處演講、放幻燈片呼籲保育觀念。有一回他們南下墾丁，在恆春國小演講，當幻燈片出現伯勞鳥誤入陷阱的鏡頭，配音發出了「唉喲！爲什麼要抓我們呀？」的慘叫聲。沒想到，臺下立時冒出幾個稚嫩無邪的回聲：「我們要吃你呀！」

雖然有更多的人支持動物保護運動，反對吃烤鳥，但偶爾也會懷疑：同是生靈，難道豬、牛、鷄、鴨就特別該殺？爲什麼野生動物就不可殺而食之？

野地西風瘦馬，夕陽西下；電影裡的西部英雄終於舉起來福槍，結束了老馬受傷的痛苦。

迪斯奈卡通影片裡的流浪狗，最怕的就是巡邏街頭的捕狗車；只見牠四處逃竄，就怕給逮個正着，捉了去「安樂死」。

「對於西方人來說，不殺死受苦中的獸類是『虐待』，但若爲了吃，而殺人們不慣吃的動物，如貓、狗和其他野生動物，則是『殘忍』，」定居曼谷的作家李簾鳳，曾在一篇文章中寫道。

西方人長久來受基督教的影響，聖經創世紀開宗明義就說：「神就照著自己的形象造人……，又對他們說，治理這地，也要管理海裡的魚，空中的鳥，和地上各種行動的活物，……」

李簾鳳指出，以這樣的思想出發，在生活中多少留下人類比一切生物爲高的觀念，認爲人必須以主人身分愛護自己的動物財產，以慈悲的態度決定財產的去留。

東方「不道德」，西方「假清高」？

這不正是西方人士，認爲東方人吃「不慣吃的野生動物」——殺老虎、烤伯勞鳥是不道德行爲而常加抨擊的原因？

若由中國人的思想出發，洋人吃牛肉，却堅拒狗肉；吃鷄胸，又對鴨頭、鷄爪退避三舍；某些保育人士吃海鮮、却不願吃野鳥的態度，不正像「紅樓夢」裡看不慣湘雲大啖鹿肉的黛玉，好生被指責的「假清高」——「……這會子腥的羶的大吃大嚼，回來却是錦心繡口。」

中國人相信「萬物有生」；孔孟、老莊等中國傳統思想，基本態度也是人與萬物並生而平等。

「雖然我們常罵人『畜牲』，」多年鑽研「中國人與動物崇拜」的藝術家楚戈以爲，那是中國人認爲人和動物一樣，除了人性，還具有獸性與神性；動物則除獸性，也具有神性與人性。所以中國民俗信仰中的動物神很多，神話故事中的動物甚至只要稍加修煉，都能以人的姿態出現。

愛鼠常留飯，憐蛾不點燈

既然「人與獸基本上生命相等」，人和獸一樣具有獸性，人自然也像萬物一樣，爲了吃而殺生；而且，萬物既相等，除非吃了傷命，又豈有某一種可食，另一種就不能食用的道理？

事實上，在中國文人思想中，因爲視萬物和人具有同等生命，對動物格外有人情味。

已故哲學家唐君毅說過，在沒有利益衝突下，中國人對動物常能見其生不忍見其死；聞其聲不忍食其肉。以至擾人的老鼠，人有時亦覺牠可憐，所以中國詩人說「愛鼠常留飯，憐蛾不點燈」。

西方人認爲該殺的野狗，中國人偏偏認爲「螻蟻尚且偷生」，多活一分鐘，便多一分樂趣；因此中國人生活的環境中，流浪的貓、狗也就特別多。只是到了工商社會，都市的居住環境不再適合狗輩「快樂的苟活」，西方的方法才被認爲比較衛生、文明。

結佛緣，仍可吃「動物」

尤其受佛教影響至深的中國民間，在慈悲爲懷的哲學下，甚至認爲人連蒼蠅、蚊子都不可殺。

佛家弟子認爲，動物是一個不得已墮入六道輪迴中「畜牲道」的生命，因此許多老一

Six or seven years ago, several scholars active in environmental protection, having just come back to Taiwan from abroad, were giving slide show lectures around the island to spread the call for conservation. At the Hengchun elementary school in the south, a slide of a shrike caught in a trap came on, with the audio having the bird saying in a wounded voice, "Why do you want to catch me?" Unexpectedly, a few innocent voices from below the stage squeaked, "Because we want to eat you!"

Although more and more people support the animal protection movement, and oppose eating roasted wild birds, nevertheless there is still doubt: All of these things are living creatures; is it possible that only the pig, cow, chicken, and duck should be specially singled out for slaughter? Why can't wild animals be killed for food?

As the sun sets in the west, the cowboy hero puts his injured steed out of its misery. In the cartoon, the stray dogs most fear the roving dogcatcher; you can see them running everywhere in fear they will be caught by this do-gooder and his "mercy killing." "For Westerners, not to kill a suffering beast is 'mistreatment,' but if one kills an unusual animal to eat, like a cat or a dog, or even wild animals, then it's 'cruelty,'" wrote author Lee Lien-feng, who lives in Bangkok, in one of her essays.

Westerners have long felt the influence of Christianity, and in the Book of Genesis it is written, "God said, let us make man in our image . . . and let them have dominion over the fish of the sea, and over the fowl of the air . . . and over every creeping thing that creepeth upon the earth."

Lee Lien-feng points out that starting from this thinking inevitably gives rise, to a greater or lesser extent, to the idea that "people are more important than animals" in daily life. This is the belief that people, as the masters, must care for the the animal property in their care, deciding the disposal of this property with an attitude of kindness.

The "immoral" Orient? The "hypocritical" Occident? Is this really the reason why many people from the West consider Asian people who eat "animals that are not ordinarily consumed" — like tigers or shrikes — to be immoral, and thus criticize and attack us?

Starting out from the point of view of Chinese, if foreigners eat beef but refuse to eat dog; eat chicken breast but recoil from duck heads or chicken feet; or if some self-proclaimed conservationists have the attitude of eating seafood, but refusing to touch wild bird. . . . Isn't this just like the famous scene in *Dream of the Red Chamber* where Tai-yu is accused of "false rectitude" because she can't accept the deer meat being consumed: "Here you are indulging in food and drink, yet turn around and try to sound above it all."

Chinese believe that all living things are on a par with each other. The fundamental view of the traditional thinking of Confucius, Lao Tzu, Chuang Tzu, and others is that man and the "ten thousand things" live side by side and are essentially equal.

"Although we often curse someone out by calling them a 'beast,'" notes the artist Chu Ko, who has for many years done in-depth research into animal worship and the Chinese, "that just shows Chinese see people and animals as being the same." Besides "human nature" there is also "bestial nature" and "divine nature." Besides having bestial nature, animals also share in divine nature and human nature. Thus there are many animal deities in popular Chinese belief, and the animals that appear in Chinese fables just need a little touch-up job and they can appear as convincing humans.

Love the rat and leave a little rice, pity the moth and do not light the lamp: Since "the lives of man and beast are fundamentally equal," man and animal both have "bestial" natures, and man is naturally like all living things — he kills to eat. And since the ten thousand things are equal, except for those things harmful to oneself, what's the logic in being able to eat one kind of critter while another one is deemed not suitable for consumption?

In fact, in traditional Chinese literati thinking, because all living creatures have life just like man, there is an exceptional feeling of "humanity" toward animals.

The late philosopher Tang Chun-yi said that, where there is no conflict of interests, Chinese often cannot bear to see them die, and are only happy when they live. They cannot eat the meat if they actually hear the animal itself. This goes even for the lowly rat: People even think that it is pitiable, so the Chinese poet has written, "Love the rat and

(左)中國藝術作品中的動物多爲寫意之作。(鄭元慶攝)

(left) The animals in Chinese art are mostly impressionistic. (photo by Arthur Cheng)

(右)西方重視理性的科學研究，往往將動物製成標本。

(right) The West emphasizes scientific research; animals are often turned into stuffed specimens.

輩操刀宰殺雞鴨時猶心存不忍，口中必念：「做雞做鳥無了時，趕緊去出世，出世富貴郎囝兒，務免各再做衆生。」

幽默大師林語堂曾對此發表意見，他以爲佛教雖然忌殺生，卻不否認人生而爲肉食者。許多信徒不吃牛肉，理由是牛的用途遠較其他動物爲大。直到現在，許多老太太不吃牛肉，也是基於「牛爲人耕作，不忍再食其肉」的惻隱之心；對無法忍受不吃肉食，又想結點善業的人，也可以定期偶爾吃蔬齋，不須長年茹素。

至於儒家，則有「親親而仁民，仁民而愛物」的看法，要完成君子的人格，仁民與愛物，缺一不可。禮記祭義篇中更有「斷一樹，殺一獸，不以其時，非孝也」的說法。

中國人看重孝道，而亂砍樹與亂殺動物，竟被認爲是不孝的大罪。顯然，在民胞物與的原則下，不是不能殺生，而是必須「取之有道」。

因此仁君打獵的態度是「春田不圍澤，不掩羣」，意指打獵絕不一網打盡，總會留一缺口做爲動物逃生之路；對自然資源（當然包括動物）的使用，是「取之有時，用之有節」。

周朝就有「國家公園」

雖然現代第一座國家公園出現在美國，在一段周朝的文獻記載中，活生生出現了現代化國家公園的管理辦法——周王保留一大片未受人爲開墾的邦田，供鳥獸有所生息，而且明令「有地而無政，則其生不能蕃息，雖有政不爲厲禁以守之，則侵地盜物……」換句話說，光劃一塊地做「國家公園」是不夠的，若要動物在此安心傳宗接代，還得設「管理處」、立「國家公園法」，對盜獵者嚴加懲罰才行。

「國家公園」範圍之外，也有禁令，農民春耕時，常有禽獸出沒危害水利、農作，因此可以設阱擭、溝瀆防止牠們侵入；但秋收之後，陷阱一定得拆除，動物才不至斷後。

國內保育人士也常引用孟子與梁惠王論及治國之道時所說的：「……數罟不入汚池，魚鼈不可勝食也；斧斤以時入山林，林木不可勝用也。穀與鼈不可勝食，林木不可勝用，是使民養生喪死無憾也……」來說明古人早有保育的觀念。

因爲孟子的教訓，正與現代保育的標的「資源能爲人類永續利用」的概念不謀而合。儒家思想由仁民愛物出發，爲其他生靈留下一片生存空間，也達到人類「利用、厚生」的目的。

養鳥不如種樹

另一影響中國文人思想極深的道家，由於酷愛自己精神的自由而「愛屋及烏」，對動物十分「放縱」。

莊子臨死前，弟子想將之以棺槨安葬，莊子卻有「在地上被烏雀老鷹吃，在地下被螻蟻食」的議論。反正死後都被動物吃，還選食客不成？

在莊子秋水篇中，他說牛馬四隻脚是天性，至於人爲的絡馬頭、穿牛鼻，則是人毀滅天機，傷害生靈。看人「虐待」動物，他渾身不舒服。

到了魏晉，文人思想精神更解放，順動物順其天性的故事也就更多。

支道林好鶴，某日有人送他一對，不久鶴翅長好想飛走，支道林捨不得，遂剪去鶴的飛羽，鶴因此不能飛走而垂頭喪氣，令支道林頗有愧意：「旣有凌霄之姿，何肯爲人做耳目近玩！」因此等鶴羽再度豐盈，就放了牠們。

清代揚州八怪之一的鄭板橋，曾在給其弟的兩封信中提到，他生平最不喜歡將鳥養在籠中，因爲「我圖娛悅，彼在囚牢，何情何理，而必屈物之性以適吾性乎！」

不願養籠中鳥，卻又愛鳥，怎生是好？「欲養鳥莫如多種樹，使繞屋數百株，扶疏茂密，爲鳥國鳥家……」鄭板橋的「上體天心」，實可被視爲現代自然保育的先驅。

現實生活無法「物我合一」

莊子曾說他心目中的至德時代，是人和禽獸相混雜而居住，不僅與萬物相處生活，還

leave a little rice; take pity on the moth and do not light the lamp.''

Where the Westerner might be inclined to put the stray dog ''to sleep,'' the Chinese believe that each minute of its life is one more minute of possible happiness. This is why there are so many wandering dogs and cats in the living environment of Chinese. It's just that with the advent of an industrial society, an urban setting is not very suitable for a ''dog's life'' of ''happy wandering,'' and the Western method is only now being accepted as more sanitary and civilized.

Tying in with Buddhism while still being able to eat "animals": In particular, the Chinese people have been deeply influenced by the philosophy of Buddhism. Under this philosophy of mercy, people believe that even the killing of flies or mosquitoes is prohibited. People who have been influenced by Buddhism believe that animals are but a form in which something has returned in its new incarnation. Thus many older people feel uneasy when slaughtering a chicken or duck, and find it necessary to say: ''There is no end to being a chicken or a fowl. Hurry up and leave this world, and come back as the scion of a wealthy family, so that you are no longer just a common creature.''

The humorist Lin Yutang has expressed his view that although Buddhism prohibits the killing of animals, it doesn't deny that people survive by eating meat. Many believers don't eat beef; the reason is that cattle are far more useful than other creatures. Even now, many older women refuse to eat beef out of natural compassion. The idea is that ''the buffalo plows the field for man, so you can't just consume it as food as well.'' Those who cannot forsake meat for life but want to do the meritorious thing can go vegetarian for short periods.

As for Confucianism, since there is the view that one should ''be close to one's family and then benevolent to the people, benevolent to the people and then loving of things,'' benevolence to the people and love for all living things are indispensable to completing the qualifications to be a true gentleman. In the *Book of Rites* there is the statement that ''to chop down a tree or kill an animal at the improper time is unfilial.''

Chinese see filial piety as the most important thing, so it is very significant that cutting down a tree or killing an animal should be seen as the major offense of ''failing to be filial.'' Obviously, under the principle of people being the equals of animals, the idea is not never to kill, but only ''to take where it is right to do so.''

Thus the attitude of the benevolent gentleman in hunting is ''not to encircle the whole field, not to entrap the whole herd,'' indicating that one should not aim for eradication in hunting, and one must always leave the animals an escape route. For all natural resources (of course including animals), the catchphrase is ''there is a time to take and a time to be sparing in use.''

Chou dynasty "national parks"? Although America was the first place to have modern national parks, records from the Chou dynasty (1122-221 BC) indicate the vigorous appearance of modern national park management methods. The Chou Emperor set aside a stretch of land as yet undeveloped by man as an animal preserve, and also stated: ''Having land but no regulations is like living without being able to reproduce; although there might be rules, with failure to strictly implement them, there will be encroachment and theft of animals. . . .'' In other words, just to map out a ''national park'' is not enough. If you want the creatures to be able to live in peace and propagate, there must be an ''administration,'' and a ''national parks law,'' with heavy penalties for poachers.

Outside the prescribed ''national parks,'' there were also prohibitions. During spring planting, often wild animals could cause damage to the irrigation systems or crops, so farmers were allowed to set traps or dig ditches to prevent them from tracking across the fields. But after the autumn harvest, the traps had to be removed, so that the animals would not be cut off from propagating.

Domestic conservationists often quote Mencius to the effect that, ''don't use a net in the fish pond, do not consume fish and turtles until they are exhausted; when cutting wood only go into the forest when it is time, and do not use the wood to exhaustion. Not allowing the exhaustive consumption of shellfish and turtles, nor the exhaustive use of the woods, enables the people to live and die generation after generation without interruption. . . .'' This clearly shows the ''environmental protection'' ideas of the ancients.

Thus Mencius' lesson fits in perfectly with the

由生物進化史來看，人類社會未形成前，人與動物相爭是極其自然的事。
(翻拍自戶外雜誌出版「臺灣三百年」)

From the point of view of natural history, conflict between man and beasts was a natural fact before societies took shape. (photo from *Taiwan: Three Hundred Years* published by *Outdoor Life* magazine)

和禽獸共同遨遊。興起，還能爬上樹去看看鳥鵲的巢窩。

表面看來這有些一廂情願，因爲鳥兒沒見到人，大概就先嚇跑了；想與猛獸同遊，無異捋虎鬚。莊子可能也無意自己的「幻想」成眞；這只是中國文人思想，確視動物與人是同具生命，而希望達到「物我合一」境界的表白。

雖然在中國文化發展歷程中，不乏對動物心存善念，但在每天爲開門七件事忙碌的眞實人生中，卻也無法人人都類似哲人，可以對所有動物都充滿情意。

北宋時代，文人雅士有所謂的「壺蝶會」，人人手攜一壺酒、一碟好菜，在百花盛開的「花朝節」這天聚在一起，飲酒賦詩，賞蝶、賞花。

如此詩情畫意的點子，其實是起於農民的「撲蝶會」，因爲蝴蝶的幼蟲專吃菜葉，如此可減少蟲害，保護農作。

美國籍的生態保育專家謝孝同說過，西方常仰慕中華民族有天人合一「莊周夢蝶」的境界；但哲學和藝術中所表現出來的天人和諧、物我合一，只是存在少數人的想法。民衆因爲要和自然搏鬥始能生存，鮮少刻意尋求他們和自然間的和諧關係。

只分涼、熱，不分「東西」

人類學者李亦園在深入探討「中國人愛以野生動物進補」的背後原因後，卻提出了另一種看法。他以爲，中國人的生命中不斷追求著個人身體、人與社會、人與大自然三種和諧與均衡。

爲求得人體內部系統的和諧，就須借助外來食物，因此將食物分成涼性和熱性，火氣大就吃退火的涼性食品；體冷則補熱食，這也正是中醫所說的「食療同源」。

強烈希望自己身體冷熱、陰陽協調的慾望，使中國人對食物只問補不補，就無所謂「慣食不慣食」了。

今天一般人已不能確實分淸何謂涼性、何謂熱性食物，何者可補，何者不能補；但「補」卻仍是中國民間的重要「文化」。

「必須要由這樣的層次，才能理解爲何中國人給人『凡動物都能補』的印象，」李亦園以爲，否則就會有「凡吃野生動物，都是化外之民」的膚淺認定。

愈稀奇的愈補？

此外，李亦園也指出，傳統中國民間對動物分類的標準，其實很偏重實用。在「科學發展的文化因素探討」一文中，他曾引用西方人類學者的說法，由於受舊約聖經的影響，古代歐洲人把一些反常動物，如鴕鳥是鳥類卻不能飛行，海馬似馬卻生活在水中，視

modern conservationist ideal that "resources are for the perpetual use of mankind." Confucian thinking also starts with the assumption people must love living things and leave living space for other living things in order to achieve the goal of "use and improving living conditions."

Raising a bird is not as good as planting a tree: Another profound influence on the thinking of Chinese people has been Taoism. Because of the deep love in Taoism for one's personal spiritual freedom, that love is also extended to those things around man, so that man's attitude to animals is quite "tolerant."

Just before his death, Chuang Tzu's followers proposed burying him in a coffin. But Chuang Tzu countered with the idea that "one is eaten by the carrion above ground, by the worms below it." In any case, one will be consumed by animals after death, so is it such a big deal to choose which ones to be eaten by?

Chuang Tzu says that the four-legged creatures like the bull and horse are natural. But when people put harnesses on horses' or cows' heads, this is a destruction of nature's intention, and is harmful to the natural order of things. Seeing people "mistreat creatures," he felt moved to the depths of his soul.

By the Wei-Tsin Era, the thought of the literati was even more embracing, and there are many stories of the desire to allow animals to give full play to their natures.

Chih Tao-lin loved cranes. One day someone sent him a pair as a gift. Not long thereafter the cranes wanted to fly away, but Chih Tao-lin could not bear to part with them, so he trimmed back the birds' wings. Because they could not fly away, they fell into deep melancholy, which made Chih remorseful: "Since they are blessed with the ability to ride the clouds, how could they be happy serving the eyes and ears of a man?!" Thus when the birds' plumage was again thick and full, he set them free.

One of the eight eccentrics of Ching dynasty Yangchow, Cheng Pan-chiao, in two letters to his younger brother, said that the thing that he most disliked in life was the raising of birds in cages, because "for my own pleasure, the other is put in a prison; what manner of logic or sentiment is this, that one must subvert the nature of the beast

爲不祥之物，付予審判後處死。

中國人對這些模稜兩可的動物，不僅不消極規避，反採積極看法。人類學者以爲，海馬、海參、燕窩、穿山甲等中藥，都因爲特殊性狀，才被中國人視爲補品。

明朝李時珍的「本草綱目」，被西方學者認爲是十六世紀最偉大的博物學，此後卻一直只停在治病、食療的應用階段。

至於中國人文精神，又重心靈體驗，對於萬事萬物不多做西方式的理性分析。二者相乘，「使得中國人對動物和對自己一樣的不了解，」李簾鳳指出，我們看大自然，往往只講究迷濛的詩情畫意，卻不曾去研究大自然究竟是怎麼一回事。中國人釣到奇怪的魚，便打主意如何去烹調；洋人釣到不知名的魚，卻忙著去做標本。雖然中國文化發展中，會養蠶，會閹鷄割牛，但是我們對獸類的生理與心理都缺乏有系統的認識。

老虎爲患，要不要保護？

而一向自認人比動物高一等的歐洲，在十九世紀達爾文提出物競天擇、適者生存的理百鳥在林不如一鳥在手？一般人很難實踐「養鳥不如種樹」的道理。

Two birds in the bush are really not worth one in the hand? It would be tough for most to realize the reasoning of "raising a bird is not as good as planting a tree" where birds will come to live.

to satisfy my own!?"

Unwilling to raise birds in cages, yet loving them nonetheless, what would be the best solution? "To raise more birds is not nearly as good as planting more trees, providing hundreds of trees for a bird homeland...." Cheng's "superhuman sensitivity" means that he could be seen as a forerunner of modern ecological conservationists.

Hard to practice "live and let live" in the real world: In Chuang Tzu's ideal era of perfect virtue, the dwelling places of man and beast would be intermingled. Man would not only coexist with the animals, but would roam with them. When his interest was piqued, he could climb into the trees and look into the nests of the birds.

It seems on the surface this is just daydreaming. Anyway, the bird would be frightened off long before it actually laid eyes on the guy climbing the tree. And if you swim with sharks, you just might get eaten. Perhaps Chuang Tzu himself did not mean that this vision could be realized. This is just a clear expression of the thinking of ancient Chinese philosophers who saw people and animals as sharing life in common, and hoped that the world of "live and let live" could be achieved.

Although people have good instincts toward animals in their hearts, the real world in which people are hectically pursuing their daily concerns does not always permit such lofty sentiments toward animals.

In the Northern Sung period (960-1127), poets and literati had a "fete for butterfly appreciation." They would drink wine, compose poetry, and revel in the butterflies and flowers of early spring.

Yet this poetic idyll in fact had its origins in the "fete for butterfly eradication" of the peasants. Because caterpillars eat plants, they could in this way reduce crop damage and protect their livelihoods.

The Chinese-American environmental scholar Hsieh Hsiao-tung says that Westerners admire the Chinese parable of "Chuang Tzu's Butterfly Dream." But the harmony between man and nature, the "live and let live" expressed in poetry and philosophy, is something that only exists in the minds of a few. Because most people must struggle against nature to survive, only a scant few deliberately seek harmonious relations with nature.

It's either "hot" or "cold," and nothing else matters: Li Yih-yuan, a research fellow in the Institute of Anthropology at the Academia Sinica, has been exploring the underlying reasons why "Chinese people are fond of using wild animals as dietary supplements," and has come to a different view.

He believes that Chinese have always sought harmony and balance in three areas: with their bodies, with society, and with nature.

Achieving internal bodily harmony requires help from aliments from without. Thus foods have been divided into "hot" and "cold" natures. When one's "fires" are high, one should eat cold foods; when the body is cold, one must compensate with the hot. This is the idea of "an ounce of prevention is worth a pound of cure" in Chinese medicine.

The strong desire for the hot and cold, the *yin* and *yang*, in one's body to be balanced means that Chinese only ask if a food source is "tonic" or "compensatory" or not, and there is no distinction between so-called "customary" and "non-customary" nutriments.

Today it is hard for most people to be sure what's "cold" and what's "hot," what is really tonic, and what isn't. But "tonic" or "compensatory" is still an important part of Chinese popular culture.

"You have to approach it from this level to understand why Chinese give the impression that 'if it moves, it's tonic food,'" says Li Yih-yuan. Otherwise people will have the superficial con-

漁獵時代的捕魚工具雖簡陋，却無一網打盡之虞。

Although the nets used to catch fish in the fishing and hunting age were primitive, at least they didn't pose the danger of wiping out a species at a single stroke.

論後，對動物科學研究更加發達。他們曾在世界各地大量獵取、收集珍禽異獸，製作標本、做研究，為此犧牲的動物也頗為可觀。但西方在明瞭人對動物迫害太過、於己並無益處之後，對動物這項人類的重要財產的保護，也是中國人望塵莫及的。

認為烏魚子可吃，伯勞鳥當然也可以吃，卻不甚了解動物這門學問的中國人，又該如何面對全世界動物數量日稀、身價日增的情形？

有個例子可茲參考。兩年前，泰國政府為響應西方生態保育，對特產老虎禮遇有加；結果老虎趁此休養生息，數量增加，還傷害不少人；泰國人想殺老虎，又被西方保育團體指責，一氣之下，以頗具挑釁的口吻回信給國際保育聯盟：「我國虎滿為患，若眞不能殺，你們帶些回去保護，如何？」此案最後以「適量獵捕」落幕。

合理對待人類最親密的伙伴

可見保護動物也應視民族文化與動物族羣的個別情況靈活運用。許多保育人士都反對吃伯勞鳥時，資深的生態保育學者、林試所集水區經營系主任柳榗就獨排衆議。

他認為，瀕臨絕種的稀有動物理當保護。至於每年大批湧到的伯勞鳥，和烏魚一樣，在經過科學調查、研究，確定不至瀕臨絕種的情況下，主管單位應當明白決定它是那一種資源，是觀賞資源或食用資源？或兩者皆可？再擬定具體的保護和使用方法。如此一來，保護有根據，使用有理由，就不怕任何人的詰問了。

隨時代改變，身為現代中國人，你是否想過，究竟應該怎樣看待我們最親密的伙伴？

S

（原載光華七十八年一月號）

clusion that "only the uncivilized eat wild animals."

Make mine rare: Further, adds Li, the classification standards Chinese traditionally used for animals tended to be extremely pragmatic. In the study *An Exploration of Cultural Factors in Scientific Development*, he employed a Western anthropological theory to say that, because of the impact of the Old Testament, Medieval Europeans thought of unusual animals — like the ostrich, which looks like a bird but cannot fly, or the sea horse, which looks like a horse but swims with the fish — as inauspicious, and thus put them to death.

As for these confused creatures, not only did Chinese not avoid them, they were seen as positive. Anthropologists think that many kinds of animals used in Chinese medicine, like sea horse, sea slug, sparrow's nest, and pangolin, were only accepted by Chinese as being especially tonic because of their unique features.

The *Compendium of Materia Medica* by Li Shih-chen of the Ming dynasty has been seen by Western scholars as the greatest work of natural history of the 16th century, but in China it was only used for medical reference.

As for the Chinese humanitarian spirit, emphasis had always been on spiritual meanings, and very few people did Western style scientific studies. Taking this into account, "this has left Chinese as unable to understand animals as themselves," points out Lee Lien-feng. When we look at nature, we often only think about its mystery and elegance, but don't try to study it to find out what, after all, nature really is. It is joked that when a Westerner catches a strange fish, he will try to classify it; when a Chinese does, he will try to figure out how to cook it. Although in the development of Chinese culture we find the raising of silkworms or the castration of oxen, we lack a systematic understanding of the biology or psychology of these creatures.

The tiger is a menace, why would you want to protect it? Yet in Europe, where people had always seen themselves as superior to animals, after Darwin produced the theory of natural selection and survival of the fittest, zoology and biology became even more developed. But they have also captured and taken precious creatures from everywhere in the world, to serve as samples or for testing, and the losses from this are also significant. But after Westerners came to realize that the harm people do to animals has gone too far, at no advantage to ourselves, they have achieved a level of protection for this valuable property of mankind that Chinese can only aspire to.

How can Chinese, who believe that if it's OK to eat caviar, then it must be OK to eat shrike, yet lack the discipline of zoology, face the dwindling number and rising cost of wild animals around the world?

There is an example for reference. Two years ago, the Thai government, responding to the entreaties of Western conservationists, gave favored treatment to their rare local tigers. As a result, tigers took the chance to recover and their numbers increased, and they even injured many people. Thais wanted to kill the tigers, but were chastised by Western environmental groups, so in anger they responded in a letter to an international conservation alliance: "We have so many tigers they are becoming a menace. If we really cannot kill them, how about you taking some back for protection yourselves?" This incident was resolved through "appropriate hunting limits."

Rationally treat our intimate partners: It is obvious that animal protection must be adjusted to each national situation. While many environmentalists oppose the eating of shrike, Liu Chin, director of the Watershed Management Division at the Taiwan Forestry Research Institute, who has long been involved in conservation work, stands out from the crowd.

He contends that, naturally, endangered species must be protected. As for the shrike, which is trapped in large numbers every year, after scientific surveys and research confirm that it would not be endangered by use, the governing authorities must specify precisely what kind of resource it is, a resource to be watched for the joy of appreciation, or a food resource, or both. Then concrete protection and use methods should be stipulated. In this way, protection becomes rooted, use is rational, and there is no need to fear the interrogations of anyone.

Times are changing. As modern Chinese, have you ever thought, how should we treat our intimate partners in this world?

(Chang Chin-ju/photos by Vincent Chang/ tr. by Phil Newell/ first published in January 1989)

從犀牛角談起⋯

Starting from Rhino Horns . . .

文・張靜茹　圖・王煒昶

在全世界一片保護犀牛的呼聲中，政府雖早在四年前就立法管制輸入犀牛角，我國卻仍是進口犀牛角最多的國家。國人到底看上犀牛角那一點？

小說裏，小男孩臉上長疔瘡，老祖母毫不猶豫地拿出蟾蜍肝，敷在瘡口上。這樣的例子，在生活中也時有可見：蟾蜍、壁虎、草魚牙，甚至蟑螂都可能被老人家搬作救兵。

有人於是說中國人無所不可入藥。只有中國人以動物爲藥材嗎？曾被用來治病的動物究竟有多少種？這些讓現代人半信半疑的動物藥材，眞的有療效嗎？

Although the government, amid worldwide calls to save the rhinoceros, passed a law to control the importation of rhino horns more than four years ago, the R.O.C. is still the world's largest importer of them. Just what is it that Chinese people see in them?

The boy in the story has a boil on his face — his grandmother, without the slightest hesitation, spreads toad liver on it. Similar examples are constantly cropping up in real life: Toads, lizards, grass carp teeth and even cockroaches are used as nostrums by old folks.

Some people say that the Chinese make use of anything and everything in medicine. Are we the only people who use animals as medicinal ingredients? Just how many different kinds are used? And do they really work?

不少犀牛因「懷璧其罪」被濫殺。(張良綱攝)
Many rhinos have been slaughtered for the ''precious jewels'' on the ends of their noses.(photo by Vincent Chang)

許多臺北的訪客，喜歡到廸化街開開眼界。除了買些南北貨、欣賞老建築，也不會錯過擺滿各種動物標本的中藥店。店面玻璃櫃裏最顯眼處，總會放著一、兩個造型像座模型小山的犀角。

根據我國中藥書的記載，犀角是極具療效的解熱涼血劑，中醫藥界也肯定它的退燒療效，至今仍有少數堅持使用中藥的人以它代替西藥退燒。

把目光由犀角移開，可能會看到黑黑一大塊像煙燻火腿的東西，它的味道也許沒有火腿可口，價錢却絕對昂貴，因爲它是另一種稀有動物藥材——「熊膽」；一根根小腿骨似的「鹿鞭」已不算稀罕，被零亂疊成一堆；桃形、毛刺刺的「麝香囊」則因「異香」撲鼻，被安置在玻璃罐中。

動物入藥，東、西皆然

藥行師傅拉開抽屜，藥盒中裝著像金龜子、蠶以及不知名的小昆蟲，偶爾也能見到熟悉的東西——蟬蛻。碰巧有人上門抓藥，八開大的藥紙上，剛烤熱的海馬和乾扁、呈大字型的蛤蚧趴在當歸、枸杞等植物藥上等待打包……，看得人嘖嘖稱奇。

以動物治病並非中國人的專利。遠溯人類老祖先最初在地球上出現，環境中沒有任何現代科技產物，很自然的只能由動植物身上找醫療材料。

中國醫藥學院藥學系教授葉豐次表示，任何民族的藥物學都起源於民間的藥草和藥用動物。我國如此，古希臘、羅馬也都有載述藥材和處方的書流傳至今；同屬東方古國的埃及人，也保存了上千年歷史、記滿植物藥材的「紙草本」。

由於植物藥比之動物藥易於取得，所有國家流傳下來的藥材都以植物較多；我國在「食療同源」、「食補」……等觀念下，營養成分較高的動物藥，中選爲藥材的機會自然極高，因此動物藥材的豐富不下植物。

是珍饈，還是奇藥？

中醫認爲，當人患病，不單是患處出了問題，而是整個身體的系統藉此亮起紅燈，發出警告。因此只要營養豐富、身強體壯，出問題的那一部分，自然比較容易痊癒。

幽默大師林語堂曾描述過卅年前上海河南路的中藥舖：「你竟難以斷言裡邊究竟是藥物多於食物，還是食物多於藥物？你在那裏可看見桂皮和火腿，虎筋、海狗腎及海參，鹿茸、蔴菇和蜜棗被並排陳列……。」

火腿、海參早成日常佳餚，是爲口慾或治病已難分野；若再把喜宴上常見的魚翅、燕窩，開刀者常被要求食用的鱸魚，和養顏美容的聖品珍珠粉、蜂王乳等等都算在內，再加上從陸上哺乳動物、海中水族、天上飛禽，到兩棲動物等眞正藥材，被中國人用過的動物藥恐怕無法估計。

吃形補形，以毒攻毒

就在中國人鑽研使用天然藥材時，西方醫學、藥學與化學的研究正如大海奔流。十九世紀歐洲人首先由天然藥材鴉片中提煉出嗎啡、在金雞納皮中抽取出奎寧。到了本世紀初，部分藥材開始以人工合成製造，不再直接食用天然生藥。今天我們服用的西藥，大都已是經過藥性分析、藥理實驗的合成藥，天然藥材的比例大爲降低。

中藥雖然也有一些人人耳熟能詳的藥理，如「吃形補形」——蛇是最靈敏的動物，故能補神經、治皮膚過敏，烏龜常年爬行水邊，不懼濕冷，因此腹殼可治風寒、補血；或「以毒攻毒」——全蠍可治瘰癧瘡毒，蟾蜍可醫疔瘡、癰毒。但到了一切講究科學、證據和分析的現代，這些原理原則實難爲現代人接受。

風水輪流轉、中藥正當令

加上許多人爲的穿鑿附會、自行引申，所謂「江湖術士」自創前所未有的動物藥，連帶的，傳統中藥也因此被懷疑。和進口犀角同樣聞名國際的「五花大綁殺老虎」，殺虎者往往將虎爺全身上下都誇成補品。被稱可壯陽的虎鞭，實際上從未在正統中藥書上出現。

Many visitors to Taipei head off to Tihua St. for an eye-opening shopping trip. Besides browsing at goods from all over China and gazing at the old buildings, they're bound to be struck by the Chinese medicine shops full of all sorts of animal specimens. At the most conspicuous spot inside the windows are invariably displayed one or two rhinoceros horns, looking like miniature model mountains.

According to Chinese medicine books, rhino horns are a highly effective ingredient for relieving fever and cooling the blood. Even today, some people who swear by Chinese medicine insist on taking rhino horn powder instead of Western medicine to reduce fever.

Looking past the rhino horns, the visitor may see a big black block like a smoked ham. It may not smell as appetizing as ham, but it costs a lot more because it too is another rare animal medicinal ingredient — bear gallstone. Deer's pizzle, jumbled together in a heap like a pile of small femur bones, is another frequent sight. Prickly, peach-shaped "musk sacs" are placed in glass jars because of their strong pungent odor.

Used in the West too: The pharmacist opens a drawer full of little medicine boxes containing desiccated ladybugs, silkworms and nameless other little insects or — a familiar sight in Taiwan — cicada shells. A customer happens by, and the pharmacist wraps up some roasted sea horse and horned toad with some Chinese angelica and wolfberry or other rare and exotic ingredients in an octavo-sized piece of paper.

Using animals as ingredients in medicine is no monopoly of the Chinese. When mankind's ancestors first appeared on the earth they naturally looked to plants and animals for medicinal ingredients.

Yeh Feng-tzu, a professor at the China Medical College, says that folk remedies around the world were originally based on medicines using herbal and animal ingredients. That is true not only of China but also of Egypt, Greece and Rome, from which ancient texts recording traditional remedies still survive today.

Being easier to obtain, plants have been used in the traditional medicines of most countries more widely than have animal ingredients. But in China, where the concepts of health and diet have always gone hand in hand, the use of nutritious animal ingredients has been much more common than in other countries and in no way secondary to that of plants.

Delicacies or wonder drugs? According to Chinese medicine, when a person is ill, the ailing part of the body serves merely as a red light or warning sign for the overall system. So long as the body as a whole is kept healthy and well nourished, the part with a problem will naturally be cured more easily.

The humorist Lin Yu-tang once described a Chinese medicine shop on Honan Road in Shanghai of decades back: "It was hard to say whether there was more medicine there or food. You could see cassia bark and hams; tiger tendons, seal kidneys and sea cucumbers; deer antlers, Chinese mushrooms, and candied dates all lined up together. . . ."

Ham and sea cucumber have long ago become common dishes at the table. If you figure in shark's fin soup and swallow's nest soup, which are commonly served at banquets; Chinese perch, which is often eaten by people who have undergone surgery; and pearl powder and queen bee extractions, which are used in cosmetics; and then throw in all the medicinal ingredients made from animals on land, fish in the sea and birds in the sky — not to mention amphibians — there is really no way to count all the animals used for health and medicine by the Chinese.

Fighting poison with poison: At the same time that the Chinese were studying the use of natural medical ingredients, Western medical science, pharmacology and chemistry were advancing like a flood tide. During the 19th century, Europeans refined morphine from opium and extracted quinine from cinchona bark, and by the early 20th century, they were making medicines from man-made ingredients rather than natural herbs. Most of the Western medicines we take today are chemical compounds that have been pharmacologically analyzed and tested. The proportion of natural ingredients has greatly declined.

Some principles of Chinese medicine are familiar to us all, such as *chih-hsing pu-hsing* (making up for a deficiency in a certain area by ingesting something rich in that area) — snakes, for instance, are the wiliest creatures, and so eating them is believed to help the nervous system and cure skin allergies. Turtles live in water and aren't afraid

國內外醫藥界不斷有人對中藥經典「本草綱目」中的藥材進行研究，但因它的機轉太複雜，很難確切分析出有效成分。

「一味人蔘，中外研究報告成千上百篇，療效的成分至今仍爭議不休，」衛生署中醫藥研究委員會主委張齊賢表示，動物藥成分比植物複雜得多，研究自然更加困難。

但目前無法解析出有效成分，却並不表示動物用藥的療效就被全盤否定，或再無發展性。

西藥在發現之初，確實對某種疾病有藥到病除之效，却往往在使用的幾年後被發現有副作用。許多人身受其害的抗生素即是一例。今日特效藥，明日致命傷，常常使得管理藥物者不得不加以限量、禁用，最後甚至淘汰。

廣尋天然藥材

近三百年來的醫療領域雖由西藥領其風騷，「但它的副作用，也使西方人回頭廣尋天然藥材，」必安中藥研究所生藥系主任張憲昌表示，至今許多天然藥材仍是人工藥物無法代替的；許多合成藥的基本原料，也多由天然藥材中淬取。

近代西方發明的荷爾蒙藥劑中，不少就由雞或羊卵子和牛鞭、鹿鞭中提取賀爾蒙。可惜國人青睞這些動物已千百年，却沒有去研究分析其中的有效成分。

中國醫藥學院藥學系教授許喬木曾舉我國唐朝的藥書「千金方」爲例，書中記載許多利用動物肝臟製成的藥劑；吃豬血能「清血」，更是所有中國人的普通常識；但由豬肝、豬血中提煉成分來治療內臟疾病的却是西方。

上一屆國建會實驗動物小組中，一位由美國來的學者就曾指出，在利用動物治療疾病上，老祖先留給我們許多「線索」，有待我們進一步探究。

我不殺犀牛，犀牛因我而死

不管科技多麼進步，我們的生存不能不依靠動物，西方因此有動物「基因庫」的建立。美國一直不斷研究目前已知唯一不會得癌症的鯊魚，希望能由牠身上找到治癌良方。即使不在動物身上找藥，今天各種藥物的發明，更需實驗動物的幫助不可。

由此看來，既然人類非得借助動物來「安身立命」，那麼中國人以犀牛角做藥材，原本無可厚非。

問題是許多中藥商因爲野生動物進出口限制愈來愈嚴格，且犀牛既然愈來愈「稀有」，就抱著「還能買到就盡量買來存放」和「即使沒人買去治病，也可留做擺飾炫耀」的心態搶購犀角。

犀角價格因此被哄抬得很高，也間接助長了生產國家違法捕殺犀牛，割取犀角。我不殺犀牛，犀牛因我而死，道義責任不能不負，政府因此立法禁止進口。

迪化街一家中藥店老闆表示，事實上犀角藥效強，每回所需的劑量極微，一個犀角可抓上千帖藥，「消費量」很有限。

除進口犀角被指責，最近大陸廣西省也傳出獵人大量捕麝，取走雄麝香囊，再剝皮剔骨，將肉當補品販賣，在國際上引起抗議。

取之原有道，殺之何太急

「何必這麼費力？」張齊賢表示，過去西藥未傳入我國，民間對動物藥材的需求量比現在多很多，因此有「藥農」專門繁殖、豢養各種藥用動物，根本不需捕殺野生品種。

直到今天，河南、山東省的蠍子，廣東省的白花蛇、遼寧的蛤蟆……，仍爲當地的重要產業；全身幾乎無一不可利用的鹿幾已成爲「家畜」。幾年前，四川馬爾康獐鹿飼養場人工繁殖麝成功，直接由活蹦亂跳的雄麝香囊中取麝香，還被視爲一項創舉。「每年定期取鹿角、羚羊角，也不必殺死牠們，」張齊賢表示。

話說回來，犀牛却不同。因爲犀牛無法人工飼養、繁殖，加上牠會攻擊招惹牠的人，盜取者往往只爲一個犀角，就把犀牛殺死。

「其實在現代，退燒藥有許多代用品，犀牛目前已瀕臨絕種，我們的確不該再進口犀角，」張齊賢強調，否則加速牠的滅亡，也

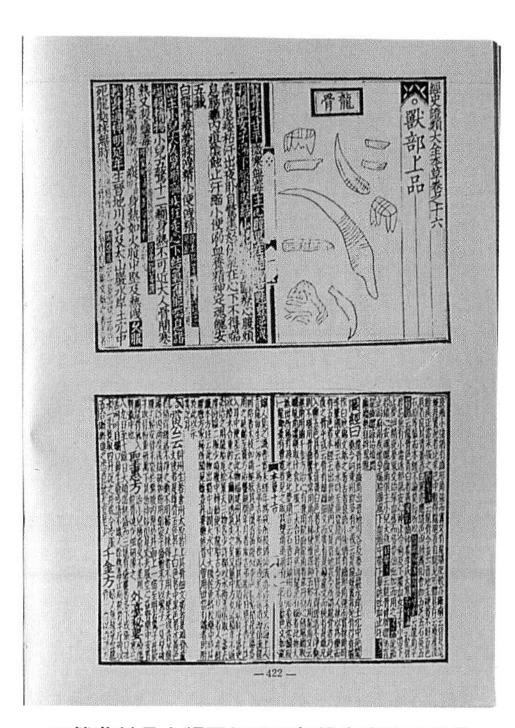
經史證類大全本草卷之十六

獸部上品

龍骨

—422—

天然藥材是人類歷經千百年親身實驗的結果。
Natural medicines are the fruits of thousands of years of experience.

of the damp and the cold, and so their undershells are believed to cure colds and build up the blood. Another principle is *i-tu kung-tu* (fighting toxins with toxins): Scorpions cure scrofula; toads cure carbuncles and furuncles. But in today's world of science, evidence and analysis, these principles are hard to accept.

What's more, some unscrupulous quacks have gone out and concocted bogus medicines of their own, bringing genuine Chinese medicine into disrepute. Tiger killers, for instance, claim that every part of the tiger has a tonic effect, while in fact the theory that tiger pizzles improve virility never appears in the orthodox Chinese medical tradition.

Chinese medicine makes a comeback: Research has long been carried out, both in China and abroad, on the medicinal ingredients described in the *Pen-ts'ao kang-mu*, or *Compendium of Materia Medica*, the great pharmacological work of the Ming dynasty (1368 to 1644), but because of the complex factors involved, their effective components have been difficult to determine.

''Ginseng alone has been the subject of hundreds of research papers, and the components that make it effective are still in dispute,'' says Chang Chi-hsien, chairman of the committee on Chinese Medicine and Pharmacy of the R.O.C. Department of Health. The composition of animal medicinal ingredients is even more complex than that of herbal ingredients, making research into them even more difficult.

But the fact that their effective components have so far eluded analysis in no way impugns the effectiveness of Chinese medicines or imply that they cannot be further advanced.

Many Western drugs that were effective in curing certain diseases have later been found to produce harmful side effects, antibiotics being a prime example for many people. The wonder drug of today may turn out to be a lethal poison tomorrow and be pulled off the market.

Search for natural ingredients: Although Western drugs have been in the forefront of medical science for the past three centuries, ''their side effects have caused many Westerners to turn back in search of natural medicines,'' says Chang Hsien-chang, chief of the Pharmacognosy Section at the Briion Research Institute of Taiwan. Even today, he adds, many natural medicines cannot be replaced by man-made drugs, and many of the basic ingredients of man-made drugs come from natural medicines.

Many hormonal supplements developed recently in the West are extracted from chicken or lamb ova and from cow or deer penises. Chinese have looked favorably on these animal medicines for thousands of years — unfortunately, we never researched and analyzed the effective elements in them.

Hsu Chiao-mu, a professor of Chinese medicine at the China Medical College, cites the tome *Chien*

琳瑯滿目的動物藥，您認得幾許？
How many of these Chinese medicinal ingredients can you name?

將使犀角這味中藥永成歷史陳蹟。

誰來研究動物藥？

近年來，有關單位不斷推動中醫科學化，也希望中藥材的研究能有突破，甚或在未來的藥劑上扮演更重要角色。國內研究植物藥的人日漸增加，研究動物中藥的人却寥寥可數。

理由之一，就是在生態保育倍受重視的今天，我國却仍給外人濫殺動物的形象，許多學者根本不願去「碰」動物藥，免得被視爲「不文明」。臺北醫學院藥學系教授楊玲玲就因此對動物藥材的前景不表樂觀。

想要中藥科學化？恐怕得由停止濫殺動物和嚴禁進口犀角開始。

（原載光華七十八年一月號）

chin fang (*A Thousand Golden Remedies*) from the Tang dynasty (618 to 907) as an example. The book records many types of medicine made from animal organs and states that eating congealed pig's blood ''clears the blood,'' which is common knowledge to every Chinese. But it was Westerners that eventually refined components from pig's liver and pig's blood for use in treating internal organ diseases.

At the last National Development Seminar, a scholar from the United States pointed out that our ancestors have left us many clues on using animals to treat disease that are just waiting for us to explore.

Not killing them but causing their deaths: No matter how advanced our technology may become, we remain dependent on animals for our very existence, one reason why Westerners have set up gene banks. Sharks, apparently the sole animal species immune to cancer, are the subject of constant research in the Untied States in the hope of finding a cure for cancer. And even if not all animals yield medicinal ingredients, their help is required in experimentation and the testing of new drugs.

Seen from this point of view, the use of rhinoceros horns in Chinese medicine should not be criticized too severely.

The problem is that as restrictions on wild animal imports become ever tougher and rhinoceroses ever scarcer, some purveyors of Chinese medicine — holding the attitude of ''stock up while you can'' and ''even if nobody buys it, you can still keep it around as a display gimmick'' — have been buying up horns as quickly as they can.

Rhinoceros horns have consequently been driven sky-high in price, thereby encouraging poachers to kill the animals illegally and cut off their horns. Not to shirk the indirect moral responsibility involved, the government has prohibited their importation.

According to the owner of a Chinese medicine store on Tihua Street in Taipei, rhinoceros horn is highly potent. Only a tiny amount is required each time, so that a single horn can be used to prepare of over a thousand doses. The quantity consumed is extremely limited.

In addition to criticism over rhinoceros horns, international protests have been raised over reports from the mainland that hunters in Kuangsi Province have been capturing musk deer in large quantities, removing their musk sacs, stripping off the hide and selling the meat as a dietary supplement.

Use, don't slaughter: ''Actually, there's no reason they should go to all that trouble,'' Chang Chi-hsien says, explaining that before Western medicine was introduced to China, the demand for animals used in Chinese medicine was so great that many of them were raised in captivity, obviating the need to hunt them in the wild.

Even today raising scorpions is an important local industry in Honan and Shantung, as are raising snakes in Kuangtung and toads in Liaoning. The deer, almost no part of which is left unused, is practically a domesticated species. ''They harvest the horns each year without having to kill them,'' Chang says.

Rhinos, though, are different. Besides facing extinction, they cannot be kept and propagated in captivity, and because they attack people who approach them, poachers often kill them just for their horns.

''Actually there are a number of drugs to relieve fever these days, and there's no reason we should import rhino horns,'' Chang stresses. Otherwise, by accelerating the rhino's extinction, we will only make its horn even rarer.

Who'll study animal medicines?: In recent years, the agencies concerned have constantly promoted making Chinese medicine more scientific in the hopes of achieving a breakthrough in research so that Chinese medicinal ingredients can play a more important role in the future. More and more researchers are studying herbs and plants, but those who study animal ingredients are few and far between.

One of the reasons is that Taiwan still has an image with many foreigners of being a nemesis to wild animals, and many scholars simply don't want to have anything to do with the subject, lest they be seen as ''uncivilized.'' Taipei Medical College Pharmacology Professor Yang Lin isn't optimistic about the future of animal medicinal ingredients for just that reason.

If we really want to make Chinese medicine more scientific, then a good way to begin is by stopping the slaughter of wild animals and prohibiting the importation of rhino horns. S

(Chang Chin-ju/photos by Vincent Chang/ tr. by Peter Eberly/ first published in January 1989)

用麵粉捏塑可愛的供羊，既不違背佛教忌殺生的原則，又符合祭祀者對神靈的虔誠敬意。（楊文卿攝）

Using flour dough to mold adorable "sacrificial lambs" does not violate the Buddhist proscription of taking life, but at the same time is suitable to express sincere respect to the gods. (photo by Yang Wen-ching)

爾愛其羊‧我愛其禮

Beastly Rites

文‧陳桂芳　圖‧王煒昶

在廟會賽豬公、祭孔拔牛毛的背後，中國人想表達的是什麼樣的心意？

鄉間小廟裡，救苦救難觀世音菩薩案前除了菓蔬糕餅，赫然出現鷄鴨牲禮。懂規矩、知禮數的人看來不免大驚：佛教忌殺生，何來葷腥供品？

可是很多臺灣民間信仰佛道不分；一些善男信女手頭闊綽，不免傾其所好以敬神，心意既在，也就不必太見怪了。

不食人間煙火？

在敬天法祖的中國人生活裏，祭祀一直扮演著重要的角色。民間對於祭什麼神，用什麼物品，也約定俗成，自然累積出一套規矩禮數。香、酒、金紙、祭品，都是祭祀中不可或缺之物；逢到年節、神誕等重要場合，又以「牲禮」顯其隆重。

中央研究院院士、人類學家李亦園指出，祭神時用不同的祭品是有相當深的含意，表達出祭祀者對不同神靈「親疏遠近」的關係與感情。

民俗儀式雖然繁複，但李院士表示，從來辨識神祇親疏遠近，有兩項基本原則——「全部」與「部分」、「生」與「熟」。

他指出，民間一般用「全」來表示最高的崇敬與最隆重的行動，而肉塊切得愈小，尊敬的程度隨之降低；以外，又用「生」來表示關係的疏遠，用「熟」來表示熟稔（見圖表）。

比方說，拜天公一定要殺豬公，且殺的豬公一定要整隻敬供，表達最高的敬意；同時以未經烹煮的生食敬奉不食人間煙火的天神，也反應了「天高皇帝遠」的遙遠關係。

大小、生熟皆有禮數

祭祀一般神明像媽祖、祖師、王爺、千歲等等，祭品可用五牲（豬頭、鷄、鴨、魚、蝦），或略作三牲（豬肉一塊、鷄、鴨各一）。但不論是三牲或五牲，大多不完整；特別是獸肉，都是一大塊，即使是鷄和魚，也不必一定全隻上供。而這些不全的供品，在祭供之前會先稍加烹煮，表示對天公以下、各種經常直接控制人間禍福的神祇較次一等的尊敬，比較天公更爲親切的關係。

Behind the slaughter of boar in temple festivals, and plucking ox hair as an offering to Confucius, what are the meanings Chinese are trying to express?

In the little rural temple, accompanying the appeals to Kuan Yin to come to relieve the suffering in this world, besides the usual fruits, vegetables, and baked goods, there surprisingly appear offerings of poultry. It seems those in the know about the ritual and the rules cannot but be somewhat nonplussed: Buddhism forbids killing living things, so where did these flesh and bone offerings come from?

But many Taiwanese do not distinguish clearly between Buddhism and Taoism. Pious and well-meaning men and women have brought these to show their respect to the gods; since they have the right intention at heart, there's no need to look askance.

Don't hang around the mundane world: In the daily life of Chinese, where it is essential to make deference to Heaven and revere one's ancestors, ritual offering has always played an important role. People have, through a long period of setting rules and cultural change, accumulated a set of rites governing which gods should be sacrificed to, and with what offerings. Incense, alcohol and paper money are all indispensable components. But at important occasions like New Year's, major festivals, or deities' birthdays, the "sacrificial rites" are especially solemn.

Li Yih-yuan, an anthropologist and member of the Academia Sinica, points out that using different gifts to worship different gods has a quite profound inner meaning. It reveals the differences in intimacy and reverence that the worshippers feel toward different spirits.

Although the rites and ceremonies are complex, Li says that there have always been two fundamental distinctions to differentiate the proximity of the divine beings — "whole" vs. "partial," and "cooked" vs. "raw."

He notes that people have always used "whole" to express the most profound reverence and solemn action. The smaller the pieces into which the meat is cut, the lower the degree of respect. Further, "raw" is used to express a more distant relationship, while "cooked" represents a warmer connection (see chart).

至於供奉祖先的祭品，則與家常菜餚無大差別。魚肉大多切成可以食用的小塊，不但煮熟了，有時還加以調味。祖宗原是自家人，煮些他們生前喜愛的家常菜，在懷念敬意中，還帶着親暱的感情。

此外，在中國人祭祀的場合中，常可在桌下或一旁小桌上看到一些米飯菜餚，加上一兩杯酒，這些則是爲過往「小鬼」準備的「便飯」。對於小鬼，自然談不上全與生，也不講究成盤整碗，備些飯菜，表達點人間溫情也就是了。

祖先眞的來吃嗎？

殺豬宰鴨、祭天拜祖，在中國人看來心誠意正，理所當然；但外國人却往往不解。

一位外國學者就曾經當面對李亦園院士提出質疑：「你們中國人眞的相信祖先神明，會來享用你們準備的食物嗎？」李院士當下笑問：「你們西方人眞的相信你們祖先會聞到墓前的鮮花嗎？」

說穿了，鮮花和牲禮，除了宗教上的意義，不過都是一份心意；而以動物爲犧牲，東、西皆然，由來已久，在中國，尤其有着千年不墜的傳統和象徵意義。

根據周禮的記載，中國人很早就用牛、羊、豬、犬等牲畜爲祭品，它們多脂肪、多肉、取得容易，又是四季皆有的動物，人們視爲飲食美味，也就將之敬奉天地神明，藉以祈求風調雨順、國泰民安。

對於牲禮的選擇，禮記的禮器篇說的更詳細——凡是天時所不生，地上所不長，君子不用之以爲禮，因爲那也不是鬼神所想要享用的。例如居高山的人用山上所不生長的魚鼈爲祭品，或是住水濱的人用水濱所不生長的鹿或野豬爲祭品，就不合宜。此外，規定祭祀不用懷身孕的母牛，也顯示了祭祀者的好生之德。

過與不及皆不足取

在古代，牲禮的選用，依奉祀者的身分地位決定，禮數嚴格，過與不及都算失禮。天子用太牢（牛、羊、豕）、諸侯用少牢（羊

祭壇上的牲羊，脖繫紅帶被架成飛奔狀，默默地等待祭典開始。（楊文卿攝）

These rams on the sacrificial platform, ribboned in red and positioned in a gallop, quietly await the start of the ceremony. (photo by Yang Wen-ching)

For example, to worship the Lord of Heaven it is essential to slay a pig, and to offer it intact, to express the deepest respect. At the same time, uncooked food is put forward for the Lord of Heaven — who remains aloof from the mundane events of this world — reflecting the remote relationship revealed in the expression ''Heaven is high, the Emperor is far away.''

There's a rite way to do everything: For sacrifices to the most common deities, like Matsu, Wang Yeh, or Chien Sui, the ''five sacrifices'' (chicken, duck, fish, shrimp and head of a pig) are appropriate, or an approximation can be made with the ''three sacrifices'' (one piece of pork, one chicken, and one duck). But whether it be the five sacrifices or the three, most of the offerings are not intact wholes. In particular, the animal meat must be cut in large chunks; even chickens and fish are not necessarily offered up as single units. These incomplete offerings can be cooked slightly before the ceremony, to express these deities who directly control the ordinary affairs of good or ill fortune in the secular world are at a secondary level of respect, more intimate with the people than the Lord of Heaven.

As for offerings for ancestors, there is no great difference with the foods one commonly keeps around the house. Fish is normally cut into small pieces suitable for cooking, and not only cooked fully, but even with a little spice thrown in. Ancestors are after all one's own family, and cooking up some of the victuals they enjoyed so much in life can add a little more personal meaning and emotion to the respectful remembrance.

Further, on sacrificial occasions for Chinese, one can often see some rice and/or snacks, with a glass of wine or two, below the table where the offerings are placed, or on a smaller table set next to it. These are ''take-out food'' for the wandering ghosts who might come by the ceremony. Naturally one needn't even mention ''whole'' or ''raw'' when it comes to these minor spirits, nor need one be meticulous about setting out a full course dinner; a little wine and vegetables express well enough the warm concern of those still in this world.

Do ancestors really come to eat? To kill a pig or slaughter a duck to make offerings to Heaven or to ancestors seems to be a perfectly natural way for Chinese to express the sincerity of

、豕）；大夫與士，有田的用小羊或小豬，沒田的行薦禮，薦禮的供品是黍麥魚雁之類的四季時鮮。

當年齊桓公宰相管仲用太牢爲祭禮，就被認爲僭越身分，甚至有篡謀之嫌；而歷事靈公、莊公、景公的齊國名相晏平仲用一塊小到蓋不滿碗的豬蹄膀祭祀祖先，當時的人又批評他太小氣了。

在如此看重祭祀的禮數下，被天子、諸侯選中的牛羊牲禮，也倍受呵護。秋天，太宰太祝必須經常探看祭祀用的牛隻，觀察牠們的毛色是否純一、肢體是否完整、所食草穀飼料是否足够，還要注意牠的肥瘦情形、量度體型、顏色、長短，務使其完美無缺。

然而，春天一來，養尊處優的「牛上牛」，終要以自己完美無缺的身子奉上祭臺。這在現代保護動物者看來，不免是個哀怨的故事。做爲一隻毛色純美、出類拔萃的牲牛，究竟是幸？是不幸？

莫作「牛上牛」？

這個千古大問要算莊子看得最透徹。楚威王聽說莊周賢能，派人以重金聘爲宰相。莊子哈哈大笑，指著特使說：「你難道沒看過郊祭裏的牲牛？牠錦衣玉食，以入太廟，到那時，想當個無名小豬，已經來不及啦！」

孔子的得意門生子貢，也對祭典中的牲羊提出過疑問。老夫子的回答是：「賜啊，你憐惜的是那隻羊，在我看來，該珍惜的，乃是其中禮數啊！」

孔夫子顯然一語中的。鷄鴨牛羊，本爲人類畜養食用；選其美者，敬天祭祖，的確不必見怪。

根據「呂氏春秋」的記載，中國上古有名的好皇帝湯，曾經因爲國內大旱，五年不收，而剪下受之父母的頭髮，以木枷十指，發願以自己爲犧牲，祈求上蒼降雨。不久，果然天降甘霖，萬民大悅。

由此看來，中國人以犧牲祭天敬祖的心意，不過是傾其最爲佳美者；逢到關鍵，也不惜以身請命，原沒有欺負畜牲的意思！

（原載光華七十八年一月號）

千斤豬公，身披紅采、口含鳳梨，以贏神靈歡心，也令過往遊客嘖嘖稱奇。

A thousand catty (about 600 kilos) pig, draped in red and mouth with pineapple, is used to win the good will of the gods; it also gives passers-by an unusual thrill.

their sentiment. But foreigners don't always see it that way.

One foreign scholar once confronted Li Yih-yuan with the following query: "Do you Chinese really believe that your ancestors and gods will come to enjoy the food you have prepared?" To which Li laughingly responded: "Do you Westerners really believe that your ancestors can smell the flowers you put on their graves?"

When you get right down to it, flowers and sacrifices, besides differences in religious connotation, carry the same heartfelt meaning. And using animals as offerings has been known in both East and West. In China, in particular, they also carry thousands of years of unbroken tradition and symbolic meaning.

According to records of Chou dynasty rituals, Chinese very early on were already using cows, sheep, pigs, and wild animals as oblations. They had more fat and more meat, and were easily accessible; moreover animals were available in all four seasons. To top it off, people find them tasty. Having all these superior traits, they are fit to express respect to the gods and spirits, to seek steady rains and favorable weather, and to pray for the safety of the country and the tranquility of the people.

As for the choice of sacrificial items, the *Book of Rites* gives refined instructions, admonishing the proper gentleman what not to use and when in making sacrifices, because certain offerings may not be what the gods or devils want to enjoy at that given time. For example, for people of the mountains to use fish as an offering, or for people of the coast to sacrifice a mountain boar, would be inappropriate. Moreover, it is stipulated that pregnant females are not to be used in offerings, indicating the moralistic respect for life of the supplicants.

Too much, too little, too late — too bad: In ancient times, the sacrificial rituals were determined by the rank and status of the supplicant. The rites were rigorously prescribed, and overdoing or underdoing it was seen to be a breach of propriety. The Emperor would use greater beasts (cattle, sheep, and pigs); dukes and marquises could use lesser beasts (sheep and pigs); landed gentry could use small sheep or pigs; those without land were stuck with oats, wheat, fish, or wild fowl, whatever happened to be in season.

At one time Duke Huan of Chi, the prime minister, used greater beasts in his sacrifice, which was seen as exceeding his proper position, and even suspected of being usurpation. But when leading nobility of the state of Chi used only a pig's thigh that could not even cover a plate as an offering to their ancestors, people at that time criticized them for being miserly.

Given the seriousness with which sacrifices are thus taken, the cattle and sheep selected for sacrifice by the Emperor or the royal household were given excellent care. In autumn, the imperial ritemeisters would check up on the care and feeding of the cattle for sacrificial use, to see whether their external color was not consistent, whether their limbs were not intact, and whether they had enough fodder to eat; they would even check their weights and measure their dimensions to insure they would be defect-free.

Nevertheless, once spring came, these "cattle among cattle" would still have to give up their perfect bodies in ritual sacrifice. From the point of view of modern animal conservationists, this seems to be a tragic ending to the tale. If you were a supreme creature of your kind, literally a cut above the herd, would you deem yourself fortunate? Or ill-fated?

Don't be so exceptional: This age-old question was seen most clearly, it seems, by Chuang Tzu. King Wei of Chu heard that Chuang Tzu was an able and virtuous man, and sent an emissary with gold to ask him to serve as premier. Chuang Tzu, a man with little regard for earthly treasures, laughed and, pointing at the emissary, said, "Haven't you ever seen the sacrificial bull? He is given the finest in food and lodging so that he might enter the imperial temple — at that moment he would rather be an anonymous pig, but by then it's too late!"

Confucius' proud student Tzu-kung asked a similar question about the animal to be sacrificed. Confucius' answer was: "In your eyes that sheep in pitiable; in my eyes what should be treasured is the rite of which it is a part."

Confucius hit the nail on the head. Chicken, ducks, cattle, and sheep have always been used as food by human beings. There should be nothing strange about choosing the finest to offer to Heaven.

According to the *Chronology of Lu*, the legendary emperor Tang was willing to offer his own hair and fingernails as a symbolic sacrifice of his own body to end a terrible drought that had lasted for five years. Not long after, as expected, Heaven delivered rain, bringing joy to the people.

From this we can see that a sacrifice to Heaven or to one's ancestors is for Chinese a most beautiful thing. To get to the heart of the matter, using a body to appeal to fate has never been in any way intended to "mistreat the poor savage beast"!

S

(Chen Kwe-fang/photos by Wei C. Wang/ tr. by Phil Newell/ first published in January 1990)

別鑽犀牛角尖

The Horns of a Dilemma

文・張靜茹　圖・張良綱

從關渡自然公園的設立，到黑面琵鷺棲息地的爭論，我們的生態保育工作，一直是在保育與開發孰重的爭論下進行。

近來國外保育團體指控臺灣為「犀牛終結者」，則為保育掀起了一場全新的保育與民族文化的戰爭。

犀牛角事件眞的只是高舉生態保育旗幟的國外環保團體，無權干涉我們的文化傳統嗎？

From the controversy over setting up the Kuantu Preserve to the spat over the black-faced spoonbill, wildlife protection in the R.O.C. has long been caught in the dilemma of "which is more important: conservation or development?" A British wildlife protection group recently accused Taiwan of being a "rhino terminator," sparking a brand-new conflict of "animal protection versus nationalism and culture."

Is the rhinoceros horn incident really just a case of "foreign environmental groups trying to interfere in our cultural heritage by flaunting the banner of ecological protection"?

犀牛的保護在廿世紀的保育工作上有重要意義，絕不只是為什麼皮草可穿，牛肉可吃，為何不能以犀角入藥的問題。

Protection of the rhinoceros has been a significant part of the wildlife protection work of this century. It's definitely not a matter of "people wear fur and eat beef, so why can't rhino horns be used in medicine?"

對大部分國人而言，犀牛角事件只讓人覺得，民族情緒與文化保存的大旗飄來揚去、威風凜凜，但到底發生什麼事，恐怕仍然「莫宰羊」。

這輩子不說沒吃過，犀牛角長什麼樣子，可能也還沒福份親眼一睹。政府官員不是說我們早就不用犀角了嗎？反正我們就是莫名其妙給人臭罵了一頓，然後呢？

保育與文化的戰爭

事實上，十五年前，附屬聯合國之下，有一百多個會員國的「華盛頓公約組織」，已將犀牛與其他三百多種瀕臨絕種野生動物，詳列在「華盛頓公約」中的第一類名單之中，嚴禁各國進行任何狩獵、交易，或利用。

偏偏近年來因石油致富的中東回教國家葉門，以有犀皮製的小刀爲地位象徵，對犀皮的需求日多。而一向以犀角爲貴重藥物傳統的東亞國家，在經濟起飛後，也有更多人開始購買犀角，造成走私市場旺盛，犀牛仍不斷遭盜獵。

五年前，華盛頓公約組織遂採取在動物保育上，史無前例的作法，要求各國銷毀國內所有犀角存量，希望能因此完全禁止世人使用犀牛相關產品，以維持犀牛族羣的延續。

但此舉卻並未解決犀牛走私的問題。

「你能把歐洲豪門宅第裡囤積的象牙、犀牛頭壁飾，全部收集來燒光嗎？」屏東技術學院森林資源系副教授裴家騏說，同樣的，要長久使用某一動物的民族，立刻改變傳統使用習慣，更不可能。

完全禁絕或永續利用？

尤其如今世界上大部分的野生動植物，都集中在工業化程度較低的貧窮國家，動植物往往是他們最重要的生存資源。

近年來聯合國所屬的保育機構，甚至各國主管保育工作的部門，也發現過去立法、採取嚴厲手段，禁止稀有野生動物的交易、使用，結果動物族羣仍然不斷減少，消費市場依然存在，保育策略無法落實，因此也不斷在修改保育方向，開始重視資源擁有地原有的使用習慣，與當地人的想法。

因此，國際間雖普遍關切犀牛，卻並不代表所有國際保育團體，就都贊成英國環境調查基金會，對我國採取的激烈做法。

對生態保育工作，保育團體間也各有不同的理念，採取的方法與手段也不盡相同。

理念不同，地位相等

今天國際上的生態保育團體，常被大分爲兩種，一是以積極的行動支持完全禁絕利用瀕臨絕種野生動物；另一種保育團體卻認爲，保育工作並非一蹴即成，因此深入當地調查、了解，與當地民間團體、政府或學術單位合作，進行長期、有計劃的保育工作。

「激怒」國人的環境調查基金會，被歸類爲前者，而來臺設立分支機構的「人猿基金會」，則屬於後者。在國際上，這兩種性質的團體並無高下之分，環境調查基金會曾要求一百多家航空公司，不再載運野生鳥類，就被視爲保育工作上的一大貢獻。

完全禁用派與長期計劃型的保育團體，其實有互補的功能，「這兩種團體一樣重要，」人猿基金會亞洲分會執行長斐馬克說。

尤其永續經營野生動物的方式，雖然逐漸受到認可，但動物穩定繁衍的數量既已被打破，並不是每一種動物都能任人隨心所欲，就妙手回春，恢復原貌，達到永續經營的理想。許多動物，如貓熊，人們費盡心機爲之配對，卻難有成果。犀牛也是一例。

調整態度或成爲館藏

犀牛出現地球四千萬年的歷史中，種類已由三百多種，衰減到今天只剩五種，總數也由上個世紀的難以計數，降到一萬頭以下。其中黑犀牛的數量，根據最新的數字顯示，更慘跌到三百以下，在保育學者所畫的圖形中，是一條直落曲線。

斐馬克舉例說，同樣瀕臨絕種，也被列爲第一危機線上的大象，由於環保人士的推動，與產象國如辛巴威的努力經營，族羣已在恢復之中。因此象牙也被地主國「翻案」，要求華盛頓公約組織，准許他們可以出售一

For most Chinese, the rhinoceros horn incident was a matter of unfurling the grand banner of cultural pride and national sentiment — without really knowing what the issue was all about. Most of us have never actually seen a rhino horn, much less ever taken any as medication. Haven't government officials said that we stopped using rhino horns a long time ago? We've been tongue-lashed for no good reason — now what?

Culture vs. ecology: In fact, rhinos, along with more than 100 other species, were listed as a category-one type of endangered wildlife species, for which hunting, trade or consumption is strictly prohibited, by the Convention on International Trade in Endangered Species (CITES), signed by more than 100 member states of the United Nations, 15 years ago.

In later years, unfortunately, the demand for rhino hide picked up in Yemen, an oil-rich Middle Eastern country where rhino hide daggers are status symbols. And in East Asia, where rhinoceros horns have traditionally been prized as a medicinal ingredient, the economic takeoff enabled more people to buy them, creating a rampant market for smuggling and poaching.

Five years ago, CITES took the unprecedented step of demanding that every country destroy any and all stocks of rhino horns within its borders. It hoped in that way to prevent the people of the world from using any rhino-related products and thereby help maintain the continuation of the species.

But that step didn't solve the problem of smuggling.

"Do you think it'd be possible to collect and destroy all the ivory decorations and rhino horn ornaments in the mansions and palaces of Europe?" asks Kurtis Pei, an associate professor of wildlife biology at National Pingtung Polytechnic Institute. It's just as impossible, he maintains, to ask a people who have made use of a certain animal for a very long time in a certain way to suddenly change their traditional habits.

Complete prohibition or sustainable use? In particular, wild plants and animals in the world today are mainly concentrated in poor, non-industrialized countries, where they may be one of the countries' most important resources for survival.

In recent years, as wildlife protection groups and agencies under the United Nations and in various countries have found that rare animals have continued to dwindle despite the strict laws and measures adopted to prohibit their trade and consumption and that markets continue to exist, they have begun to revise their approach and attach greater importance to the habits and thinking of people in the areas concerned.

As a result, even though concern about rhinoceroses is shared around the world, that doesn't mean that all international protection groups approve of the extreme measures adopted towards the R.O.C. by the Environmental Investigation Agency (EIA) of Britain.

Wildlife protection groups differ in concepts and methods, and the measures they resort to are not all the same.

Different in thinking but not in status: Wildlife protection groups around the world today generally come in two kinds: One takes radical action to support the total prohibition of trade in endangered species; the other believes that wildlife protection cannot be accomplished in a single step and tries to study and gain an understanding of local conditions and works with local agencies or scholars in order to carry out long-term, systematic protection.

The EIA that infuriated Chinese in Taiwan can be categorized as belonging to the first type, while the Orangutan Foundation, which has set up a branch in Taiwan, belongs to the second. Neither of these two types is considered higher or lower in status than the other. The EIA once demanded that more than a hundred airline companies stop carrying wild bird species, and its success is considered a major contribution to wildlife protection work.

In fact, the two types of groups supplement each other in effect. "Both are equally important," says Marcus Phipps, Asian regional director of the Orangutan Foundation.

In particular, even though the validity of wildlife husbandry for sustainable use is gradually being affirmed, herd populations are too low in some species for stable propagation, and not every species can be restored as desired to its original state. Rhinos are an example.

Let's change our attitude or they'll end up in a museum: Rhinos have a history on the earth of 40 million years, yet the number of species

中醫使用犀角治病的量極其微小，學者認爲嚴禁走私犀角，但開放國內庫存量供中醫使用，是旣保護犀牛，也保存我們傳統文化的最好方法。
Only a tiny amount of rhinoceros horn is used in preparing Chinese medicine. Scholars believe that strictly prohibiting the smuggling of rhino horns while deregulating the use of domestic stocks in medicine is the best way to protect the rhinoceros and to preserve our traditional culture.

定量的象牙。但黑犀牛就沒有大象來得「幸運」。

保育工作的推動，確實需要考慮文化的差異，否則很難進行。但今天人們也需要考慮到，人口的劇增，已使人類對動物的威脅，到了進而威脅自身、就連文化也可能隨著動物消失。別說以犀牛角入藥救命，未來恐怕大夥只能帶著兒孫輩，排隊進博物館看犀角，再指手劃脚描述犀牛了。因此今天各民族也需要重新考慮本身傳統使用資源的態度與方法。

這顯然已經不是別人也吃牛肉、也穿皮草，爲何我們不能以犀角入藥的問題了。

犀牛保衛戰

我們的情況也不能與其他開發中國家等同而論。例如去年六月地球高峯會議中，馬來西亞爲自己國內熱帶雨林的使用，與西方國家大開論戰，原來飽受抨擊的馬國總理馬哈迪，也儼然成爲第三世界與西方保育戰的代言人。

森林是馬國的重要經濟來源，且有一套自己的經營、管理方式，能得到已開發國家某種程度的理解與同情。犀牛卻並非我們自己的資源，又非關國人生計。根據調查，臺灣地區百分之九十以上的犀角走私，只是極少

today has dropped from more than 300 to only five, while the total population has declined from countless numbers the last century to fewer than 10,000. The population of black rhinos has fallen below 300, a straight-line drop in the charts drawn by wildlife experts.

Phipps says that herds of elephants, also listed as a category-one endangered species, have been recovering, thanks to the efforts of wildlife protectionists and of ivory-producing countries such as Zimbabwe. As a result, there have been "revisionist" calls among owner countries, demanding that CITES permit them to sell a certain amount of ivory. But the black rhino hasn't been so lucky.

Wildlife protection work must indeed take cultural differences into account or else it will be very difficult to carry out. But people also have to consider that the rapid increase in the human population has reached a stage where we threaten the existence not only of wild animals but even of parts of our own culture. Not to mention the fact that rhinoceros horns may no longer be available for use in medicine to save lives, a day may come where people will have to gesticulate to describe a rhino to their children or line up at a museum if they want to show them a rhino horn. People around the world need to reconsider their traditions, methods and attitudes about using resources.

Clearly, it's not a question of, "other people eat beef and wear fur, so why can't we use rhino horns for medicine?"

Rhino protection battle: Nor can our situation be discussed in the same breath with that of other developing countries. At the Global Summit in June, for instance, Malaysia engaged in a fierce debate with the West over its use of rain forests. Prime Minister Mahathir, roundly attacked, clearly became the spokesman for the Third World in a dispute over ecological protection.

Forests are an important economic resource in Malaysia, and Malaysia was able to win a certain amount of understanding and sympathy from developed countries for having its own methods of operation and management. Rhinos, however, are not a resource of ours and do not involve our livelihood. Surveys show that more than 90 percent of the rhino horns in the Taiwan area that have been smuggled in have been bought as ornaments for vain display or stockpiled to sell at a higher price in the future.

Most countries that are home to rhinos are very poor compared with the countries that consume them. Civil war, disease and famine make local people willing to risk their lives to hunt them. The poachers have weapons, and the police charged with managing and protecting the animals are sometimes murdered, while areas from which rhinos have disappeared are often overexploited and became barren.

Pity the poor rhino: Local governments once tried drastic tactics — cutting the horns off living rhinos — but the horns grew back by about six centimeters a year, and the rhinos continued to fall prey to covetous poachers. Others have suggested sending rhinos to protected sanctuaries, but based on past experience 20 percent die in the moving process. In fact, both methods are inhumane and run counter to the very meaning of wildlife protection.

The fight to protect the black rhino holds an important symbolic significance in the annals of wildlife protection in the 20th century. Smuggling and trading in rhinos is now treated as a crime in most parts of the world, as evil as trafficking in drugs. If all of us understood this background and were better informed of what has gone on in the rest of the world, we might realize why a wildlife protection organization could adopt extreme measures to stop rhino horn smuggling.

Marcus Phipps, who belongs to a wildlife protection group himself, says that even though the EIA is based in Britain, that neither means that only British environmental groups are concerned about the use of rhino horns in Taiwan nor excuses explaining away the dispute as racial discrimination or a conflict in cultural differences.

"Wildlife protection groups in almost every country around the world are concerned about rhinos just as deeply," he says.

The seeds of the dispute: There's no smoke without fire, it's said — if something has gone wrong there must be a reason. Taiwan began to control the importation of rhinoceros horns in 1985, and in 1989 it implemented the Wildlife Conservation Law, which extended controls to all the category-one species listed by CITES. But rhino horn smuggling continued unabated.

The government has burned smuggled animal

人口有增無減，自然被大量開發，保護野生動物變成奢侈的事情，因此今天靠國際力量支援生態的保護，已成爲共識。（鄭元慶攝）

With unabated population growth and the destruction of nature, protecting wild animals is an unaffordable luxury, and so nowadays the consensus is to rely on international help to support wildlife protection. **(photo by Cheng Yuan-ching)**

數人買來當擺飾炫耀，或貯存以備未來缺貨時賣到好價錢。

目前存有犀牛的國家，相對消費國又是極爲貧窮的國家，內戰、疾病、飢荒，使許多人不惜拼命盜獵。盜獵者更擁有武器，使得管理、保護犀牛的警察常遭殺害，犀牛消失的地區，也立卽遭到濫墾，而成不毛之地。

悲情犀牛

當地政府曾使出「殺手鐧」，試圖將犀牛去角圖存，但犀角每年仍會生長約六公分，依然遭人覬覦。也有人建議把犀牛全部遷移至保護區，但過去的經驗是，遷移過程中會有百分之廿死亡。事實上，不論去角或遷移，也都違反人道，失去保育的意義。

黑犀牛的族羣保衞戰，在廿世紀的保育工作上，有重要的象徵意義。今天犀角的走私、交易，國際上就將之等同於國際販毒一樣罪大惡極。放眼天下，如果大家都能得到充分的國際資訊，理解這樣的背景，就可以了解，爲何一個保育組織，會針對犀角走私地採取激烈的手段了。

針對犀牛角事件，同屬國外保育團體成員的斐馬克認爲，環境調查基金會總部雖然在英國，卻不代表只有英國環保團體關心臺灣地區使用犀牛角的情形，大家不需要將之解釋成種族對立或文化差異的衝突。

「因爲幾乎全世界每個國家的環保團體，對犀牛都有相同深度的關切，」他說。

無風不起浪，事出必有因。臺灣地區在民國七十四年開始管制犀角的進口，七十八年「野生動物保育法」實施，也將華盛頓公約組織中第一類的動物列入管制。但犀角走私卻未曾中斷。

早種禍因

五年來，政府雖然焚燒過五次含犀角在內的走私獸體，但從未抓到過任何走私客。反而南非政府不斷在當地截獲寄往臺灣的走私犀角；前年更抓到兩個臺灣走私客，並取出寄往臺灣的包裹中，藏著一百多隻犀牛角。

包括環境基金會在內的許多國外官方與非

remains including rhinoceros horns five times in the past five years, but it has never caught any smugglers. South Africa, on the other hand, has frequently intercepted smuggled rhino horns destined for Taiwan. In 1991, two smugglers headed for Taiwan were caught with more than 100 rhino horns in their luggage.

Many official and unofficial protection groups from overseas, including the EIA, have come to Taiwan hoping to gain a better understanding of the smuggling and rhino horn situation here. Local officials have repeatedly stressed that we have a wildlife protection law, but the visitors still see horns displayed in many Chinese medicine shops. Chi Wei-lien, member of the board of directors for the Asian region of the Orangutan Foundation, says frankly, ''The seeds of today's rhino horn dispute were planted early on.''

When the members of CITES met in March in Japan, some countries called for a boycott of Taiwan products to protest the continuation of rhino horn smuggling here. The suggestion was finally vetoed, but it represented a strong protest by the world community, and local scholars who attended the conference delivered a serious warning on their return, a warning that unfortunately didn't receive much popular attention or media coverage.

In November, the activist group EIA finally launched its campaign in Britain. After the incident exploded, Chinese were wrapped in a haze of nationalist sentiment, and the causes and effects of the matter were lost in the fog.

The last straw: It's true that it was Western hunters who created irreparable damage by decimating rhino herds in the first place. But today the West is reflecting on its actions of the past. The time may be a little late, but when rhinos have reached a stage of near extinction, ''do we want to be the final executioners?'' asks Fang Chien, president of the Green Consumer Foundation.

The EIA hasn't aimed its barbs only against Taiwan. Radical wildlife protection groups have taken up the fight to save endangered species in places around on the globe, including their own countries. Examples are efforts to stop the killing of whales in Japan, of sea turtles in France and of seals in Canada.

Beyond endangered species, fox hunting is frequently condemned in England. Even more extreme

官方保育團體，都曾來過臺灣，希望了解國內犀角的情形。國內保育官員一再強調，我們已經有了保育法，但他們仍然見到許多中藥店擺設犀角。人猿基金會董事祈偉廉不諱言地說：「今日的犀角事件，早就種下禍因。」

去年三月，華盛頓公約組織在日本開會，就有國家針對臺灣仍有人從事犀角走私，要求會員國對此進行抵制臺灣產品。雖然提議最後被否決，但等於是國際上共同對我發出嚴重抗議。參加會議的學者，回來後曾對此事提出嚴重警告，可惜並未受國人重視與媒體青睞。

十一月，積極的保育組織環境調查基金會，終於在英國展開攻勢。事件雖爆發，國人卻瀰漫在民族情緒之中，前因後果依然墜在五里雲霧。

壓死駱駝的最後一根稻草

過去西方在非洲的狩獵行爲，確實對犀牛造成不可彌補的傷害，族羣大傷元氣，但今天西方也在反省自己過去的行爲。也許爲時稍晚，但在犀牛已到了油盡燈枯的階段，「我們是否還要做最後的劊子手呢？」綠色消費者基金會負責人方儉說。

環境調查基金會也絕對不只針對臺灣而來。今天地球上幾乎不時都有激進的保育組織，以瀕臨滅絕動物爲議題，在世界各地，當然也包括他們自己的國家，掀起稀有動物保衞戰。例如日本有殺鯨事件、法國的海龜事件、加拿大的海豹事件。

除了瀕臨絕種的動物，像英國的獵狐行爲，也常遭致嚴厲的抨擊，更激烈的是英國動物權組織，反對動物實驗，甚至企圖謀殺動物實驗科學家。

有人指出，但是這些保育組織不會拿其他國家的外貿產品開刀。

「因爲他們必須針對當地國的情形，做最有效的訴求。對臺灣，這一招最能達到效果，在其他國家，可能就得換招數，」一位保育人士解釋說。

正如臺大大氣科學系主任林和所說，今天我們就像「壓死駱駝的最後一根稻草」。所有人的目光都放在最後一根稻草身上，只能說我們很尷尬，「恰好在犀牛快絕種的這段時間很富有，恰好國際上生態保育思潮已然成熟，恰好又是以外貿爲導向的國家，」他說，是很寃。

欲速則不達？

其實並非沒有國際保育團體，考慮到國人使用犀角的傳統習慣和文化背景。

針對犀牛角的走私與使用，分支機構遍佈世界各洲的「國際動植物交易調查委員會（Traffic）」，也在四年前開始與我國保育團體接觸，並和國內生態學者，共同進行中醫藥界犀角存量的調查。

獸醫師祈偉廉正是這份調查主要成員之一，他表示，現階段，犀角的走私絕對應該禁止；但對於現有的犀角存量，從事調查的學者，卻不認爲應該如國外環保團體所說的，即刻焚毀殆盡。

殺頭的生意有人做，一旦存貨燒光，恐怕立卽會有大批犀角走私進來，塡補這一眞空市場。對呈半飢餓狀態的犀牛產地，不可否認，也會有人爲此賣命，反而助長犀牛被殺戮。

今天化學合成藥物帶來許多副作用，使西方也在回頭尋找、利用傳統的動、植物藥材；接受調查的中醫藥師，也沒有人願意犀角這一古老的藥方消失，因而表示願與學界合作。

「牽涉走私的人是少數，政府甚至可以釋出沒收的犀角來降低市場價格，讓中醫藥界合法使用，也延長了國內庫存量的使用時間，」祈偉廉說，既然連法律禁止都無法斷絕走私，不如先不要禁止存量的使用，而和中醫師們合作，了解國內的存量之後，才容易進行管理。

這樣的態度和做法，正是學界認爲兼顧物種與我國文化保存的最佳方法。

也因此，當環境調查組織以激進的手段，掀起犀角事件之後，農委會卽刻宣布，犀角及其相關產品，由過去的管制使用（必須向

are some animal rights groups that oppose vivisection and have even tried to murder animal lab scientists.

Some people say, however, that these groups don't try to use another country's exports as a weapon. ''You've got to make the most effective appeal depending on the local circumstances. In the case of Taiwan, that's the ploy that works most effectively. In the case of another country, maybe you'd have to use another tactic,'' a wildlife activist explains.

It's just as Lin Ho, head of the department of atmospheric sciences at National Taiwan University, has said: What's happening now is that Taiwan has become ''the straw that broke the camel's back.'' Everyone focuses attention on the last straw, putting us in an embarrassing situation. ''We just happen to be prosperous at a time when the rhino is facing extinction. The wildlife protection movement just happens to be in full flood. And we just happen to be the number one nation in terms of foreign exchange reserves,'' he says, sounding wrongly abused.

Haste makes waste?: In fact, there are a few wildlife protection groups that do take into consideration our cultural background and traditional use of rhino horns. Aimed at rhino horn smuggling and consumption, the Trade Record Analysis of Flora and Fauna in Commerce (TRAFFIC) of the World Wildlife Fund began to get in touch with local experts and wildlife protection groups here four years ago to carry out a survey of rhino horn stocks used in Chinese medicine.

Veterinarian Chi Wei-lien, one of the main members of the survey team, says there is no question that smuggling should be stopped. But the scholars don't believe that current stocks should be burned, as called for by CITES.

There are always a few unscrupulous people where there's money to be made. If the stocks are burned, the scholars are afraid that large batches of horns will be smuggled in to fill up the vacuum. People in the half-starved countries that are homes to the rhinos will undoubtedly risk their lives to supply the horns, and the effect will be further slaughter.

None of the pharmacists surveyed wanted to see rhino horns -- this ancient medicinal ingredient — disappear, and all expressed a willingness to work with the experts and scholars. ''Very few people are involved in smuggling,'' Chi says. ''The government should legalize the use of stocks by practitioners of Chinese medicine. It could even release the horns it has confiscated to lower market prices.'' Since criminalization has failed to stop smuggling, it would be better, he says, to legalize the use of stocks and work with the Chinese medical community to gain a better understanding of the situation and manage stocks more easily. That kind of attitude, scholars believe, is the best way to look after both the animals and the preservation of Chinese culture.

When the EIA set off the controversy with its radical action, the Council of Agriculture promptly announced that the use of rhinoceros horns and related products would be completely banned rather than allowed on the basis of a prior application, as previously. Chi worries that that approach will only force the practice underground, making future control even more difficult.

Even if no hurt feelings were involved, looking at it strictly from the standpoint of wildlife protec-

如果人們可以將黑熊養得肥胖慵懶，大量繁殖，就是將之當成豬一般宰來吃也無妨。但人與動物的關係就僅止於吃的層次嗎？

Even if we could raise black bears as plump as pigs and slaughter them for the dinner table, does the relation between man and animals stop at the level of pleasing the palate?

農委會申請許可），進而為完全禁止使用。祈偉廉懷疑，此舉是否會讓國內犀角的使用完全走入地下，使未來的管理更加困難。

即使不牽涉任何情緒，就保育論保育，「我也看不出環境調查基金會的保育策略在哪裏，」裴家騏直言。

認養犀牛？

事實上，這一份調查報告雖由外人提議進行，主要調查人卻是我們的學者。主管保育工作的農委會也間接得知，並曾提議做經濟上的支援。祈偉廉認為，政府可以大大方方的表示，我們讓國外環保組織在此設機構、做調查，還願意支持這樣的報告，難道我們管理犀牛的誠意還不够？

今天國際上已了解到，要求貧窮的非洲國家獨立保護自然資源，並不公平，過去已開發國家使用、享受最多資源，今天地球上僅剩的自然資源，既然是全人類共有的維生系統，那所有的人都有義務盡一分責任。

地球高峰會議之後，這樣的作法更成為各國的默契，被視為「經濟動物」的日本國民，就是捐款、認養熱帶雨林最多的國家之一。我們未嘗不能這樣做，政府就可以委託一個基金會募款，來援助犀牛的保育工作。

保育也要長大

環境調查基金會的做法與態度，雖遭同為保育團體的人士批評，「但要證明他們錯了，還是要靠我們自己的表現，」環保團體負責人方儉說。

臺灣在經濟上的表現令人刮目相看，又急於重返國際舞臺，別人自然也會睜大眼睛，看我們有什麼樣的成績，能不能遵守國際舞臺的遊戲規則？

在保育上，我們的確需要再加一把勁，迅速吸收國際資訊，也把我們自己的文化傳統、保育誠意和實際行動，傳播出去，做到良性的交流，以免犀角事件重演，既傷了和氣、耗了元氣，還冒下一味傳統中藥可能從此消失的危機。

（原載光華八十二年一月號）

tion, "I don't see the point in EIA's strategy either," Kurtis Pei says frankly.

Adopt a rhino? In fact, even though foreigners suggested the survey in the first place, its main members are scholars from Taiwan. When the Council of Agriculture, which is in charge of wildlife protection work, found out about the survey, it suggested providing economic assistance. Chi believes the government should come right out and say that it will allow foreign protection groups to set up branches here and is willing to support the study — or aren't we really sincere about rhino horn management?

Today the world realizes it isn't fair to demand that poor African countries protect natural resources all on their own. Developed countries consumed and enjoyed the bulk of them in the past, but since the resources that remain make up the common maintenance system of mankind, all of us have a duty to do our part.

Following the Global Summit, that sort of practice has become a tacit understanding. Japan, home of the "economic animal," has donated the most money to adopt and save tropical rain forests. We could do something similar. The government could commission a foundation to raise money to support the protection of the rhinoceros.

Wildlife protection also has to grow up: The EIA's methods and attitude may not be employed by other wildlife protection groups, but "if we want to prove they were wrong, we still have to rely on our own performance," Fang Chien says.

Taiwan has drawn international attention for its economic success and is eager to return to the international stage. People naturally look closely at our record to see whether we can play by the rules of the game.

We do need to do more in wildlife protection. We need to create a beneficial exchange, bringing in more information from abroad and putting out the story of our own cultural heritage, the sincerity of our efforts and our practical actions, to avoid a repeat of the rhino horn incident, with its hurt feelings, wasted energy and the risk that a traditional ingredient in Chinese medicine may disappear.

(Chang Chin-ju/photos by Vincent Chang/ tr. by Peter Eberly/ first published in January 1993)

你知道自己也不過是萬物中的一種嗎？ Do you realize that we're animals, too?

天生萬物以養人？

Creatures vs. Comforts

文・張靜茹　圖・張良綱

緊接著犀角之後，魚翅、燕窩這兩味中國人傳之久遠的補品，也被外人提醒，不能再這樣吃下去了；殺老虎、喝虎骨酒，更被環境調查基金會視爲攻擊我們濫殺動物的下一波目標。

以後是不是除了家禽、家畜，其他動物就此一概碰不得了？

「人類使用動物，實在沒什麼不對，」生態學者裴家騏表示，在自然界，動物本來就互相利用、共生，除了人類以外的其它動物也會使用、改變自然環境與資源。只不過萬物彼此制衡，維持在一個平穩的情況下。

要馬兒好，又要馬兒不吃草

但今天人口劇增、自然環境的開發壓力如此之烈，人們一方面要吃、要用，一方面又拚命開發（侵佔？）動物的棲息地。像大象、犀牛、鯨魚這種大型動物，所需要的棲息地最廣，於是成了人類文明開發下的第一波受害者。

又要馬兒好、又要馬兒不吃草，的確不可能。除了大型動物，魚翅、燕窩的來源——某些種類的燕子、鯊魚，數量確實在減少，老虎在野外的族羣也正告急。

除非我們覺得殺光、吃光就算了，管他後代子孫不後代子孫，否則就只好節制使用或根本暫停使用。

但禁止使用，又常牽涉到人們原有生活習慣的改變，也等於要人們大幅度改造原有的文化。「改變文化的保育方式，確實很困難，」裴家騏說。

禁用犀牛就是最好的例子，要改變非洲人民獵食犀牛與東方犀角入藥的傳統，等於同時改造兩個民族的傳統文化，引起的反彈自然很大。

與自然共存

事實上，許多地方資源的改變，主要的罪魁禍首並非當地人，而多是外力的入侵，才在當地造成文化與生態的浩刼。像早年白人入侵北美洲，對美洲野牛展開大屠殺，當時還產生了有名的西部牛仔「水牛比爾」，可以不停發射子彈，一天槍殺四千頭野牛。歐美人士在非洲獵殺各種野生動物，也都是俯拾可得的例子。

相反的，在當地落地生根、長治久安的民族，為了種族的綿延，反而不會對自己的資源做掠奪式使用，而與當地環境維持穩定關係，以達到永續經營、利用的平衡狀態。

今天人們已查覺各大洲許多原住民族的生活方式，都遵循著這樣的法則，而中國古人也深諳此理。孟子梁惠王篇中說道：不要以網口太細的魚網捕魚，以免仔魚也被捕撈殆盡，如此「用之有節，取之有時」，正是最好的永續利用範例。

建立世界性的交易市場

辛巴威的大象則是今天典型的例子。

大象被列在華盛頓公約第一類絕種的動物名單上，明令禁止利用，結果造成五、六個大象產國威脅要退出華盛頓公約組織。後來國際上就同意，如果大象族羣可以恢復穩定，就分配給當地人照顧，並規定限額，可以出售給國外動物園，和進行象牙交易。結果大象在辛巴威的族羣，反有增加的趨勢。

不提必需改變使用習慣，今天，對貧窮國家而言，保護生態是很「奢侈」的事。像非洲一些國家公園，面積有半個臺灣那麼大，許多當地人只會覺得，保護起來給富人觀光，所得利潤又微不足道，根本沒有誘因去保護。

中研院經濟研究所副研究員蕭代基就說，最新的資源經濟學認為，既然世人公認這些資源是不應該滅絕的公共財，就建立世界性的交易市場，讓世人都來支援、保護牠。當地人有利可圖，動物反而能繁衍，無滅絕之虞。

而原有使用這些動物的民族，也無需改變他們的使用習慣。

在歐美國家，許多地方政府出售狩獵執照，然後把收入拿來保護、繁衍動物，供人狩獵，也被視為一種永續利用的作法。

僅止於吃的層次？

換句話說，如果人類眞能成功地養殖犀牛，使無絕種之虞，甚至大量繁衍；那麼，別說以犀角入藥，就算將之當鷄、鴨般宰來吃，也可以理直氣壯；魚翅、燕窩的情況也一樣。事實上，如今市面上的魚翅，一部分根本來自人們養殖的魚類，完全禁止使用並無太大道理。

Following hard on the heels of the rhino controversy, foreigners are also reminding Chinese to stop eating those two traditional Chinese tonic foods, shark fins and sparrows' nests. And the killing of tigers and the consumption of tiger bone wine are being targeted for the next wave of criticism of Taiwan by the Environmental Investigation Agency (EIA). Will it be that in the future, except for basic poultry and livestock, other animals will be completely untouchable?

"In fact, there's nothing inherently wrong with utilizing animals," says ecologist Kurtis Pei. In the wild kingdom, animals constantly use one another as well as coexist. Other animals besides humans will also utilize resources and alter the natural environment. The only thing is that "the ten thousand things" balance each other and sustain each other under stable conditions.

You want to have a pony, but you don't want the pony to eat the grass: But today, with the dramatic expansion of the human population and the pressure on the natural environment so intense, on the one hand people want to have these animals around to eat and utilize, but at the same time are developing (invading?) the habitats of these animals in a helter-skelter way. Large animals like the elephant, the rhino, and the whale need the largest habitats, so they have become the first wave of victims of the spread of human civilization.

You want the pony, but you don't want the pony to eat the grass. That, of course, is a pipe dream. Besides the larger creatures, the number of sharks and swallows which provide the materials for shark fin or bird's nest soup are also on the decline. The number of tigers in the wild has also reached a crisis point.

Unless we say that it makes no difference whether we kill them all and eat them all, and to heck with future generations, then there must be restrictions or an outright stoppage on their use.

But completely prohibiting their use involves altering established patterns of human behavior, and is equivalent to asking people for a major change in their own culture. "Conservation methods which imply altering cultures are difficult indeed," concludes Kurtis Pei.

Banning the use of rhino horns is the best example. Trying to change the African tradition of hunting the rhino and the Asian tradition of turning the horns into medicines is the same as trying to simultaneously alter the traditional cultures of two peoples, so naturally there has been a tremendous backlash.

Peaceful coexistence with nature: In fact, the greatest offenders in the alteration of resources in many places are by no means the local residents themselves, but external forces which create a local cultural and ecological disaster. For example, when the white man first went to the American west, he massacred the bison in droves, and even produced "Buffalo Bill" who could kill 4,000 head of wild bison in a day. There are also countless examples of the hunting to extinction of wild animals in Africa by European or American "sportsmen."

On the other hand, for the local people who have set down roots and settled in a place, in order to insure their lifestyle is sustainable, they are not likely to make exhaustive use of their resources. The local ecology thus retains a stable relationship in order to reach a condition of a balance of management and utilization.

The Chinese of old were more than familiar with such logic. In the "King Liang Hui" chapter of *Mencius* it is written: Don't use too-fine nets to catch fish in order to avoid trapping the young ones and wiping them out. In this way, "use with restraint, take only in its time" became just the formula for sustainable use.

Create a global exchange market: The elephants in Zimbabwe today make a classic case in point. The elephant was listed in the Washington Treaty as an endangered species, and its use was forbidden. The result was that five or six countries with native elephant populations threatened to withdraw from the treaty. Thereafter there was agreement that if the elephant population could be kept stable, these would be turned over to the local people to care for, and quotas would be set for numbers that could be sold to foreign zoos or for the trade in elephant ivory. The result is that in fact the herds in Zimbabwe are increasing.

Besides altering living habits, today for poor nations, protecting the environment seems like an "extravagant" matter. For example, some of the national parks in African nations can be as

至於虎鞭、虎骨，由於貓科繁殖能力強，今天很多動物園裡的貓科動物，如獅子、老虎，在吃好、睡足，又無事可做的環境下，數量暴增。許多動物園根本不敢再收外人「餽贈」的貓科動物，甚至連熊也因過度飼養、繁殖，只好做「人道」結紮、閹割。

就有學者認爲，營養過剩的國人若眞的還敢吃虎鞭、喝虎骨酒，或吃熊掌、以熊膽入藥，反倒能爲動物園解決煩惱。

但是如果人們一方面大吃、大喝，一方面又不維護動物族羣，別說道義上說不過去，最後就眞的除了雞、鴨、豬、牛，想嘗嘗鮮、換換口味也難。而人與動物的關係，又豈僅止於吃的層次？

誰來扮演上帝？

人類由與自然共存，到剝削自然，如今開發殆盡之後，又希望走回頭路。但環境的改變，已使今天「利用、厚生」的做法，和過去不同。

過去人類是在順自然之性下，進行所謂的永續利用。不捕捉幼鹿、不釣仔魚，如中國人所說的「春田不圍澤，不掩羣」。但自然環境漸失，原有的一點人道情懷，在今天也變成理想高調，無法說服大衆。永續經營，也往往只能在有限的空間下，加入人爲的手段進行。

國外曾經有例子是，由於經營野生鹿羣供人狩獵，結果鹿隻大量繁衍，反而威脅了其他動物的生存。

今天大家其實是把動物的價值，建立在人類認定的價值上，動物已變成人類意志控制下的動物。有一天，可能我們周圍的某些野生動物，雖沒有滅絕之虞，卻都變成了寵物、實驗動物或家禽、家畜，否則也只能在人們劃定的「保護區」內活動，這恐怕是生態學者最不願見到的。

永續經營、利用，可能是人類目前所能找到的、最適合與動物共存的一套方法。但人們眞的願意萬物的存在，只是爲了人類的實用？

（原載光華八十二年元月號）

非洲象是現代人永續經營、利用生物的最佳範例。（鄭元慶攝）
The African elephant is the best example of a case of human management for sustainable use. (photo by Cheng Yuan-ching)

large as half the size of all Taiwan. Many local residents can only think that these areas are protected for tourism for the wealthy, while the profits are minimal, so there is no real incentive for conservation.

Hsiao Tai-chi, an associate researcher in the Institute of Economics at the Academia Sinica, states that the most recent versions of resource economics argue that since there is already world recognition that these resources are common assets that should not be driven to extinction, then a global exchange market should be set up to allow people from all over the world to lend support and protect them. Local people thus have a source of income, while the animals can flourish, and there are no worries about eradication. And the people who originally utilized these animals have no need to change their ingrained habits.

In countries of Europe and North America, many local governments sell hunting licenses and use the income for conservation and to propagate life. This provides hunters with their prey yet insures that there can be perpetual use.

More than just a meal: In other words, if people could really successfully raise rhinos so that there would be no fear of annihilation, and perhaps they might even flourish, then don't even mention using them for medicinal purposes — you could even go right ahead and slaughter them without any more thought than people today give to chickens or ducks they will eat. The situation for sharks' fins or swallows' nests is the same. In fact, some of the fins on the market today have been raised in captivity, so a complete ban on their use wouldn't make much sense.

As for tiger penis or tiger bone, because creatures of the cat family have the ability to propagate quickly, the numbers of big cats in many zoos — given conditions where they eat their fill, get enough sleep, and have little to do — is growing rapidly. Many zoos simply will no longer accept gifts of large cats from outsiders. Indeed, even bears can be subject to "humanitarian" sterilization or castration because they are overly pampered and are reproducing too quickly.

So some scholars suggest that if people in Taiwan, who are already overnourished, still have the desire to consume tiger penis, drink tiger bone wine, eat bear paw, or turn bear gall into medicine, then this could help zoos resolve their problem.

But if people continue to eat and drink avariciously, and on the other hand do nothing to maintain the animal populations, then — not to mention the humanitarian considerations — in the end there really will be nothing left to eat but chickens, ducks, pigs, and cows. It would then be next to impossible to try something a little different for variety's sake. Even then, is the relationship between people and animals limited to eating?

Who should play god? After having moved from living with nature to exploiting nature to developing it to the point of destroying it, people now want to go back to square one. But the alteration of the environment means that a "utilization and propagation" method must be quite different than in the past.

In the past, when mankind had to adapt itself to nature, there was an inclination to so-called sustainable use. People lived with simple rules like "don't hunt fawns," or "don't catch small fry." But today, it has become a case of mouthing ideals which fail to convince the mass of people. So sustainable management can often only be implemented in limited areas with human intervention and techniques.

There is an example of this from abroad. Because there was management of wild deer in order to insure supplies for hunting, the result was that the number of deer exploded, threatening the existence of other creatures.

Today many people are determining the value of animals on the scale of human values, so that animals have already become things subject to the will of humans. One day, although there will perhaps be little fear of the extermination of the wild animals around us, all will have become essentially pets, experimental animals, or livestock, or otherwise will only be able to live in "reservations" marked out by humans, which is something that ecologists most wish to avoid.

Sustainable management and use are perhaps the methods for coexistence with animals that are at the moment most accessible to and most appropriate for people. But do people really only want to maintain the existence of, as Lao Tse called them, "the ten thousand things," only for our own use?

(Chang Chin-ju/ tr. by Phil Newell/ first published in January 1993)

內神通外鬼？

International Ecologists: Foreign Meddlers and Local Conspirators?

文‧張靜茹

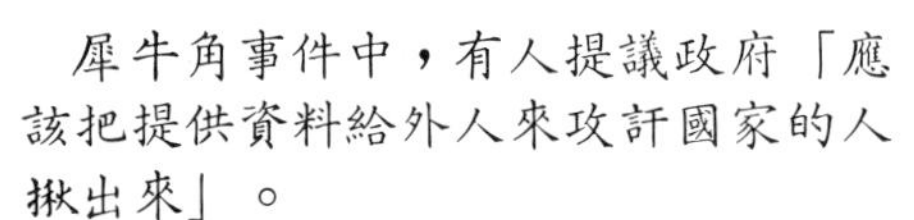

犀牛角事件中，有人提議政府「應該把提供資料給外人來攻訐國家的人揪出來」。

我們的環保工作為什麼得「挾外人以自重」？借助外力一定是壞事嗎？

多年來，經濟一直是我們的發展旗艦，原來就佔去最多資源。對大部分埋頭苦幹、在資源匱乏的情況下，創造出經濟奇蹟的大衆而言，生態保育聽來理想堂皇，但比起醫療、福利、教育、文化，既非切身急事，也絕對比不上常常被拿來與保育「送作堆」、統稱「環保」的環境污染與公害來得有賣點，被作為議題討論的機會因此少之又少。

不只國際資源的維護，本土的生態保育，也常陷入經濟考量的泥沼。規劃了七年的關渡自然公園，就卡在「關渡是大臺北最後一塊可以大展商業鴻圖的地區」，而遲遲無法定案。

「我們整天呼籲保育，其實是沒什麼人聽的，」大部分的保育學者與團體都有相同的感觸。的確，從犀牛、紅毛猩猩到黑面琵鷺，似乎都多少借助了所謂的「外力」，或根本是國外先主動表示關懷，才炒成新聞事件，引起大家關注。

借力使力，不得已也

我們可以說，生態保育在國際上也是新的思潮、一般大衆尚未認知它的重要性；也可以說以前很窮，那裏想得到這些事。但臺灣的富裕，如今也不是一天兩天的事了，在學者看來，答案簡單，國人就是太重視經濟開發，相對的環境教育少受重視。誰又曉得生態保育，其實還是為了人類自己長久的利用、厚生。

事情推不動，國際上又有人關心，保育人士只好借力使力了。落腳臺南縣七股鄉的瀕臨絕種鳥類黑面琵鷺，棲息地被劃為工業區

黑面琵鷺棲息地的保護，在世界各地鳥會紛紛來信聲援後，才廣受國人知曉。(鄭元慶攝)
The need to preserve the habitat of the black-faced spoonbill only came to the attention of local people after a letter-writing campaign by bird associations around the world. (photo by Cheng Yuan-ching)

During the rhino horn incident, people suggested that the government ''ought to ferret out those who helped provide the foreigners with information.''

Why has our country's environmental work ''come to rely on the support of foreigners''? Is it necessarily bad to rely on help from abroad?

For many years now, the economy has been the engine of our development, and it has been absorbing most of our resources. To the people of Taiwan, who have struggled to bring about an economic miracle despite a lack of resources, ecological conservation may sound grand, but it pales in comparison with the importance of medicine, social welfare, education, culture — or even pollution and environmental hazards, which, despite being lumped with conservation as environmental issues, more directly affect people's lives. As a result, ecological conservation's chances of being a focus of attention are smaller than small.

Besides foreign wildlife, ecology in Taiwan itself is often sacrificed at the altar of economic considerations. Planned for seven years, the Kuantu Preserve can't get final designation amid protest that it is ''the last area available for commercial development in greater Taipei.''

''We shout all day about ecological preservation, but no one listens'' is the common complaint of ecological scholars and activists. Indeed, from the controversy over rhino horns to those over the orangutan and the black-faced spoonbill, if foreigners weren't lending a hand, they were the ones whose protest originally attracted the media reports that brought the issue to people's attention in the first place.

Relying on foreigners can't be helped: Conservation is a new concept even abroad, and most people don't understand its importance. You can say, how could people be bothered with these matters before when they were so poor? But Taiwan's affluence didn't arrive overnight. As scholars see it, the problem is simple: Chinese put too much importance on economic development and correspondingly too little on ecological education. As a result, no one understands that over the long term ecological preservation will actually improve people's living conditions.

With stagnation at home and with outsiders who share concerns about these problems, conservationists have had no choice but to draw on outside support.

Hoping to save the black-faced spoonbill from extinction, the R.O.C. Wild Bird Society has exhausted itself protesting plans to make the spoonbill's habitat in Chiku Rural Township of Tainan County the site of an industrial park. Yet not many people care much about the birds or have any understanding of the matter.

The attitude of the Tainan County government is that the county is big: The birds can go anywhere — why must it be the Chiku Industrial Park? Ecologists argue that industry hasn't made full use of the land it has already been allotted. Why must an industrial zone be set up here, on this irreplaceable habitat of birds?

The Wild Bird Society was forced to write letters requesting support from bird organizations all around the world before the question of a home for the black-faced spoonbill finally came to the attention of Premier Hao. ''If it weren't for that,''

，中華民國野鳥學會早就聲嘶力竭，爲琵鷺請命，但鳥人鳥事，又有多少人會想多關心、了解一點，是怎麼回事。

臺南縣政府的反應是，臺南那麼大，鳥兒到處可棲，爲何偏偏要選在七股工業區預定地？保育人士說，工業用地多到閒置，爲何又一定要在這樣一個不可取代的鳥類棲息地設工業區？

鳥會只好寫信求助世界各地鳥會來信支援，黑面琵鷺落脚何處的問題，於是受到行政院關注。「若非如此，臺南縣政府早已開發了七股工業區，」農委會官員說。

無關內神通外鬼

經濟學者蕭代基看來，借助外力作保育並無不妥。過去，我們的各種經濟政治利益已成一穩定平衡的狀態，很多政策是在目前的政經利益狀態之下協調出來。不論在民間或行政部門，生態保育是新領域，很難打破舊體系爭取資源，這時候，如果能聽聽不同的聲音，未嘗不是好事。

除了可以打破已僵化的平衡，媒體的報導、輿論的爭議，也給民衆很好的教育。犀牛角事件之後，現在大概除了掌管保育工作的農委會之外，其他相關單位像抓走私的海關、管理中藥的衞生署，甚至主管外貿的單位，也都會警覺到保育已是國際性的事件，不能等閒視之。

在保育工作上，很多事其實已經無法區分政治上的疆界。

這次，國內一保育團體由於協助國際環境調查基金會在臺灣舉行記者會，曾被誤解成「內神通外鬼」。實際上，國際間的保育團體互相支援，極爲平常。

善用國際壓力

只是國際壓力可以善用，但萬一用的不好，讓地方百姓討厭，也可能使得原本的立意適得其反，而走上回頭路。犀角事件多少就因爲雙方的隔閡，使原本可能「面對面作溝通」的機會，演發成一場外貿抵制與民族尊嚴的混仗。

政治學者劉必榮看來，由於國際環境的丕變，議題的改變，許多問題單靠一國一地無法解決，例如販毒、環保，世界愈來愈小，國家與地區間的關係也愈趨緊密，如果你的問題會影響世界生態，他國當然有權表示意見，甚至出面協商。

因此只要干預是在各國簽定的公約，或遵守某些原則、程序，而非強國對弱國做不合理要求，態度不是帶着強烈種族優越感，各國也逐漸能接受某種特定程度的「干預」。

也有人不平，我們的保育事件，怎麼都是「強國在告訴我們要怎麼做」？綠色消費者基金會負責人方儉表示，如果我們願意，當然也可以去告訴別人該做環保。如今我們就該趕緊昭告天下，我們不買犀角了，也該去告訴非洲國家：「我們不買你們的犀角，不要再盜獵犀牛了」。

靠人不如靠己

事實上，針對環保事件，除非事情已成熟、嚴重到形成國際共識，例如臭氧層破洞，國際共同起而立約，消滅禍首氟氯碳化物的使用量，我們不去遵守都不行；否則卽使有外力介入，保育是否有成果，主動權還是在我們身上，借助外力畢竟是消極的。

但是正如劉必榮所說，隨着國際議題的改變，環保的國際干預會越來越多。「一旦你的事要由別人來告訴你，那是很可悲的，」方儉也強調，不管外力是溫和的或激烈的，都表示我們自己失去了自我療傷的能力，才需要別人來告訴我們該怎麼做，「我們若能做好，根本不需要等別人來干預，」他說。

有人開玩笑，規劃了七、八年的臺北關渡候鳥保護區，如果也能借助一點外力，也許這座國內首座鄰近人煙密集市區的候鳥保護區，早就落成了。但是，能以工業革命發源國的資歷，在倫敦市區中心享受大片綠地的英國，會千里迢迢來告訴我們一座自然公園，對一個像臺北這樣的城市有多重要嗎？如果眞有那一天，那犀角事件恐怕眞的只是個開始，而不是結束。

（原載光華八十二年一月號）

says an official of the Council of Agriculture, "Tainan County would have already developed the Chiku Industrial Park."

They're not meddlers and traitors: Hsiao Tai-chi, an associate researcher at the Institute of Economics of the Academia Sinica, contends that there's nothing wrong with relying on this support from abroad for conservation efforts. Our economic and political situation has reached stability and balance, and in this period many government policies have been worked out cooperatively. For both the people and the government, conservation is a new realm, and it's very difficult to smash the status quo and fight for resources. At such a time, it might be well advised to hear different voices.

Besides shaking things up, media reports and public debate also educate people. Now, besides the Council of Agriculture, which is responsible for ecological matters, other government units like customs, which heads up anti-smuggling efforts, the Department of Health, which oversees Chinese medicine, and even those responsible for foreign trade will all be alerted that conservation is an international matter of great importance.

Ecological issues transcend national boundaries.

Because it once helped the Environmental Investigation Agency to hold a press conference in Taiwan, the Green Consumer Foundation was once unfairly accused of "conspiring with outsiders." In reality, international conservation groups commonly provide mutual support. Yet while international interference can be of benefit, it might backfire if used improperly, making people resentful. Because there was no meeting of minds over the rhino incident, a chance for "face-to-face communication" turned into a melee of trade boycotts and wounded honor.

Making the best use of international pressure: As political scientist Steve Bih-rong Liu sees it, because of changing international circumstances, such problems as the drug trade and environmental protection cannot be resolved in one country or one place. The world is growing smaller. National and regional links are tightening. If one country's problems affect the world ecology, other countries certainly have a right to make their opinions known — and even show up to negotiate.

Hence, if the interference involves what has been covered in international treaties or if certain principles and procedures are respected — so strong countries don't place unreasonable demands on weak ones or adopt an attitude of racial superiority — all nations will begin to accept a certain amount of "outside interference."

But some are shouting foul about these ecological incidents. Why is it, they complain, always "the developed countries telling us what to do?" Jay Fang, president of the Green Consumer Foundation, says we can of course tell other nations to do ecological conservation if we so desire. Today, we ought to declare that we won't buy rhino horns, telling African countries, "We're not buying them; don't illegally hunt them any more."

Self-reliance is best: In reality, if an environmental situation is serious enough to engender common international concern, such as the hole in the ozone layer for which the nations of the world have signed a pact reducing consumption of fluorocarbons, we have no choice but to go along. In other cases, with or without outside support, we've got to shoulder responsibility if conservation efforts here are to be successful. Relying on outsiders is too passive.

But as Liu says, just as the focus of international attention changes so too will international interference in environmental matters increase. "It's a tragedy when others have to tell you about your own problems," Fang stresses. It doesn't matter if the outsiders are diplomatic or boorish — in either case it shows that we have lost the ability to solve problems ourselves and that we need outsiders to come and tell us what to do. "If we do it well ourselves," he says, "there simply won't be cause for outside interference."

People joke that if Taipei's Kuantu Preserve, which has been planned for seven or eight years, had been the beneficiary of a little outside interference, then perhaps the sanctuary, the first for migratory birds near an urban area in Taiwan, might have been built long ago. Will we learn from England, the birthplace of the industrial revolution, where large tracks of green land were nonetheless preserved in central London? Will it teach us the importance of parks to a city like Taipei? If there really is such a day, the rhino incident may just be the start of things.

(Chang Chin-ju/tr. by Jonathan Barnard/ first published in January 1993)

從吸血鬼到蝙蝠俠

From Vampire to Caped Crusader

文・張靜茹

（龍祥電影公司提供）
(photo courtesy of Long Shong Pictures Ltd.)

（美商華納兄弟公司提供）
(photo courtesy of Warner Bros.)

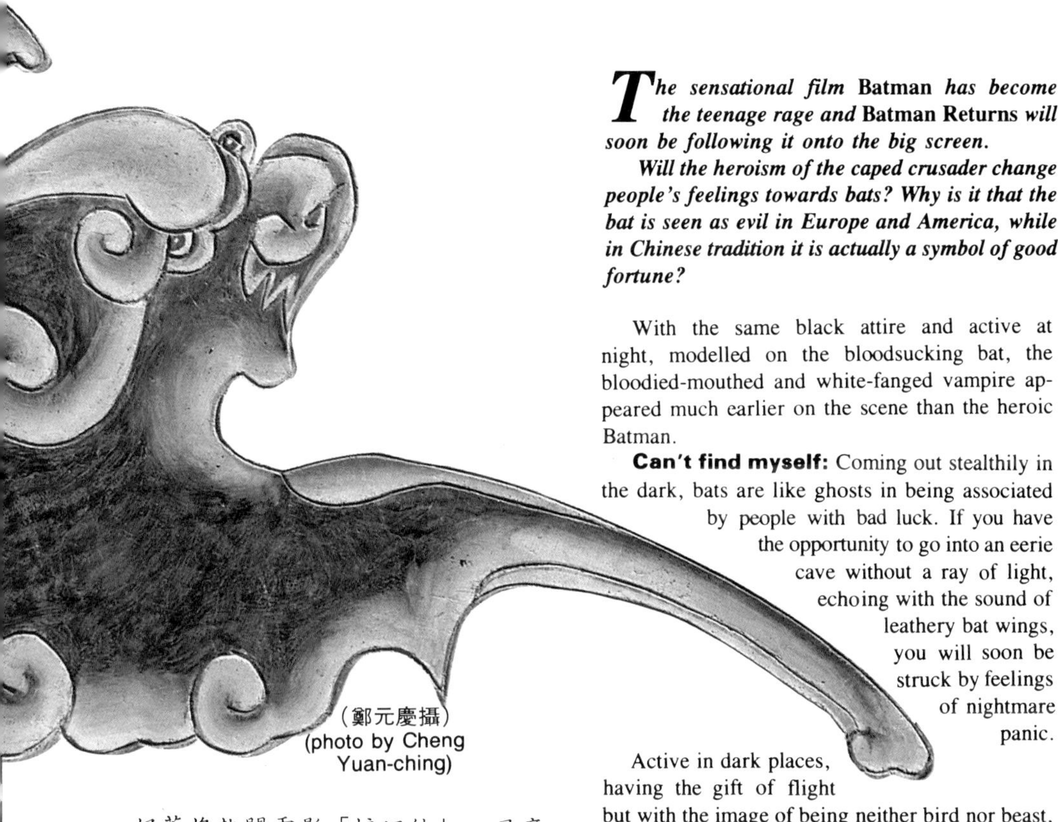

（鄭元慶攝）
(photo by Cheng Yuan-ching)

好萊塢熱門電影「蝙蝠俠」，風靡青少年；暑假將臨，蝙蝠俠第二集打鐵趁熱，即將連線上映。

只是，人們對蝙蝠的觀感，能夠因爲鋤奸扶弱的「Batman」而改變嗎？爲什麼在歐美是邪惡化身的蝙蝠，在中國傳統裡卻是吉祥象徵？

同樣披著黑色風衣、喜愛在黑夜活動，脫胎於吸血蝙蝠、血口白牙的吸血鬼，可比高大挺拔、身穿金屬防彈衣的蝙蝠俠「出道」早多了。

蝙蝠總是從黑暗中悄悄出現，像幽靈般引起人們不祥的聯想。

The sensational film* Batman *has become the teenage rage and* Batman Returns *will soon be following it onto the big screen.

Will the heroism of the caped crusader change people's feelings towards bats? Why is it that the bat is seen as evil in Europe and America, while in Chinese tradition it is actually a symbol of good fortune?

With the same black attire and active at night, modelled on the bloodsucking bat, the bloodied-mouthed and white-fanged vampire appeared much earlier on the scene than the heroic Batman.

Can't find myself: Coming out stealthily in the dark, bats are like ghosts in being associated by people with bad luck. If you have the opportunity to go into an eerie cave without a ray of light, echoing with the sound of leathery bat wings, you will soon be struck by feelings of nightmare panic.

Active in dark places, having the gift of flight but with the image of being neither bird nor beast, in Western culture bats have become a symbol of uncleanliness and danger. In Aesop's Fables the bat plays the role of the one who cannot find his own identity.

In medieval European paintings, the Prince of Darkness and his underlings are shown with bat wings. In the seventeenth century, as the conquistadores colonized South and Central America, tales of attacks by bats soon came back to Europe so that people's impression of the bat became even worse. With the blood-sucking vampire bat being a carrier of rabies, the distance between humans and bats increased even further.

Hung up on vampires? With the appearance of vampire books all kinds of European vampire legends came to be portrayed even more vividly. Horror stories also love to use scenes with bats to

「找不到自我」的生靈

如果有機會走進因凹凸岩塊阻擋，不見一絲陽光的洞穴，陰森的岩洞盡處，往往迴響起蝙蝠拍打牠乾燥皮翼的聲音，惡夢驟醒的驚悸感隨之而來……。

在暗處活動，擁有飛行天賦、卻不屬鳥類的「非禽非獸」形象，蝙蝠在西方成爲不潔和危險的象徵；在希臘伊索寓言中，蝙蝠是找不到自我的角色。

在歐洲，中世紀畫師就把畫中凶殘的魔王及手下，如數添上蝙蝠翼。十七世紀，西班牙軍人在中南美洲遭蝙蝠攻擊的消息傳回，歐洲人對蝙蝠印象愈加嫌惡；而吸血蝙蝠吸食家畜血液，成爲狂犬病的散播者，更令人對所有蝙蝠退避三尺。

與吸血鬼掛鈎？

藉著吸血蝙蝠出現，歐洲各種吸血鬼傳說更被描述的如影如繪；恐怖故事也總愛用蝙蝠串場，才顯得出陰森詭譎。於是「黑夜裡，結滿蛛網的古堡，飛出幾隻兩眼發著燐光的蝙蝠」，每被用來爲各種不可告人的陰謀拉開序幕，蝙蝠聲名狼藉的形象，至此只有聖經中的蛇可與之相比了。

直到「科學昌明」的廿世紀，蝙蝠仍然不得翻身。

黑白電影時代，美國將羅馬尼亞「吸血魔

(右圖)年畫中，喜氣洋洋的中國童子與蝙蝠「嬉耍」。(張良綱翻拍)
(right) Auspicious New Year pictures feature children frolicking with bats. (photo by Vincent Chang)

(左圖)臺中摘星山莊牆上的五蝠圖。雖然古人視善終爲第五福，但在避諱談死的近代社會，五蝠圖常被省略成四蝠。(鄭元慶攝)
(left) Because the fifth bat in ancient designs symbolizes a happy death it is often left out in today's death-shy society. (photo by Cheng Yuan-ching)

create a weird and strange atmosphere. How many scenes open with: "On a dark night, in an ancient cobwebbed castle, out flew some bats with flashing eyes," as a means to add suspense to the plot? All this means that the notoriety of the poor bat is comparable only to that of the biblical serpent.

By the beginning of the "enlightened scientific" twentieth century, the bat had still not been rehabilitated. In the era of black and white movies the Transylvanian Count Dracula hit the silver screen. When people saw a real person portraying the evil vampire carrying out his wicked deeds wherever he went, the speechless bat was even more condemned to hell and unable to make a comeback.

Ten years ago many people in the United States and Europe would clear out bat colonies located near their homes. Take the United States as an example, which likes to see itself as the most conservationist country in history: The bat colony in Austin, Texas, that draws so many tourists today, ten years ago was being chased out by the residents!

It is only natural that today the centuries of deep-rooted myths have still not been dispelled. In China it is the same, with the symbolic significance of the bat still lingering on over the ages of history.

Bad luck in the West, good in the East: When still being seen as evil by European and American artists, the image of the bat was flying into the homes of the Chinese people. Very moving, their bright and brilliant shapes became an omnipresent part of daily life.

No matter whether it be window lattices, clothes,

王卓久樂伯爵」的傳說搬上銀幕，且乘勝追擊，一續再續。人們頭一次見到眞人演出的吸血鬼四處行凶；蝙蝠無言，卻因此更被打入地獄，不得超生。

不過十年前，歐美許多出現在住家附近的蝙蝠羣，還時遭人們惡意驅逐。以自稱從事生態保育歷史最悠久的美國爲例，德州奧斯丁吸引觀光客的龐大蝙蝠羣，在十年前初出現時，當地人就提議趕走牠們呢！

百年神話，根深柢固，至今陰魂不散是很自然的事。正如在中國，蝙蝠所象徵的意義，同樣歷久不衰。

西方懼蝠，東方接「蝠」

當歐洲和早期美國藝術家還以蝙蝠影射邪惡時，蝙蝠身影，早飛入中國尋常百姓家。大量動人、絢爛的蝙蝠造型，遍佈人們日常生活。

不論在花窗、服飾、寢具、椅背或年畫上，總不乏五隻蝙蝠圖，以代表健康、長壽、富貴、美德、壽終（平靜而自然死亡）等五種人生最大的幸福。蝙蝠和桃子、石榴擺在一起，則象徵多子、多孫、多福氣。

親朋好友往來的信箋上，也常見童子仰視蝙蝠飛來的圖案，這若發生在過去的西方，童子大概早嚇跑了。中國童子卻全無懼色、笑臉相向。迎春「接福」，誰不喜歡呢？

爲什麼蝙蝠在歐美是邪惡之兆，在中國卻是吉祥象徵？

中國蝙蝠不寫實

故宮博物院古器物專家楚戈表示，要探究根源，必須由中國古代的宗教信仰追溯起。在比商、周更早的年代，蝙蝠就是中國境內某一民族的族徽、守護神，也就是受族人崇拜的圖騰。

中國各民族拜物的意念和其他民族有很大不同。西洋源流的希臘、埃及文化，崇拜的是寫實的動物與人，中國人卻認爲視覺經驗的事物不值得拜，「老虎再凶，也能克服，」楚戈說，因此人們拜的是龍、鳳等各種世間沒有的幻獸。

而存在世上的動物要成爲物神，則必須具有一些特別的條件，或經過人文觀念加以改造，人們拜的不是實物本身，而是牠代表的觀念與背後隱藏的意義。如此想法的承襲，更深遠的影響是中國美術重寫意、不重寫實的精神。

兼具「禽、獸」二性、生活「日夜顚倒」的蝙蝠，也因此經過轉化成爲守護神。而早期巫術信仰，更迷信對於守護神，看其圖像、念其名字，就能得到與其名相關的東西。

飛入尋常百姓家

蝙蝠展翼形狀與來自「雲紋」的如意相似，是爲吉祥一；晝伏夜出，過去被稱爲「伏翼」，與「福」同音，是爲吉祥二。因此春秋戰國圖騰崇拜日衰，漢朝文化統一後，蝙蝠就由一族的信仰對象，統一爲全民族共同的吉祥符號。

過去僅存於統治階級的崇拜儀式，也日益成爲民間信仰，留在民族意識與日常生活、語言習慣中。尤其到清朝末年，國家積弱，民生凋蔽，人們特別缺乏安全感，求好日子的心更急切，代表福氣的蝙蝠，出現在建築、家具、器物、服飾，與各種民藝品中的機會也就更多了。

不僅配合吉祥話的蝙蝠圖案應有盡有，蝙蝠也由長相凶惡、以驅趕不祥物進門，到化繁爲簡，以幾筆線條簡單鉤勒的寫意造型。

到了現代，由於建築形式改變，製作百物的材料、技術更新，民俗手藝如剪紙、中國結、刺繡的沒落，蝙蝠圖案在日常生活中才逐漸減少。而來自西方的蝙蝠象徵，亦逐漸步入中國社會。

金庸武俠小說「倚天屠龍記」中就出現了一個性情怪異、必須長期吸食人血來療傷的角色——明教四大法王之一的青翼蝠王韋一笑。他輕功獨步天下，受害者遠遠聽到他磔磔的笑聲，脖子已然出現兩個血洞。和西方吸血魔王稍有不同的是，他只攻擊奸邪小人，亦邪亦正，可說是蝙蝠中、西合璧的混血兒。

近來，隨著電影「蝙蝠俠」的上演，印有

bed clothes, the backs of chairs or New Year decorations, a bat design would never be lacking as a representation of good health, longevity, wealth, virtue and a peaceful and natural death — the five most important things in a person's life. The bat, the peach and the pomegranate together symbolize many descendants and good fortune.

Personal note paper for intimate letters would also often feature images of children looking at flying bats. If such a thing had happened in the West, the children would probably have run away in terror as soon as possible, whereas the Chinese children do not show the slightest bit of fear, but are smiling. Who could fail to like these bringers of good fortune?

Why is it that bats are seen as evil in America and Europe but as lucky symbols in China?

Chinese bats are not realistic: Chu Ko, an expert on ancient artifacts at Taiwan's National Palace Museum, says that if you want to get to the bottom of this then you must trace it back to ancient Chinese religious beliefs. In an age before the Shang and Chou dynasties, the bat was the ancestral symbol and guardian spirit of a race living in the area of China, and was thus worshipped as a clan totem.

The significance of the objects worshipped by the Chinese people is very different from that of other races. In ancient Greece and Egypt, the sources of Western civilization, it was realistic animals and people that were worshipped. Chinese people, however, thought that visible objects were not worthy of worship. "The ferocity of the tiger can be overcome," says Chu Ko, because of which people worshipped all kinds of mythical and unworldly beasts, such as the dragon and the phoenix.

For real animals to become spirits there had to be special conditions, or some cultural transformation so they would not be seen as the original objects but rather their underlying ideas and significance. This kind of thinking was most far reaching in its influence on the way that the spirit of Chinese art has been idealistic rather than realistic. With the bat having the nature of bird and beast, and turning day and night upside down, it became transformed into a guardian spirit. Moreover, early shamanism was even more superstitious about guardian spirits, believing that looking at their images and chanting their names could bring about things that were related to their appellations.

Flying into the everyday home: The first reason that a bat shape is auspicious is that the shape of a flying bat's wings is similar to the lucky cloud-scroll pattern of decoration; the second reason is that, sleeping in the day and coming out at night, it was called the "sleeping flyer," which is a homonym of the character for "wealth." In the Spring and Autumn and Warring States periods totemism was on the decline; when Chinese culture was united under the Han dynasty, from being the object of worship of one clan, the bat became the shared lucky symbol of the whole race.

From being a kind of worship that was the preserve of the ruling class, the bat daily became a more popular object of belief that was a part of daily life and language. This was especially true during the late Ching dynasty, when the country was very weak, living conditions bad, and there was an acute lack of security. With the pressing desire for a better life, the bat as symbol of good fortune appeared everywhere on architecture, furniture, tools, clothes, and whenever there was an opportunity for it to appear in the objects of folk art.

Not only was the auspicious bat design seen everywhere it was required, but from a ferocious form that was meant to chase away any bad luck that might come through the door, it became simplified to a bold design of a few lines.

Today, with architectural forms having changed and with new materials and technologies, folk crafts such as paper-cutting, knot-tying and embroidery are on the decline and the bat design appears less in everyday life. Moreover, the Western image of the bat is entering Chinese society.

In one famous Hong Kong martial arts novel there appears a weird character called Green-Bat-Winged Wei I Hsiao, one of the four Zoroastrian lords, who must drink human blood to heal wounds. He stalks the earth and when his victims hear his spine-chilling laugh in the distance two bloody holes appear on their necks. But there is a difference from the Western Dracula in that this one only attacks bad people. Both good and bad — it could be said to be the result of combining the Western and Chinese bats. More recently, Batman T-shirts, bags and stickers have become the bewitching fashion.

The rise of the Western bat: The recent

蝙蝠俠的T恤衫、背包、貼紙，亦大行其道，滿街招搖。

西方蝙蝠「出頭天」

自廿世紀下半期自然保育風氣興起以來，近代動物學領域中，動物生態習性的研究大行其道，西方對蝙蝠在生物鏈中的角色已有了新的認知，發現蝙蝠不但不主動攻擊人，在人類生活中更佔有重要角色。保護蝙蝠的呼聲遂四起，一九八八年還是歐洲的蝙蝠年，「我愛蝙蝠」的貼紙處處可見。

一個和蝙蝠相關的全新形象——蝙蝠俠也適時出現，粉碎了蝙蝠只能和吸血鬼掛鈎的神話。這和西方人對蝙蝠印象在逐漸改觀不無關係。

已成立十週年的美國蝙蝠保護協會(BCI)，更以中國的蝙蝠吉祥圖案做爲該會的會徽，他們認爲，「中國人對蝙蝠的喜愛，能鼓勵西方善意正視蝙蝠。」該協會更在刊物中指出，中國人每將代表長壽、多子孫的桃子與蝙蝠擺在一起，是有生態意義的。桃樹是五千年前最早爲中國人栽種的植物，桃樹依賴蝙蝠傳花授粉，沒有蝙蝠，桃樹也就無法繁衍「子孫」。

「多子多孫多福氣」的圖案，代表了中國老祖先早就察覺到自然萬物間，彼此有著密切不可分的微妙關係呢！

由三千年前中華民族的族徽，到今天成爲西方蝙蝠學會的會徽，蝙蝠終於可以完全釋然、揚眉吐氣。

祝福滿人間

走過擺著花盆的街道，盆沿赫然印滿蝙蝠，雖然不夠精緻，卻不失樸拙；牆上的一張中國年畫，走近瞧瞧，裏面可不是藏了隻五彩繽紛的蝙蝠；在路邊吃碗麵，把湯喝淨，碗底竟浮現出五隻手牽手、圍成同心圓的蝙蝠……。現代生活中，蝙蝠仍然一隻隻飛向我們。

展開笑顏迎接，說不定你眞會收到滿滿的祝福。 S

（原載光華八十一年七月號）

rise of conservationism has meant that ecological research has become fashionable. There is a new awareness in the West about the ecological role of the bat, with it now being realized that it does not attack people but plays an important role in the life of human beings. There are now widespread calls to protect the bat and 1988 was even the European "Year of the Bat," with "I love bats" stickers appearing everywhere.

It was at this time that a new image connected with bats appeared — Batman. Shattering the myth that bats are connected with vampires, this cannot be said to be unrelated to the gradual change of the image of the bat in the Western mind.

Ten years ago the American Bat Conservation International was established and decided to use a lucky Chinese bat design as its logo. It is their view that "the love and affection of the Chinese people for the bat can encourage a good and proper view of the bat in the West." They also point out that the Chinese view of the bat and peach together as a symbol of fertility has ecological significance: The peach tree was first cultivated in China about 5,000 years ago, and the wild tree relied on bats for the dispersal of its seeds; without the bat, the peach tree had no way to spread its "descendants." The "many descendants and good fortune" bat and peach design is thus an ancient Chinese representation of the close relationship that exists between all natural things.

From a Chinese clan symbol of three-thousand years ago, to today's logo for Bat Conservation International, bats can finally take a breather.

Bringing luck among us: As you walk down the road you can see that the flower pots arranged around you have impressions of bats on their sides, perhaps not very refined but at least not having lost their naivete; take a close look at the New Year picture on the wall and it contains five bats; have a bowl of noodles at the roadside and, when you have finished drinking the soup, at the bottom appears a ring of five bats . . . .

In modern-day life the bats are still flying at us from all directions. Welcome them with a smile and you might bring yourself a whole lot of good fortune. S

(Chang Chin-ju/tr. by Christopher Hughes/ first published in July 1992)

蝙蝠正傳 Bats — The True Story

文・張靜茹　圖・徐仁修

甫結束的地球高峯會議上，已開發和開發中國家吵吵嚷嚷，爲了誰該付出更多心力、金錢保護熱帶雨林而爭議不休。動物學者卻氣急敗壞地疾呼：熱帶雨林保育計劃的第一步是保育蝙蝠。爲什麼？

千百年來，中國人以蝙蝠象徵吉祥，西方人卻藉蝙蝠以示邪惡，但人們可曾好好認識過蝙蝠本身？你又知道蝙蝠在臺灣的現況嗎？

這不是時下流行的腦筋急轉彎——

牠們擁有雙翼，能够飛行，卻不是鳥類；牠們眼不盲，卻以聲音在「看」東西；牠們極爲脆弱，但其中有些份子，卻以大型動物的血液爲食。

牠們是——蝙蝠。你猜對了嗎？

艱辛的生存策略

有人形容，蝙蝠乃是動物世界中的一首「變奏曲」。

牠們是唯一會飛行的哺乳動物。雖然有少數飛鼠能在空中滑行，但哺乳類中只有蝙蝠具有眞正的「翅膀」。和鳥類不同的是，牠們的翅膀沒有羽毛，而是由前肢演化而成，四根細而長的手指間，長著一片可伸縮的薄膜，就是牠們的飛行利器。細小的姆指則單獨生長，用來保持儀容整潔——清理皮翼和寄生蟲。

蝙蝠倒掛高處棲息，使牠們不像飛機需要跑道助飛，也不需要大力抖翼才能騰空而起；只要鬆脚、落下，就身在空中了，十分省力。

但雌蝠懷孕時，可不比懷胎十月的人類輕鬆，牠們必需大腹便便，負擔胎兒的重量飛行、覓食，還要倒吊著入眠，與地心引力「抗衡」，極爲艱辛。因此牠們每回繁殖多是「一胎化」，但也有少數雙胞胎和三胞胎的情形。

地面生活競爭激烈，蝙蝠不畏萬難，「飛」上朗朗青天，但鳥族掌握了白晝的領空權，牠們只好另謀策略，選擇競爭對手較少的黑夜活動。沒想到一頁歷經辛酸才定型的生活史，卻使蝙蝠成爲人們心中的「夜魔」。

只吃蟲、不吃血

不過，蝙蝠的生存策略也許是正確的，牠們種類多達一千種，佔哺乳動物的五分之一強，僅次於老鼠。地球上，除了寒冷的極區，可說蝠踪遍天下，但仍以高溫多濕地區種類較繁盛。地處熱帶與亞熱帶交界的臺灣地區，就有廿二種蝙蝠。

雖然，此地的蝙蝠全數和吸血蝙蝠同宗，屬體型較小的「小翼手目」，但和大部分宗親一樣，仍保留著已有廿五億年傳統的飲食習慣，以昆蟲爲主食。也有少數發現了花蜜、花粉和蟲兒們一樣含高蛋白，且更「秀色可餐」，而改變了食性。

獵物改變，有少數蝙蝠的飛行技巧也跟著改良，牠們能像蜂鳥一般，騰空停在花蕊面前，把纖長的舌頭伸進花苞中收集汁液；美洲的墨西哥食魚蝙蝠，能在水面上捕魚；也有一些蝙蝠能由空而降獵捕青蛙、蜥蜴。

掌握了飛行訣竅，使蝙蝠獲益匪淺。但要在夜間更精確發現、捕食飛行獵物，食蟲蝙蝠發展出一套比視力更精密的「導航系統」——利用音波來指示方位。牠們在大洋中的親戚海豚、地底活動的地鼠，也有和牠們相同的音響設備。

聲波導航系統

蝙蝠以嘴或鼻發出頻率極高的聲音，音波會碰觸到牠面前的東西，再反彈廻響進耳際。藉這套音波定位系統，牠可以準確無誤的判斷出獵物的位置。就像雷達自動系統，只不過雷達採用的是電波，牠們用的是聲波。

比雷達聰明的是，「牠也能藉此測知獵物的質地是細、是粗，合不合胃口，」碩士論文是「臺灣東亞家蝠活動模式之研究」的盧道杰說。蝙蝠中的「漁翁」墨西哥食魚蝙蝠，還能藉聲波察覺出水面一根流動的頭髮和一絲絲水紋。

接收到回音後，牠們可以很快刹車捕食，

During the continuous disputes at the Earth Summit in Rio de Janeiro between the developed and developing nations about who should put in most effort and bear the brunt of the costs of conserving the tropical rain forests, why did one zoologist exclaim: "The first step in conserving the rain forests is to have a plan to protect the bat"?

Over the centuries the Chinese have seen bats as creatures symbolizing good luck, while in the West they have been seen as evil. How can we get a proper understanding of bats?

They have wings and can fly, but they are not birds; their eyes are not blind, but they use sound to "see" things; they are fragile, but some live on the blood of large animals. They are bats — did you guess right?

A difficult strategy for survival: Some people still describe bats as being an "improvisation" on the theme of mammals.

They are the only mammal that can fly; although there are a small number of rodents that can glide through the air, it is only the bat that has real wings. The difference from birds is that bats have no feathers but have evolved their forelimbs so they can fly using a thin membrane that stretches between their four long, thin fingers. Their small, thin thumbs have also grown particularly long so that they can be used for grooming.

Bats roost hanging upside down in high places. Unlike airplanes, they do not need a runway for takeoff, and do not need to expend much energy to get airborne. All they have to do is relax their feet, drop, and they are in the air — most economical.

When the female bat is pregnant, however, she cannot be as relaxed about it as human beings with their nine-month gestation period; she must fly when heavily pregnant, search for food, sleep upside down, fight against gravity — it is really tough. Most of them therefore only have one offspring, although there are cases of twins and triplets.

With the struggle for survival on the ground being fierce, bats took to the skies. But this is the sovereign territory of the birds, so they adopted an even better strategy and chose to go out at night when there is less competition. The bat never thought that with the passage of time this would make it a "demon of the night" in the eyes of mankind.

Bats eat insects, not blood: But the bat's strategy for survival is also correct. There are more than a thousand species of bat, which accounts for a fifth of all mammals, second only to rodents. Apart from extremely cold areas of the globe, you can say that bats are found everywhere, although the numbers of species is most diverse in warm and humid regions. Situated between the tropical and subtropical zones, Taiwan has 22 species of bat.

All the bats on the island of Taiwan have the same ancestors as the vampire bat, belonging to the Microchiroptera group, characterized as being comparatively small, with small eyes, complex ears and the ability to echolocate. But, as with most of their relatives, they have preserved their culinary habits of more than 2.5 billion years by feeding mainly on insects. There is also a small minority that has discovered that foods such as nectar and pollen have as much protein as insects but are perhaps rather more appetizing.

With changes in their prey, some bats have changed their flying techniques accordingly so that they can hover like humming birds in front of a flower's stamens and lick out the nectar using their long tongues. Then there is a type of Mexican fish-eating bat that can catch fish on the surface of water, and types of bat that can dive out of the sky to catch frogs and lizards.

Having unlocked the secret of flying, the bat has profited a lot. But so as to more accurately locate and catch its night prey, the insect-eating bat has also developed a sonar navigation system which is even more sensitive than the power of vision. Whales in the deep seas and rodents that are active underground use similar systems.

Sonar navigation system: The bat emits high-frequency sounds from its mouth or nose which rebound off objects in front of it and return to the bat's ears. Such a system enables the bat to determine the position of its prey with pinpoint accuracy. It is just like a radar system, except that radar uses radio waves while the bat uses sound waves.

What is even more clever than radar about the bat's system is that it can tell whether or not the prey is edible, says Lu Dao-jye, whose master's thesis was "A Study of the Activity Pattern of *Pipistrellus abramus* in Chutung." The Mexican

食果蝙蝠長相似初生的幼狐，又稱狐蝠。
Fruit-eating Flying Foxes derive their name from their striking resemblance to foxes.

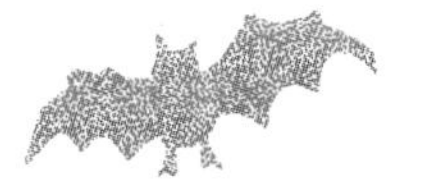

或迅速移位，做三百六十度的轉向，去追捕試圖逃逸的獵物。「如果鳥類是戰鬥機，蝙蝠則是靈活機動的直升機，」盧道杰比喻。

蝙蝠的聲納系統，恐怕是早期西方對牠唯一不排斥之處。科學家認爲，若能發明一套類似功能的裝置，盲人也能辨識障礙，行走自如，減少失去視力的不便。

十八世紀西方就有人嘗試解開蝙蝠的聲納之謎，可惜人類至今仍無法完全了解它的運作。當人們闖入蝙蝠洞穴，成千上萬的蝙蝠會不斷繞飛，但牠們倒底以何種方法，在如一團拚命衝飛的火球當中，不致碰撞，訊號回收也不會互相干擾？

貌似稚犬的食果蝙蝠

爲了一邊享受美食，一邊還可以發出聲波，蝙蝠克服困難，將鼻子發展成「擴音器」，臉部因此長出許多發聲組織。比如用來矯正聲音，形狀如枯葉、釘矛的構造，由於捕食對象不同，和爲了分辨不同距離，又有各種不同組合。由人類眼光來看，牠們臉上的特殊造型，極爲「醜怪」，人們創造惡魔時，因此常借牠們的形象助威。

事實上，在「族繁不及備載」的蝙蝠世界中，有許多大眼睛、長吻的狐蝠，就像一臉稚氣、剛學會蹦跳的小狗，足以讓人們對蝙蝠容貌的刻板印象完全改觀。

只生長在綠島的「臺灣狐蝠」就是其中之一。臺灣狐蝠與分佈在熱帶與亞熱帶的兩百多種狐蝠一樣，屬「大翼手目」。小翼手目蝙蝠中，體型最小的只有人的小指頭一般長，最大型的狐蝠，雙翼展開可以長達一·五公尺，就如狗兒當中的吉娃娃與聖伯納。狐蝠飛行姿勢優雅，緩慢、規律的拍翅，能一口氣飛上七十公里，與黃昏後盤旋在天空上快速拍翅覓食、常令人誤以爲是隻小小鳥在學飛的小型蝙蝠大異其趣。

狐蝠只吃成熟、誘人的水果，因此又被稱爲「果蝠」。牠們不具聲納系統，擁有一雙大而圓的眼睛，在熱帶森林，入夜後，總可以見到成千上百如落入凡間、閃閃發光的羣星，正是一羣羣棲息林間的狐蝠。

fish-eating bat can even use sonar to distinguish between a hair and a ripple on the surface of water.

Once the sonar signal has been received back, the bat can either catch its victim or change course with a 360 degree turn in pursuit of more elusive prey. "If birds are fighter planes, then bats are nimble helicopters," says Lu Dao-jye.

The bat's sonar system is probably the only thing about it that has not been condemned in the West. Scientists think that if something like it could be recreated then it could be of great benefit to the blind.

People in the West began trying to unlock the secret of the bat's sonar in the eighteenth century, but even today its workings are still not fully understood. If someone disturbs a bat cave so that thousands of them start to fly about, how is it that they do not bump into each other and get confused by each other's sound signals in the chaos?

Innocent dog-faced fruit-eating bat: The bat has overcome the difficulty of being able to enjoy its food while still emitting sonar signals by using its nose as a "transmitter," and developing all kinds of sound-emitting instruments on its face. To emit a full, clear sound bats have faces like dried leaves with stud structures on them. Because each has its own kind of food and to keep a distance from other kinds of bat, each is incompatible with all others. In the eyes of human beings, their special features look very ugly, so that when people were creating demons they would often use their appearance for an exaggerated effect.

In fact, in the variegated world of the bat, there are also many large-eyed, long-nosed Flying Foxes. Looking like innocent little dogs, they are enough to change anyone's stereotype vision of the appearance of the bat.

The Formosan Flying Fox that only lives on Green Island is one of them. Just like the more than 200 species of Flying Fox distributed throughout the tropical and subtropical regions, the Formosan Flying Fox belongs to the "Megachiroptera" group of bats, which are characterized by having large eyes, simple ears, and, in most cases, no echolocation system. The smallest bat in the Microchiroptera group is only as long as a person's little finger, while the biggest Flying Fox can have a wingspan of 1.5 meters — like the difference between a chihuahua and a St. Bernard. The flight of the Flying Fox is slow and graceful, being able to travel 70 meters with one flap of its wings; very different from the rapid dusk fluttering of smaller bats that can lead people to mistake them for fledgling birds learning how to fly.

The Flying Foxes only eat ripe, enticing fruit and are often called Fruit Bats. They do not have a sonar system but two large, round eyes instead. If you go into a tropical forest at night you will see thousands of clusters of glittering stars in the darkness — the eyes of colonies of roosting Flying Foxes.

The female's "sperm bank": Summer is the mating season for bats, and the Taiwanese Pipistrelle family of bats mates in September. This often does not result in pregnancy, however, because the female can store sperm in her ovaries until winter has passed. This is especially the case with bats in temperate zones, where methods such as delayed sperm reception, ovulation or slow fetal growth enable the females to hibernate. Researchers say that such prolonged pregnancies ensure that the young are born in spring, when there is plenty of food about. With every country striving to establish animal gene banks, this ability of the bat to control the source of life has become the object of research second only to its sonar system.

Of even more significance for human beings than this research into the bat's physiology, was the call made during the arguments over preservation of the rain forests during the Earth Summit by a British zoologist, who demanded: "The first step in protecting the rain forests is to have a plan to protect the bat."

If you want to save the rain forests, first save the bat: Such a statement sounds very strange to city dwellers who, just like bats that have lost their sonar systems, have lost their ability to accept that people share a very close relationship with animals. Unless we do not eat, drink or breathe, we cannot ignore our relationship with the bat. This is because the bat plays a pivotal role in preserving the supply of food for human beings, and maintaining the world's atmosphere.

Take the fruit-eating Flying Fox as an example. Although we are still not very clear about the exact ecological position of the Taiwanese variety of this creature, we can know from its condition in the rest of the world that it is of great benefit to human

雌蝠是「精子銀行」

夏季是蝙蝠的求偶季節，臺灣東亞家蝠在九月交配，但雌蝠往往不受孕，而將精子儲存在子宮，度過寒冬。尤其在溫帶地區的雌蝠，有如「精子銀行」，以延遲受精、延遲着床或是延遲胚胎發育等方法，度過冬眠期。研究人員認爲，牠們延長孕期到大地回春的理由，是此時食物較豐盛，更利於哺育幼兒。

生命由精、卵結合開始，蝙蝠卻可控制生命的進展，在世界各國極思建立「動物基因庫」的情形下，蝙蝠的生理、生殖機制，遂繼聲納之後，成爲人類爭相研究的對象。

蝙蝠對人類更重要的意義，卻不只是生理上的研究與利用。當世界各國在巴西地球高峯會議上，還爲誰該保護熱帶雨林爭議不休，英國動物學者史考林則大聲疾呼：「熱帶雨林保育計劃的第一步是保育蝙蝠。」

救雨林，先救蝙蝠

此話聽來蹊蹺，因爲生活在都市的人們，就像聲納系統故障的蝙蝠，已喪失接收「動物和人類關係密切」訊息的能力。除非不食、不飲、不呼吸，否則人們和蝙蝠還眞脫不了關係。因爲提供人類食物來源，與維持地球大氣穩定、使人們能暢然呼吸，蝙蝠是其中一顆不可或缺的螺絲釘。

以吃水果的狐蝠爲例，雖然我們尙不淸楚臺灣狐蝠在生態中的地位如何，但由世界各地狐蝠的情況可知，牠們不停在爲人們創造「利基」。

芭樂、桃子、芒果、芭蕉、鳄梨……，世界熱帶市場上的七成水果，大量依賴蝙蝠傳花授粉。以日落後才開花的榴槤爲例，花朶在幾個鐘頭內就凋謝，而昆蟲體型太小，花兒又等不及白天，待鳥兒來靑睞。這時候只有食花粉蝙蝠能適時光臨，採蜜之餘，又扮演媒人傳花授粉，果農一年的豐收也有了保障。

即使一些人工果園不需要免費的媒介，但保證人工品種能長期維持優良品質，則必需靠大量保存野生品種來不時改良，因此蝙蝠至今仍是「東南亞果農豐收的關鍵生物」。

全世界有卅多種蝙蝠發現到人類的家比野外更適合牠們安享天倫。圖爲臺灣產的渡賴氏鼠耳蝠。（盧道杰攝）
There are more than 30 kinds of bat in the world that live in human habitations, which are suitable as secure and peaceful homes. (photo by Lu Dao-jye)

一夜吃三千隻昆蟲

平均而言，你我呼吸的新鮮氧氣，有百分之七十來自熱帶雨林，那也是目前地球上百分之九十動植物的家。美國蝙蝠保護協會的人員就發現，在熱帶雨林地區，許多伐木後的空地，因爲食果蝙蝠在鄰近覓食，不到廿分鐘，無法消化的種子，在飛行中就排出、掉落，童山濯濯之地不久又開始出現綠意。

食蟲蝙蝠則在控制夜行性昆蟲上，有重大貢獻。臺灣大部分蝙蝠的食物中都不乏蚊蟲；食蟲的灰蝙蝠一晚上可吃掉三千隻昆蟲。泰國有一洞穴棲息著近六百萬隻蝙蝠，每年吃下三千五百萬噸的蚊蟲，對當地夜間蟲相，具有平衡穩定的作用。

由生理到生態，蝙蝠果然如中國傳統所言，能爲人們帶來「福」祉。但隨著人們改造環境的技術與速度日益猛進，克服各種困難才成爲哺乳類第二大家族的蝙蝠，卻也在逐漸沒落。

蝙蝠喜歡成羣聚集，成爲牠們最易被傷害的弱點。果蝠常數千隻結伴，收束翅膀倒掛樹上，遠看像一片結滿纍纍果實的樹林。食蟲蝙蝠則喜歡居住在不見天日的樹洞、岩洞；在人類創造了「家」之後，也有蝙蝠搬進屋簷、磚瓦間，分享家庭「溫暖」，但洞穴則是蝙蝠數量最多之處。

在婆羅州加瑪洞穴，當黑夜來臨，成羣蝙蝠飛離棲所，奔騰如源湧成帶的河流，平均每分鐘數千隻由洞中飛出，再分散到森林、沼澤地覓食。全數飛出洞穴，需要兩、三小

beings.

Many of the main varieties of tropical fruits largely rely on bats for pollination. Take the durian, which only opens its flowers after sunset, as an example. Its flowers wither within a few hours, most insects are too small for it, and it cannot wait until daytime and the coming of the birds. Only pollen-eating bats can play the role of carrier, thus guaranteeing a good harvest.

Some cultivated fruit plantations might not need the free services of this pollinator. But if in the long term the good quality of cultivated fruits is to be guaranteed, then they will still have to rely on the occasional improvement from wild varieties. It is thus that the bat remains an animal of crucial importance for the fruit harvest of Southeast Asia.

Eating 3,000 insects an evening: On average, 70 percent of the oxygen we breathe comes from the tropical rain forests, which are also home to 90 percent of the world's animal and vegetable species. Personnel of the American Bat Conservation International have discovered that in a number of areas where the rain forest has been felled, within twenty minutes nearby fruit-eating bats have excreted indigestible seeds in their flight and greenery begins to appear on the disturbed, bare ground.

Insect-eating bats also make a big contribution to the control of flying insects. Mosquitoes and gnats are not lacking from the diet of most of Taiwan's bats. The insect-eating Grey Bat can eat 3,000 insects in an evening. There is a cave in Thailand that is the roost for nearly six million bats which, every year, eat 35 million tons of mosquitoes and gnats, thus being of great use in the stabilization and balancing of the nocturnal insect population.

From physiology to ecology, bats are finally bringing great fortune to people, just as the Chinese traditionally believed. However, having overcome all kinds of difficulties to become the second most successful mammal on earth, following the increasingly rapid advances of environment-changing technology, the bat is gradually disappearing.

The gregariousness of bats is their weakest point. Fruit bats often hang upside down from trees in flocks of thousands, from a distance looking like a forest of fruit. Insect-eating bats like to live hidden away in holes in trees and in caves. When people set up a home, then bats will also come and live under the eaves so as to enjoy the "warmth" of home. But caves are still what they love best.

As night draws close at a cave in Indonesia, a huge swarm of bats leaves the cave at a rate of thousands every minute, spreading out over the forest and marshes to look for food. It takes two or three hours for the whole lot to get out of the cave.

Although Taiwan's bats cannot be seen in such spectacular numbers, as soon as you get out of the suburbs, from north to south there is never a lack of bat caves to attract sightseers.

時。

臺灣蝙蝠數量雖然沒有如此壯觀，但一出郊區，由北到南，總不缺名為蝙蝠洞的旅遊地點。

蝠去樓空

研究蝙蝠棲息地的臺大動物系研究生陳怡文表示，蝙蝠羣居洞穴的理由，可能是洞穴數量較少，不允許個個自立門戶；羣居也可互相警戒，提防蛇、老鷹等天敵的偷襲。

但更重要的理由，卻是互相取暖提高周遭的溫度，以安然度過寒冬，這也是雌蝠成功哺育幼蝠的重要條件。由於體型小的溫血動物，散熱極快，無法維持體溫恆定。有研究發現，蝙蝠在覓食前後，體溫可以差上好幾度。

嚴冬萬物蟄伏，不易覓食，為維持體溫，又必須大量捕食，溫帶地區的蝙蝠就靠冬眠捱過冬季。即使氣候暖和、號稱四季皆春的臺灣，最冷的一、二月，傍晚一樣難見蝙蝠逐蟲而飛。

休眠中的蝙蝠若受到驚嚇，四處奔竄，會造成過度消耗體內貯存的熱量，無法度過冬天。太多旅客湧入洞中，也會改變洞穴的溫、濕度，逼迫牠們不得不棄巢離去。

臺灣的蝙蝠洞就因為開路、遊樂區開發，和人為干擾而逐漸無蝠可看。以基隆八斗子的摺翅蝠石灰洞穴為例，北部濱海公路通車後，因觀光客絡繹於途，如今蝠去樓空，只見蟑螂四處爬行。

在臺灣野外四處跑的生態攝影者劉燕明就說，臺灣蝙蝠洞的命運如出一轍：「叫蝙蝠洞的地方都沒有蝙蝠。」否則就要到人煙罕至之處，像南臺灣墾丁一些名符其實的蝙蝠洞，也是蛇踪最愛出沒的地方。

失去蝙蝠又如何？

農委會保育科技士盧道杰，在他的論文裡清楚列出，目前臺灣地區的大蹄鼻蝠、小蹄鼻蝠、臺灣葉鼻蝠、臺灣鼠耳蝠、綠島的狐蝠，都因棲地破壞，面臨族羣劇減的危機。

隨著綠地、森林、沼澤，甚至四合院建築的消失，不只是洞穴蝙蝠受威脅；兩年前竹東軟橋一戶民宅，還有近千隻家蝠，如今民宅左側多了一家大型KTV，屋後則有工地施工，準備蓋「販厝」。前往當地作調查的盧道杰便說：「光是噪音就足以嚇跑蝙蝠，」果然，當晚出現的蝙蝠只剩十幾隻。

失去蝙蝠，又會怎麼？

「如果是指臺灣的情況，恐怕只能說不知道，因為我們對臺灣蝙蝠的了解實在不够，」盧道杰說。

再求諸國外的經驗吧。美國關島有三種蝙蝠，二次大戰後美軍上岸，蝙蝠成為打獵取樂的對象，加上大量的開發，已有兩種在島上消失。當地植物種子原來有百分之四十靠蝙蝠傳播，如今失去重要媒介，數以萬計的果實花朵竟成孕育果蠅的溫床，直接威脅果農生計。

應徹底作分佈調查

西方過去由於對蝙蝠的厭惡而惡意撲殺，也是全世界蝙蝠減少的理由之一。但今天蝙蝠的神秘面紗已被生態學者拉開，牠們與人類之間的互動漸被得知，英、美、澳洲也都成立了蝙蝠保護協會，雖然和「挽救鯨類」、「海洋哺乳動物保護協會」等團體比起來，仍是很年輕的組織，但後起之秀卻拚勁十足。

他們不僅對自己國家蝙蝠的分佈與習性深入的調查，更在世界各地呼籲立法保護蝙蝠，嚴禁人們進入蝙蝠洞干擾蝙蝠。

為鼓勵人們接近、了解蝙蝠並不是那麼可怕，也規劃賞蝠據點，人們傍晚時可以在洞口之外觀賞一場活生生的自然景觀——蝙蝠離巢；也教人們製作蝙蝠食臺，餵食都市近郊的蝙蝠，讓牠們樂於留在人們生活環境之中。

何時再見蝙蝠？

國內動物生態研究起步較晚，夜行性動物又是較難掌握的研究對象。因此至今仍缺乏對蝙蝠實際分佈地點與停留季節的調查。又

Going to roost: Chen Ti-wen, a zoologist at Taiwan University who is researching bats, says that the reason why bats live in cave colonies could be that the small number of caves makes it impossible to have individual roosts; there can also be mutual protection within a colony from snakes, eagles and other predators.

But what is even more important is that the bats can raise the surrounding temperature so as to safely pass the winter. This is also a very important condition for the females to successfully suckle their young. Because small-bodied warmblooded creatures rapidly lose heat and have no way to maintain it, research has revealed that there can be a difference of many degrees in the body temperature of a bat before and after feeding.

In the cold winter it is hard to find food, and to maintain body temperature it is necessary to eat a lot. This is why bats in temperate regions depend on going into hibernation for the winter. In warm Taiwan, where there is no extreme difference between the seasons, it is still rare to see bats on January and February evenings.

If hibernating bats are disturbed and take to the air, they will use up too much of their stored heat and have no way to get through the winter. If too many sightseers go into a cave, the temperature and humidity might change so that the bats will be forced to leave. In Taiwan, due to road building, theme park development and general human interference, it is gradually getting harder to see bats. Take Keelung's Patoutzu Japanese Long-Winged Bat lime cave as an example. Since the northern coastal road was opened to traffic, the endless stream of sightseers has meant that today there are no longer any bats, just cockroaches crawling around all over the place.

Liu Yeh-ming, a photographer who has photographed all of Taiwan's wild places, says that the fate of Taiwan's bat caves seems to have left the tracks: "None of the places that are called bat caves have any bats." On the other hand, if you go to some places near human habitation, such as Kenting in southern Taiwan, where there are real bat caves, they are also places with a lot of snakes.

Where would we be without bats? The master's thesis of Lu Dao-jye, a conservation technician at the Council of Agriculture, spells it out: The Formosan Greater Horseshoe Bat, Formosan Lesser

臺灣蝙蝠分佈圖
Distribution of Bats in Taiwan

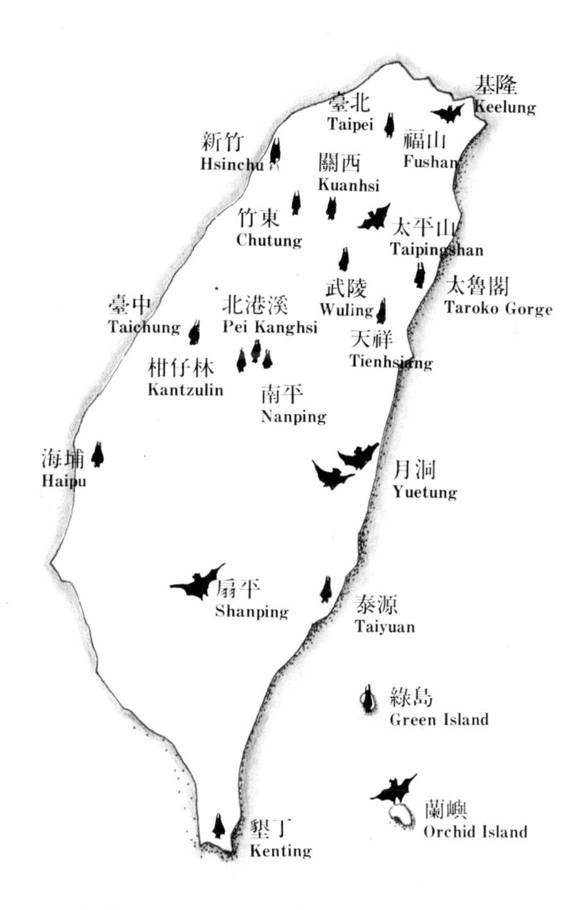

要保育蝙蝠，先要做一次「蝠」口普查，以蝙蝠爲研究對象的盧道杰，將自己見過蝙蝠的地點大略標示出來。
（盧道杰製）（李淑玲繪圖）
If you want to preserve the bat, first do a "population survey." These are the places where Lu Dao-jye has observed bats.
(map by Lu Dao-jye)
(drawing by Lee Su-ling)

以水果爲食的果蝠，體型比食蟲蝙蝠「壯碩」。(卜華志攝)
Fruit bats have a more robust physique than their insect-eating cousins. (photo by Pu Hua-chih)

因爲沒有岩洞管理辦法，「想保護，或想設觀光據點好好經營管理，都無從下手，」農委會保育科技士盧道杰說，目前最急迫的是徹底做一次蝙蝠分佈調查。

除日據時代有過全島野生動物調查，與五〇年代美國海軍在東南亞從事流行病學研究，追踪過蝙蝠之外，針對臺灣蝙蝠較深入的研究，就只有三年前臺大動物系對東亞家蝠活動模式的一份研究報告了。

近幾年來，動物學者仍不時發現新種蝙蝠。師大生物系畢業生杜銘章曾在武陵農場發現皺鼻蝠，是臺灣第一次發現，也是目前唯一的一次，至於牠們是「原住民」，或遷徙途中路過本島，答案未知。

一年前，日本國立科學博物館研究人員，前來臺灣中央山脈從事小型哺乳類調查，在二千五百公尺的鞍馬山森林附近，發現模樣奇特、耳朵與身體幾乎一般長的新紀錄種長耳蝠。

正指導研究生做蝙蝠論文的臺大動物系副教授李玲玲就擔心：「有些品種可能連我們都還不知道，就消失了。」

我們不會願意看到有一天，臺灣的蝙蝠只飛翔於器物、服飾之上，卻消失在大自然中吧？ S

（原載光華八十一年七月號）

Horseshoe Bat, Formosan Leaf-Nosed Bat and Formosan Mouse-Eared Bat are all facing extinction because their roosts have been destroyed.

Following the loss of green areas, forests, marshes and even the traditional four-walled courtyard architecture, it is not only cave bats that are under threat. Two years ago, at a house in Chutung's Juanchiao, there was a colony of nearly a thousand Pipistrelle Small House Bats. Now on one side of the building there is a large karaoke parlor, and at the back there is a building site. Lu Dao-jye, who used to do research there, says: "It is just the din that has scared off the bats." Now you cannot see many more than ten bats leaving in the evening.

"If you are talking about the situation in Taiwan, then we can probably only say that we do not know, because we do not understand enough about Taiwan's bats," says Lu Dao-jye.

Take a look at the experience of other countries. The American island of Guam once had three kinds of bat. When the United States Army landed after World War Two the bats were hunted for entertainment. With large-scale development added to this, two of these bat species have already become extinct. 40 percent of the flora on Guam depended on the bats to spread their seeds and pollen and they have now lost an important carrier, and many fruit plants have become breeding beds for pests that are a direct threat to the fruit industry.

There needs to be a comprehensive survey: The western hatred and suppression of the bat has been another reason for its being under threat of extinction. But today ecologists have already lifted the superstition surrounding the bat. Now that people know they have a mutual relationship with the bat, Britain, the United States and Australia have all established bat protection societies. Although these organizations are still fledglings compared to long-established campaigns to save other animals, they are going to flourish in the future.

Such organizations not only carry out surveys into the spread and habits of bats in their own countries, but also call for legal protection for bats all over the world so as to stop people going into their caves and disturbing them. To encourage people to participate and understand that bats are nothing to be afraid of, they set up bat-appreciation spots. These are places where people can go near the mouth of a cave in the evening and see one of nature's most spectacular and lively sights — bats leaving the roost. The societies also teach people how to build bat feeders to attract suburban bats and allow them to happily live in a human environment.

Our own ecological research began rather late, added to which nocturnal creatures are a difficult object of research. Because of this, we lack knowledge about the spread of bats and their roosting seasons. Moreover, with there being no method of managing caves, "If we want to conserve, or properly manage observation posts, we haven't even made a start," says Lu Dao-jye. What is of paramount urgency now is to carry out a thorough survey.

When will the bats return? Apart from a comprehensive survey of Taiwan's fauna carried out during the Japanese occupation, and some research carried out by the United States Navy into contagious diseases in Southeast Asia, the only in-depth research into Taiwan's bats has been a research report on behavioral patterns that was produced by the zoological department of National Taiwan University three years ago.

In recent years there have been occasional finds of new species. Tu Ming-chang, a researcher at the National Taiwan Normal University, once found a Wrinkle-Nosed Bat at Wulu Farm, the first in Taiwan and so far the only. It is not known whether this was an indigenous bat or was just in transit.

One year ago, a zoologist from Japan came to Taiwan to undertake research into Microchiroptera bats in the the central Taiwan mountain range. Near the forests of the 2500 meter Anma mountain, he discovered an unusual type of Long-Eared Bat whose ears were actually the length of its body. Because such a case had never before been recorded, he took it back to Japan for further research.

At present supervising doctoral research into the bat at National Taiwan University, Associate Professor Li Ling-ling worries: "Some species that we still do not even know about could become extinct." Are we willing to see the day when bats in Taiwan can only be seen flying on ornaments and clothing, but have disappeared from the state of nature? S

(Chang Chin-ju/photos by Hsu Jen-hsiu/ tr. by Christopher Hughes/ first published in July 1992)

你了解吸血蝙蝠嗎?

Do You Understand Vampire Bats?

文•張靜茹

月不明、星稀，吸血鬼出來獵食了。受害者痛苦掙扎、臉型扭曲，牙齒開始變形，成了另一個吸血鬼。故事看多了，吸血蝙蝠會不會也吸食人血，讓人染上吸血的惡習?!

人生不如戲，這些寃寃相報的念頭是人類想出來的。蝙蝠吸血？——混個溫飽罷了！

凡是蝙蝠，都會吸血？

其實，全世界只有三種蝙蝠以血爲食，牠們全數分佈在中、南美洲，其中兩種已近乎絕跡，人們對牠們的行爲一無所知。

吸血蝙蝠攻擊人？

人類不須往自己臉上貼金，吸血蝙蝠的「菜單」上，向來沒有「人」這一道食物。當然，就像蛇一樣，若蓄意招惹牠，也有可能成爲「點心」。

吸血蝙蝠的獵物包括生活於智利北海岸的海獅和鵜鶘，在接近人煙之處，牠們則以生禽、家畜爲主食，雞、牛、馬、豬這些在晚上不甚活躍的動物，是牠們較中意的對象。據說狗的耳朵可以接收音頻極高的聲音，因此躲避有方，不易被吸血蝙蝠攻擊。

被吸過血的動物會失血致死？

一隻蝙蝠每晚可消化卅公克到四十公克（卅到四十 c. c.）的血液，失血的對象不致因些微的出血而癱瘓。不過，由於牠們特殊的超聲波常引領牠們向同一羣牲畜索食，遭反覆侵襲的動物，健康狀態會日漸衰弱。但遭吸血蝙蝠嚙咬後眞正的危險，並非喪失血液，而是經由牠做媒介，所帶來的狂犬病菌。事實上，大部分的蝙蝠若感染狂犬病，在還沒有機會傳播給其他動物時就已死亡。

吸血蝙蝠常造成農害？

帶有狂犬病源的吸血蝙蝠，確實曾造成農產大損失，但是「原罪」恐怕不在牠們。研究人員指出，在哥倫布抵達中南美洲之前，吸血蝙蝠生活在未受干擾的自然環境，族羣很小。直到四百年前，歐洲殖民者陸續來到，帶來牲畜，提供了牠們新的食物來源。人們在此史無前例地伐林，以便能養更多牲畜，吸血蝙蝠遂能無限制膨脹，在某些地域成爲農業之害。

大量的吸血蝙蝠傳播疾病、抑制牲畜數量，過去卅年，拉丁美洲曾展開大規模控制蝙蝠數量的計畫，但很不幸，許多以昆蟲、水果爲食，對人類有益的蝙蝠，遭到池魚之殃，也被大量撲殺。

(吳鴻富畫)

(cartoon by Wu Hung-fu)

On a moonless, starless night the vampires come out to hunt for food. The faces of their victims twist as they struggle in pain, their teeth changing shape as they themselves become vampires. Vampire stories are many. But is it really true that bats suck human blood and pass on the bad habit to people? Life is not so dramatic and such ideas are products of the human imagination. Bats sucking blood? It's just their way of getting by and surviving!

Are all bats bloodsuckers? In fact, there are only three kinds of bat in the world that feed on blood, and all of them are found in Latin America. Two of them are already practically extinct, and very little is known about them.

Do vampire bats attack people? People should not flatter themselves: ''Human'' is not an item that appears on the bat's menu. Of course, just like snakes, if you disturb a vampire you could become a kind of light snack.

The prey of the vampire bat is so extensive as to include sea lions and pelicans that inhabit the regions of coastal northern Chile. Near human settlements, however, it prefers to feed on a variety of domesticated animals such as chickens, cows horses and pigs, which do not move about at night. It is said that dogs, being able to hear very high-pitched sounds, have a kind of defensive system and are not easy prey for the vampire bat.

Is the bite of the vampire fatal? A vampire bat can digest up to 40cc of blood in one evening, an amount so small that the victim will not feel any ill effects. However, because vampires have a good memory and return to feed on the same herd of animals again and again, their prey can gradually be weakened. Yet the real danger from the bite of the vampire is not loss of blood but rabies, although most bats that contract rabies die before they can pass it on.

Are vampires an agricultural pest? As a carrier of rabies, the vampire bat can actually cause high losses for ranches. But the blame cannot really be laid at their door. Researchers point out that before Columbus reached the Americas, vampire bats existed only in small numbers. The arrival of European settlers with livestock brought a new supply of food, and the following unprecedented deforestation that occurred allowed the raising of even more cattle. The result has been an unchecked expansion of the vampire bat to a degree where it has actually become an agricultural pest in some areas.

With large numbers of vampire bats spreading disease and causing heavy losses in livestock, over the past 30 years in Latin America there have been large-scale campaigns to restrict its numbers. Unfortunately, this has meant that many insect and fruit eating bats which are beneficial to people have been mistaken for vampires and been killed in large numbers.

Are vampire bats as vicious as Dracula? Zoologists think that vampires are the most fragile

吸血蝙蝠生性凶惡？

動物學者認爲牠們是蝙蝠族類中最脆弱的一羣。近千種蝙蝠中，只有牠們獵食的對象，體重比牠們大上千倍、萬倍。牠們花很多時間在地上，悄悄接近獵物，卻因此常被踩死，死亡率高達百分之五十以上，尤其是那些初學覓食的年輕蝙蝠。

想由大型動物身上取食，正如老虎嘴上拔毛，危機四伏，看來蝙蝠的日子也不是那麼好過。

蝙蝠像蚊子一樣「吸」血？

其實吸血蝙蝠並不像人們想像的是直接吸血，而是以極爲快速的動作「舔」血。

由於血液不需咀嚼，牠們的牙齒比其他蝙蝠少，但前排牙齒已衍化成兩片鋒利的三角刃，可以極精準地割開睡覺的動物皮肉，而不被察覺。

牠們會尋找背、頸等血管細微、血液流量大的地方下手，再利用舌下及下顎凹型管狀構造接收血液。由於牠的唾液中含有抗凝血成分，因此可以痛快暢飲，血液不致凝結。目前醫學界已設法在合成牠們唾液中的抗凝血素，因爲它比現代醫學上的任何一種抗凝血劑都來的「進步」。

吸血蝙蝠每回飲血時間持續卅分鐘左右，一次可吃和身體等重的血液。由於血中含有百分之八十的水，牠們有高效率的腎臟可以在攝取血液時，就一邊吸收血中的蛋白質，把剩餘的尿液排泄掉，不用擔心吃太飽會飛不動。吸血蝙蝠可以快速吸收血液中的養分，也是醫學界極力想要參透的奇蹟。

在蝙蝠世界裡，吸血蝙蝠是「比較進化」的一羣，牠們極有社羣和互惠意識。在一個有七、八十隻蝙蝠的社羣中，成熟的雄蝙蝠會爲成羣的雌蝙蝠守衛。更令人驚訝的是，由於找不到獵物的機會很大，有些蝙蝠會將吸食到的血液分給餓肚子的兄弟。

你對吸血蝙蝠有新的看法了嗎？ Ｓ

（原載光華八十一年七月號）

species of all the bats. Out of nearly a thousand species, only the vampire hunts a prey that is up to 10,000 times its own size. They spend much of their time on the ground so as to creep up on animals, so they often get crushed to death. The mortality rate is more than 50 percent, and is especially high among young vampires going out for their first bite. Choosing to feed on such large animals makes the lot of the vampire not an easy one.

Do vampire bats "suck" blood? In fact, the vampire bat does not actually suck blood as humans would like to believe. It licks it up with very rapid movements.

Because there is no need to chew blood, the vampire's teeth are actually smaller than those of other bats. Its front incisors are large and razor-sharp, however, enabling it to penetrate the skin of a sleeping animal without it feeling a thing.

The vampire can find areas on the back and neck with concentrated capillaries and a good supply of blood, which will flow up special grooves on the underside of its tongue. Its saliva contains an anticoagulant so it can drink quickly without the blood clotting. Researchers are at present working on recreating this anticlotting agent, which is far more advanced than anything that modern medicine has come up with.

The vampire bat can feed for up to 30 minutes, and absorb as much as its own weight in blood. Because blood is 80 percent water, the bat has a highly efficient system that, at the time of feeding, can absorb proteins and excrete excess water. No need then to worry about being too full to fly. This remarkable ability to rapidly absorb nutriments is also a miracle that the medical world is striving to understand.

In the world of bats the vampire is one of the more advanced species and has a high degree of social consciousness. In a colony of 70-80 bats, the brave, mature male bats will act as guards for the weaker females. What is even more surprising is that, when the opportunities of finding food are scarce, vampires will actually regurgitate blood to give to their hungry relatives. Has your view of vampire bats changed? S

(Chang Chin-ju/tr. by Christopher Hughes/ first published in July 1992)

蝙蝠的心事

Matters of the Bat's Heart

文・張靜茹

圖・邱瑞金

臺大動物系碩士盧道杰與研究生陳怡文，在竹東軟橋從事有關東亞家蝠的研究，使臺灣蝙蝠的基礎研究邁出一大步。上脚環、秤體重的當兒，蝙蝠又在想什麼呢？

National Taiwan University zoologist Lu Dao-jye and researcher Chen I-wen undertake research at Chutung's Juanchiao into the Pipistrelle Small House Bat and make a great contribution to basic research.

有人研究蝙蝠的生理結構，有人研究蝙蝠的生活行爲，蝙蝠的心裏又在想什麼？

當天邊僅剩一抹殘存的夕陽，「噗！噗！」有蝙蝠掉進長長的捕網了。

上網的蝙蝠如怒氣沖沖的火爆浪子，張牙舞爪，不停發出「吱——吱——」的威嚇聲。另一頭，又一隻蝙蝠落網，卻嚇得縮成一團，直打哆嗦，和人們印象中的「凶惡」蝙蝠，實在差太多！一樣米養百樣人，莫非蝙蝠跟人類一樣，也是鐘鼎山林，各有天性？

餓昏頭了？

過去幾年，臺大動物系在竹東軟橋一戶有近千隻東亞家蝠寄居的民宅作調查，以了解東亞家蝠的活動模式，和棲息地改變對牠們的影響。

爲建立研究對象的基本資料，必需給蝙蝠上脚環、記錄性別、量體重。工作人員遂趕在天未黑前，在屋簷下架網，捕捉蝙蝠。

對擁有「超音波偵測裝備」的蝙蝠，網子有效嗎？農委會保育科技士盧道杰的經驗之談是，入夜後，有些蝙蝠餓了一天，出門覓食往往行色匆匆，也會急得忘了使用聲納而誤蹈捕網。就像人有眼睛，若不專心，一樣會「視而不見」。若等牠們酒足飯飽，腆着肚子姍姍而回，就可以見牠們慢條斯理、輕而易舉地越網而過。

但更耐人尋味的，卻是人爲干擾引起的情緒反應。

那膽小的，在研究人員忙著給牠秤重量、上脚環時，只見牠不停戰慄，頭低低，似極爲懊喪。

有「小生怕怕」、有「捍衛戰士」

工作結束，「去吧！」研究人員將之平擺在手上說，牠卻仍緊縮四肢，不敢稍有動彈，也彷彿猶豫著：「這不會又是個陷阱吧！」輕輕爲牠掰開雙翼，再推推牠，「別怕，走啦！」牠小心翼翼在手上繞走了幾步，才展翅而去。

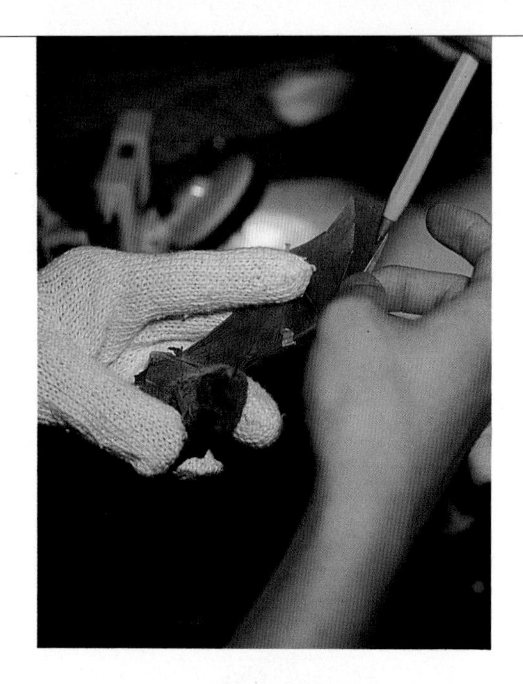

至於那不停作勢要攻擊人的「捍衛戰士」，一上網即齜牙咧嘴猛咬網線。研究人員戴手套躡手躡脚將牠由網上取下，牠則如噴火般、身體不停往前伸，試圖咬人，雙翼且不時掙扎，未曾有過片刻安靜，臨走時則氣急敗壞，彷彿在咒罵：「有够衰！」

兩年內，連續在此觀察蝙蝠卅幾次的盧道杰記得，有一次一隻蝙蝠趁大夥忙著爲其他蝙蝠作紀錄，偷偷蠕行到「安全距離」後，奮而飛走，逃脫成功，眞冷靜的可以。

心事誰人知？

人類研究蝙蝠的聲波構造、生殖系統，希望找出可茲利用的原理；動物學者調查牠們的分佈，研究牠們的棲息地、行爲模式，對如何有效保護蝙蝠，有重要意義。

至於蝙蝠的內心世界到底如何？自然狀態下，牠們也有喜怒哀樂嗎？如果牠們知道千百年來，人們一直以有色眼光看牠們，牠們會作何感想？蝙蝠又怎麼看人類呢？或者牠們根本「目中無人」？

蝙蝠的心事，誰人知。

（原載光華八十一年七月號）

***S**ome people carry out research into the physiological structure of the bat, others research its behavior. But what does the bat really want at heart?*

It is dusk and there is a muffled noise as a bat falls into a trap. It struggles furiously in the net, baring its teeth and claws as it incessantly squeaks. Another is caught, but this one curls up and trembles with fear — how different from the impression of the vicious bat that exists in most people's minds! The same rice will raise a hundred different people, and it is just the same with the great variety of bats.

Empty stomachs: Over the past few years zoological students from Taiwan University have been carrying out a survey at a house in Chutung's Juanchiao where there is a colony of more than a thousand Pipistrelle Small House Bats, so as to understand their behavioral patterns and how they are affected by changes in habitat.

To establish the basic research data for this it is necessary to attach tags to the bats' feet and record their sex and weight. So, just before dark, the researchers hurriedly place nets under the eaves of the house as traps.

How can nets be effective with sonar-equipped bats? Lu Dao-jye, a conservation technician at the Council of Agriculture, explains that when night falls and the bats have been hungry all day, they come out in such a hurry that they forget to use their sonar system and fall into the traps. It is just the same as when people do not concentrate and look without seeing. If you wait until the bats have had their feast and come back with full bellies, then you can see how easily they get round the traps.

What is even more intriguing is the reaction of the bats to human interference.

The timid and the brave: The small bat does not stop trembling as it is weighed and tabbed by the researchers, lowering its head as if in a state of extreme dejection, so that the researchers cannot avoid feeling as though they have committed some kind of crime.

When the work is over, the bat is held out on the palm of a researcher's hand for release. But it is doubtful and curls up, not daring to move, thinking ''Isn't this another trap?'' It needs encouragement as its wings are gently opened and it is given a gentle push. ''Do not be afraid, go on!'' It flutters a little and takes a few steps before finally taking off.

As for the little warrior that wants to attack the researchers, as soon as it hits the net it bars its teeth and bites at the strings. Now the researchers put on gloves and take the bat from the net, but the little firebrand does not stop trying to stretch forward and bite people, all the while struggling with its wings. Not having rested for an instant, as the time comes for the bat's release, it lets out an angry sound as if to curse: ''Tough luck!''

Having been an observer here more than thirty times in two years, Lu Dao-jye recalls how on one occasion the researchers were making records when a bat stealthily wriggled itself to a safe distance and flew off. So, if you keep your cool, escape is possible.

Who understands affairs of the heart? Human beings carry out research into the bat's sonar and reproductive systems in the hope that they can discover something useful; zoologists carry out surveys on their spread, habitat and behavioral patterns, all of which are significant in finding out how best to protect them.

What then about affairs of the bat's heart? In the state of nature, do they also feel love and hate, sorrow and joy? If they knew that over the centuries people have viewed them through tinted glasses, then what would they think? And how do bats see human beings? Or do we simply not exist for them? Who really knows? **S**

(Chang Chin-ju/photos by Diago Chiu/
tr. by Christopher Hughes/
first published in July 1992)

飛鳥入畫
——中國畫家與鳥
Flying Into Painting — Chinese Artists and Birds

文・張靜茹　圖・國立故宮博物院提供

宋朝李安忠的竹鳩圖，也就是今天的伯勞鳥，圖中竹鳩上喙微彎，遮蓋下喙，白色的過眼線與翅膀、體色層次分明，與畫家何華仁所畫的伯勞鳥(左圖)對照，栩栩如生，唯畫家特意強調了牠的渾圓體型。

In this painting of a ''bamboo pigeon'' (shrike) by Li An-chung of the Sung dynasty, the upper part of the bird's beak is slightly curved, covering the lower part, while there is white coloring over the eye and on the shoulders and layers of coloring on the body are distinct. It seems as vivid as the one painted by Ho Hua-jen (left), except that the Sung dynasty painter chose to emphasize the round body shape.

兩個月前加拿大畫家蘭士登在國內舉行「中國珍禽」畫展，對他描繪中國鳥類與其生態細膩、生動的手法，包括生態學者在內的許多人都認爲：「中國人從未好好畫過眞實的鳥類生態，這是以後可努力的一個領域。」

對中國繪畫而言，這種說法恐怕未必實在。

在保育風氣興起、賞鳥人口日增的今天，連帶畫鳥的風氣也日盛。與臨摹歷代古畫、畫譜爲主的花鳥畫，尤其不同的是，畫者以臺灣野外的各種野生鳥類爲對象，親自深入臺灣高山觀察，並記錄鳥類生態，作爲畫畫的藍本。

花鳥畫的新道路？

比較起來，缺乏深入觀察的傳統花鳥畫，總讓人覺得欠缺「眞實感」，即使以工筆描繪得再細膩，也稍嫌死板。當愈來愈多的賞鳥人逐漸都能辨識臺灣野生鳥類的名字和習性後，就更無法接受傳統花鳥畫。許多人遂認爲，生態式地畫鳥，爲一味臨摹古人的工筆，開了一條新路。

因此，前年郵政博物館發行畫家楊恩生所畫的臺灣野生鳥類郵票，廣受喜愛；去年畫家何華仁展出「臺灣鳥木刻版畫展」也頗得好評；如今以野生鳥類爲主題的卡片、月曆、筆記，也四處可見。加拿大畫家耗費四年完成的卅幾幅「中國珍禽」圖畫在國內展出，更造成轟動。

一般以爲，近代圖繪野生鳥類的風氣始於西方，因此今天畫生態鳥類者大都受到西方的啓發。但觀察生態，以眞實的鳥類生態作畫，眞是西方的傳統嗎？

有近廿年賞鳥資歷，也是鳥會資深鳥友的故宮書畫處長林柏亭可不以爲然。別人賞鳥是爲了享受自然之趣，對於林柏亭，賞鳥還附帶幫助他更了解中國歷代花鳥畫家作畫時的心境，也能進一步辨識古畫的眞僞。

發生在前故宮副院長李霖燦身上的例子，能爲他說明原因。

古人畫伯勞，今人烤伯勞？

卅年前李先生護送故宮一批國寶到美國舊金山展出。一位觀衆在宋朝花鳥畫家李安忠的「竹鳩圖」前流連了將近兩個小時，終於打起勇氣，告訴李霖燦有事與之相商，並邀請他到家裏作客。李心中雖納悶，但盛情難卻，只好拉著故宮同仁同去「壯膽」。到了他家，房裏一個大籠子中兩隻鳥兒，和李安忠竹鳩圖中的竹鳩竟一模一樣。

這位外國朋友得意地說：「兩位中國朋友，你們說，你們那張畫是照著我的鳥來畫的呢？還是我的鳥照著你們的畫長的？」三人相視大笑。

這隻古人稱爲竹鳩的鳥兒，今天賞鳥人卻有不同稱謂，牠正是每年過境臺灣、人人耳熟能詳的伯勞鳥。墾丁的「烤伯勞鳥」曾是當地人招徠遊客的佳餚。十年前國內生態保育意識初起時，保育人士曾力呼墾丁居民勿再捕捉伯勞鳥，使牠成爲第一個被保育人士要求保護的野生動物。

今天的鳥友一眼見到竹鳩圖，立刻能叫出「灰背伯勞嘛！」畫中伯勞獨踞高枝、遊目四方，正是伯勞鳥最顯著的特性。在李霖燦看來，圖中的竹鳩「舉目四顧，志在八方，不可一世的神氣模樣，豈是照相匣子所能拍攝出來的？」可惜觀察入微的李安忠沒有料到，今天捕鳥者以鳥仔踏捕捉伯勞，就是利用牠居高臨下，獨占鰲頭的特性。

謙卑的生態觀察

「中國的第一代畫家，就像今天的科學家一樣，對自然做過非常科學性的研究，」在中國美術史的課堂上，前東海大學美術系主任蔣勳如此說道。過去他總覺得中國畫家只寫胸中意氣，是寫意的，事實上是很大的誤解。特別是在宋代，理學盛行，畫家受到格物窮理的思想薰染，作畫的基本精神發自格物，因此不僅觀察物的結構，也去分解物的肌裡。

中國畫家用他們的眼睛觀察自然萬物如石

Two months ago a Canadian artist, J. Fenwick Lansdowne, held an exhibition in Taiwan on the theme of Chinese rare birds. His lively and fine depictions of Chinese birds and their habitats has led many people, including some ecologists, to exclaim: "The Chinese have not been good at giving realistic renderings of birds and their habitats, which is something they should strive hard to do in the future." Words that were not perhaps entirely fair to Chinese painting.

Today the rise of environmental awareness and the rapidly increasing numbers of bird watchers has been accompanied by a flourishing of the art of bird painting. What distinguishes this painting above all from the classical Chinese genre of bird and flower painting is that artists today take the various species of Taiwan's wild birds as their subjects and venture deep into the high mountains to make firsthand observations of birds and their surroundings to make the sketches for their paintings.

A new path for bird and flower painting? By comparison, the lack of detailed observation in traditional bird and flower painting has always tended to leave people with the impression that it lacks "realistic feeling." It is as if the extreme refinement of brush work in such paintings leaves them like dead specimens. With the growing number of bird watchers gradually getting to know the names and ecology of Taiwan's wild species, people are less ready than ever to accept such traditional bird and flower painting. Now many people have come to the opinion that the ecological style has opened a new path for Chinese bird painting, which used to consist mainly of detailed brushwork imitations of the ancients.

Two years ago, the Chinese Postal Museum thus issued a set of stamps featuring the wild bird paintings of Yang En-sheng which have been very warmly received, and an exhibition was held last year of prints of Taiwanese bird species by Ho Hua-jen, which received much critical acclaim. Cards, calendars and notebooks on the theme of wild birds are everywhere. The exhibition on Chinese rare birds by Lansdowne, which was four years in the planning, also created quite a stir.

In general, the recent observation of wild birds and painting styles began in the West, which is therefore where many ecological bird painters received their inspiration from. But is ecological observation and the realistic painting of birds and their environments really a Western tradition?

Lin Po-ting, acting curator of the department of painting and calligraphy of the National Palace Museum, with nearly 20 years of bird watching experience behind him and a veteran bird lover and fellow of the Bird Watching Society, thinks this is not necessarily so. Whereas other people watch birds so as to indulge their interests in nature, Lin Po-ting sets out with the added purpose of using it to help him get a better understanding of the sentiments enjoyed by the painters of Chinese bird and flower paintings through the ages and make progress in his knowledge of how to tell authentic from fake works of art.

The ancients paint it; the moderns bake it: An incident that happened to the previous deputy director of the museum, Li Lin-tsan, can clarify the reasoning behind this.

Thirty years ago, Li Lin-tsan sent some great masterpieces from the National Palace Museum to San Francisco for an exhibition. At the exhibition there was one spectator who stood revelling in front of the painting *Bamboo Pigeons* by the Sung dynasty artist Li An-chung for more than two hours. He eventually got up his courage and invited Li to his house because he wanted to discuss something with him. Although Li was a bit dubious, it was hard to turn down such an offer. On arrival he was confronted by a cage containing two birds which were the spitting image of those in the Li An-chung painting.

The American proudly pronounced: "My Chinese friends, you say your painting is based on my birds? Or are my birds based on your painting?" At which they fell about laughing.

For the bird watchers of today, what the ancients called "bamboo pigeons" are in fact the familiar shrike that passes through Taiwan every year. In fact roast shrike was once a delicacy served to guests in the Kenting area. Ten years ago, when environmental awareness was just picking up in Taiwan, environmentalists put much energy into appealing to the residents of Kenting to stop trapping this bird, making it the first animal for which demands for protection were made.

Today as soon as bird lovers catch sight of the bamboo pigeon painting they will exclaim, "the

頭、山脈等的結構，再以毛筆勾勒出各種卷雲皴、披麻皴等捕捉自然物的最佳技法，「不論東西方，後來都很少看到畫家會像科學家一樣，去觀察石頭的結構，」蔣勳說。

卅年前因宋朝花鳥畫家李安忠的竹鳩圖所衍生的古今對照，故事有趣。而北宋畫家馬遠的故事，更凸顯了宋代畫家對萬物心存謙卑的觀察態度。

北宋亡國後，馬遠一路由北方奔向南方，抵達西湖，察覺到以原來畫北方乾燥、雄渾山水的技法，來畫南方水融融的景物，有了很大的障礙。馬遠於是四處旅行，觀察南方山水：海浪碰撞石頭反彈起來是什麼模樣，水經過土壤時的線條如何……，然後畫了十二段水與岸關係的「水圖卷」。

西畫中的配角

由於中國人的哲學觀，從未高高在上地排斥其他的自然生命，自然萬物在中國人的民族情感中佔有重要地位。因此含括鳥類、游魚、猿猴等動植物在內的花鳥畫，一直與山水、人物並列爲中國三大繪畫，無論質、量均極爲可觀。

反觀同時期，西方的宗教畫至高無上，「動物只是西畫中的配角，」畫家何華仁說。當時，幾乎沒有具代表性的動物畫可以進入

great grey shrike!'' The most obvious of the shrike's special characteristics are that it stands alone in the high branches of trees and its eyes roam widely in all directions. In Li Lin-tsan's view, as for the bamboo pigeons in the paintings, ''How could their grand, imperious, imposing manner ever be captured by a camera?'' Unfortunately, Li An-chung, who carried out detailed observations, could not have foreseen that today's trappers would catch the shrike, exploiting its particular habit of perching up high in its position of great preeminence.

Humble ecological observation: ''The first generation of Chinese painters were like today's scientists and carried out very scientific research into nature,'' says Chiang Hsun, former director of the fine arts department at Tunghai University. In the past he always felt that Chinese artists just portrayed their inner sentiments through a kind of expressionism, but this was actually a big mistake. This was especially the case in the Sung dynasty, when artists were influenced by the empiricist zeitgeist of a flourishing rationalism, which led them not only to observation of the structures of things, but also to an analysis of their anatomy.

Chinese artists used their eyes to observe the myriad things of nature, such as the structures of rocks and mountain formations, and took their brushes to catch nature's most magnificent objects through a variety of intricate techniques of the brush. ''No matter whether you are talking about

自然萬物一直是中國畫家的繪畫對象，然而他們卻從不只囿於描寫實物，也要追求隱藏在形體之後的精神。圖爲明朝孫龍所畫的戴勝，簡單的墨塊卻充分表現了鳥兒神氣十足的神情。左爲攝影者郭智勇在野柳拍到的戴勝。

Nature's myriad manifestations have always been subjects for Chinese painting, although artists have never been confined to realistic portrayal but have pursued the spirit behind appearances. In this work by Ming dynasty artist Sun Lung, simple patches of ink fully capture the lofty air of the hoopoe. The photograph left shows a hoopoe caught in the lens of photographer Kuo Chih-yung.

西方藝術史被廣泛討論。

直到十六世紀西方汲取自希臘、羅馬的藝術母汁已發揮到飽和，加上海運大開，東方美術中，尤其源自中國的日本花鳥畫曾受西畫家青睞，注入西方繪畫大河之中，但基於東西方哲學觀念的差異，激起的漣漪太小。

而中國花鳥畫，由於古人對自然經過長時間的認知與觀察，更促進後世花鳥畫在中國的重要地位。

好鳥枝頭亦朋友

故宮博物院登記組長佘城表示，花鳥畫由唐代孕育於自然思想，至五代急速昌盛後，便一路朝寫實的路子——在表現上追求自然的模擬——邁進，而寫實需要藉助高超的技藝，因此精熟的技巧始終也是畫家追求的目標。

早期的花鳥畫風中，善於設色、華麗的工筆花鳥畫派，開宗祖師是五代西蜀的宮廷畫家黃筌，他留給兒子黃居宷的「寫生珍禽圖」，就如今天的圖鑑一般，圖上的主角們，一隻隻站立紙上，彼此沒有關係，像等著供人辨識。雖然鳥數不多，美術史學者卻認爲此畫是宋代畫家充滿觀察自然精神的最佳證明，也可以說是中國最早的鳥類圖鑑。

今天的鳥友們仍可由寫生珍禽圖中，鳥類的特徵判斷出鳥種，臺灣高山上成羣的白額畫眉、白頰山雀，冬天過境臺灣、停留西部海岸河口覓食的白鶺鴒，都出現在千年前的這幅「鳥鑑」上。只不過它是道地的圖鑑，只有鳥類圖像作爲後人的描摹對象，不似今天的鳥類圖鑑，已經加上文字，介紹鳥兒的基本生態、習性與外貌。

多才多藝、詩文書畫俱能的南宋徽宗，也是個花鳥畫寫實論者。某一次宣和殿前種植的荔枝結實纍纍，適有孔雀徜徉樹下，他急忙召來畫工對景寫生，但所畫的孔雀皆舉起右脚站立。徽宗告訴畫工，孔雀上墩爬升時必定先舉左脚，驗之果然。

保存在臺北故宮，出自徽宗畫院的「梅竹聚禽」，在美籍中國繪畫史學者高居翰看來，是「真實感底下深藏著精密而耐心的觀察

十九世紀德國畫家伍爾夫所畫的臺灣朱鸝，發表在當時英國的鳥學雜誌「朱鷺」上。近代野生鳥類繪畫的風氣，始於西方畫者描繪生物學家由世界各地收集來的動植物。(張良綱翻拍)

The maroon oriole painted by the nineteenth century German artist Wolf appeared in the English ornithological journal *Ibis*. The modern style of painting wild birds began with depictions by Western painters of the flora and fauna collected by biologists from all over the world. (photo by Vincent Chang)

East or West, there were very few later artists who were so like scientists and made such observations of the structure of rocks,'' says Chiang Hsun.

It is interesting that some thirty years ago Li An-chung's *Bamboo Pigeons* led to comparisons being made between the ancients and moderns. People have also been moved by the story of the Northern Sung painter Ma Yuan, who even more conspicuously reveals the humble attitude taken by Sung artists in their observations of nature.

Following the conquest of the Northern Sung, Ma Yuan fled from north to south China where he settled at Hangchow's West Lake. He was soon confronted by many obstacles as he tried to use the techniques he had garnered from observing and painting the arid northern mountainscapes to portray the warm scenery of the south. This led Ma to travel widely, observing the mountain scenery of the south and such phenomena as the shapes created by the action of the sea on stones and the patterns made when water passes over land. He went on to produce 12 albums on the relationship between water and coast.

A supporting role in Western art: Because Chinese philosophy has never put people in a superior position, the myriad things of nature occupy an important position in the sentiments of the Chinese people. Thus birds, fish, apes and other animals in flower and bird painting have always gone to make up one of the three main themes of Chinese painting, along with mountain landscapes and people.

Looking on the other hand at the unsurpassable religious paintings of the same period in the West, ''animals just play supporting roles,'' says bird artist Ho Hua-jen. It seems there are just not any representative works of purely animal art that can be discussed in Western art history.

By the sixteenth century the influence of Greco-Roman classical art in the West had reached saturation point. With the opening up of sea routes, Japanese flower and bird painting that originated from China came to be highly regarded by Western painters and entered into the mainstream of Western painting. However, due to the different philosophical outlooks of East and West, the ripples it made were still too small.

As for China, the long process of familiarization and observation gone through by the ancients enabled bird and flower painting to secure an important place in its later art.

A bird in the bush worth two in the hand: Hsu Cheng, director of the registration department at the National Palace Museum, says that flower and bird painting started growing from thinking about nature in the Tang dynasty, and after a rapid blossoming in the period of the Five Dynasties took to the path of realism, seeking to imitate nature in its mode of representation. With such realism requiring high artistic skills, perfect technique became the objective sought after by artists.

The earliest flower and bird paintings had excellent colors and beautifully delicate and detailed brush work, as can be seen in the work of the court painter Huang Chuan of the Western Hsu in the period of the Five Dynasties. The album he left for his son is just like today's illustrations with the subjects scattered on the paper with no apparent relationship, as though waiting to supply people with information. Although this work does not include a lot of birds, art historians believe such art is the best proof of the great spirit of natural observation possessed by the Sung artists, and it can be said to be the earliest example of bird illustration in China.

Bird lovers today can judge the species of birds from the special characteristics shown in paintings. The great tits that once flocked in Taiwan's mountains, and the white wagtails that pass the winter in Taiwan settling in the estuaries of the west coast, can all be seen in the bird illustrations of a thousand years ago. Although these are real illustrations, they were only meant as models for later people to imitate, unlike today's illustrations to which are added inscriptions introducing details of a bird's ecology, habits and appearance.

With much talent and artistic ability, the able poet and painter Emperor Tsung of the Southern Sung was also a realist when it came to painting birds. On one occasion the lychee trees in front of his palace were laden with fruit and the peacocks wandering to-and-fro beneath made a perfect scene which the emperor quickly summoned his artist to capture. However, in the resulting picture the birds were shown with their right feet raised. Emperor Hui told the artist that when a peacock rises up it should first lift its left leg, and when the artist

前年郵政博物館發行畫家楊恩生的臺灣溪澗野生鳥類郵票，頗受歡迎。要使科學與藝術結合，畫家的挑戰更大。

Two years ago,the issue of stamps by the Chinese Postal Museum showing the wild bird species of Taiwan's mountain streams by Yang En-sheng was widely welcomed. The challenge for artists wanting to unify art and acience is a big one.

。藝術家知道竹是怎樣抽枝生葉的，也了解到鴿子細長羽毛和鶉的尖銳羽毛是不一樣的。」

越過寫實，進入寫生

北宋承襲五代遺緒，「繪畫技法仍然是寫實第一，」佘城說，宋人描繪技巧的能力比唐人進步，對寫實漸能掌握之後，品評層次更進而觸及形而上的境地。在浸淫繪畫陶冶性情之餘，又喜尋繹事物的內在，因此，成爲北宋繪畫主流的花鳥畫，除了追求精細、正確的外形描繪，更注重對象內在精神的表現。

林柏亭也認爲，對宋朝畫家而言，畫得逼眞已經不够，因此不再將「寫實」放在第一位，而是要將自然生命畫出來，也就是寫生。林柏亭在「寫生在畫史上之轉變」文中就指出，今人常用的美術詞彙「寫生」，總被誤解爲就是面對實景描繪作畫，清末以來，傳統繪畫受西洋衝擊，許多企圖有所作爲者提倡畫家應拿起畫筆重新面對自然，或現實的事物，寫生遂屢被提起。事實上，寫生二字最早見於宋人評品花鳥畫作品的文章，宋人的寫生是重在「生」字，而非今人所謂的「寫實」。

林柏亭以他認爲最具觀察精神，又超脫形體、捕捉到萬物氣韻的北宋崔白「雙喜圖」爲例，畫中兩隻山喜鵲對著路過其領域的野兎做驅趕狀，動物一下、一上，喜鵲展翅作勢嚇兎，羽毛因風翻飛，野兎轉身愕然，一臉迷惑，身體鬆軟如眞，讓人欲伸手撫摸。回頭來看今人所觀察記錄的山喜鵲生態，確是具有強烈護衞領域的鳥種。

高居翰在其「中國繪畫史」書中，形容雙喜圖之畫者觀察、捕捉到了自然的天性，且「對出現在自己畫中的生命，有了同情和了解。」

迷霧森林十八年？

欲捕捉自然的內在精神，達到自然生命形、神兼具，觀察的功夫需要更深，於是花鳥畫中被視爲畫風野逸、賦色淡雅，以五代南

went to see for himself this turned out to be the case.

In the eyes of art historian James Cahill, a painting from Hui Tsung's painting academy of plum, bamboo and animals kept in the National Palace Museum reveals how, "Under the feeling for realism there is concealed concentrated and patient observation. The artist knew how bamboo grows and sprouts leaves and understood the difference between the fine long feathers of a pigeon and the pointed plumage of the quail."

Surpassing realism, entering expressionism: She Cheng explains that when the Northern Sung dynasty took over the artistic heritage of the Five Dynasties, realism was still of primary importance for painting technique. Techniques in the Sung had made progress over those of the Tang, and after they had gradually been able to catch reality, the level of connoiseurship advanced to touch the area above appearance. Apart from sheer indulgence in the pleasure of painting, art also came to be moulded by a love of searching out the inner essence of things. Flower and bird painting entered the main stream of Northern Sung painting, and apart from pursuing detailed and accurate outer description it came to place even more emphasis on expression of the inner being of the subject.

Lin Po-ting thinks that the portrayal of descriptive reality was just not enough for the Sung painters. Realism was no longer of paramount importance; it was rather the essential life of the subject that had to be brought out.

Lin Po-ting takes the Northern Sung artist Tsui Pai's *Double Happiness* as an example of art of both superseding appearances and catching the tone of the myriad things of nature. In this work, two magpies face off a hare that has crossed their territory; the birds stretch out their wings to scare the hare, their feathers ruffled by the wind; the hare turns in astonishment with a look of bewilderment, its body pliant with a realistic softness that makes you want to stretch out your hand and stroke it. Looking at what has been recorded today about the magpie, it is in fact a bird that defends its territory very fiercely.

In his history of Chinese art, James Cahill describes how in *Double Happiness* the artist's powers of observation and his ability to capture the essence of nature, "reveal a sympathy and understanding of life in the painting."

An 18-year fog in the woods? In capturing the inner spirit of nature and conveying its living manifestations, the linkage of spirit and matter requires much hard observation. Thus flower and bird painting has been seen to be a refined pursuit using a simple elegance of color, as with the Hsu Hsi school of painting of the Southern Tang of the Five Dynasties period which managed to integrate its art with nature to such a great degree. I Yuan-chi of the Northern Sung once went deep into the mountains to live with the apes so as to observe their ecology. Digging a pond for flowers, letting waterfowl gather there and concealing himself in dark places to observe their activities, his spirit of investigation was not so far removed from that of today's ecologists.

Because the ability of the Sung artists to work from nature was built on the foundation of observation, when you share their ornithological experience many problems are revealed in what are faked Sung paintings as it is possible to observe whether they have been done from life or are just imitations. Lin Po-ting explains, "Usually later people imitated the Sung paintings with great skill, but they had not been through the process of observation, so they could often unwittingly reveal their weak points."

This is especially so with the Ming dynasty, when artists also worked from life, had great ability when it came to realistic portrayal and adopted a more florid style than the Northern Sung. Many works from that time were stamped with seals counterfeiting the Sung paintings. But if you look carefully you can discover that what they ultimately lacked was actually that special quality of thinking developed by the Sung after its engagement in studying the anatomy of objects.

Lin Po-ting says that one cannot go so far as to say that everyone in the Sung was possessed by the empiricist spirit, but at that time it was certainly a prerequisite for any artist who wanted to set about painting birds. It was this flourishing of research into the nature of birds that gives Sung paintings their degree of rationality when looked at from today's perspective.

Finding the self in contemplation of nature: The basic nature of art is to evolve according to the principle of moving from simplicity to complexity, then returning back from over-

唐徐熙爲首的畫派，更融入自然。畫家北宋易元吉，曾入深山，與猿猴爲伍，動則累月，觀察猿猴生態與林石景物。也曾在長沙家中鑿池植花，任水禽自然羣聚，再藏身隱處，觀察其動靜遊息，與今天的生態學者從事生態調查的精神不相上下。

由於宋人的寫生能力是墊基在觀察功夫上，因此若能與宋人一樣擁有觀察鳥類生態的經驗，可以藉此看出很多宋畫仿作的問題，也較能够判斷係屬寫生或臨摩而來。林柏亭解釋賞鳥對從事美術史研究的助益何在：「通常後人臨摩宋畫惟妙惟肖，但並未經過觀察，常會不經意地露出破綻。」

尤其到了明朝，畫家也崇尚寫生，極盡寫實之能事，華美富麗更甚於宋畫，許多當時的作品常被後人添上僞款，誆做宋畫欺世。但若細心觀察，可以發覺仿作終究缺少宋人發自思維、悟自物理後具有的特質。

他再舉鶺爲例，鶺翅尾端黑色，因此收起雙翅時，尾巴看來是黑色的，但有些人畫飛行的鶺尾部仍呈黑色；同一個作者應該不會在兩幅畫上出現一對一錯的情形，如此就值得再進一步細究。林柏亭表示，不敢說所有宋人都具備格物精神，但在當時這是畫家畫鳥之前必須下的首要功夫，研究鳥類天性的風氣極盛，因此今天看來，宋畫表現通常比較「合理」。

萬物靜觀皆自得

藝術本身具有由簡入繁，繁極返簡的演變法則，繪畫發展至南宋，寫實主義逐漸發展到盡頭，筆墨轉趨簡化的畫法代之而起。佘城指出，南宋畫家在勾勒花鳥的輪廓上，揚棄過去工整圓勁的細線，改用粗獷奔放的粗線，用色也由濃彩敷設改爲淡彩漬染；對於對象則力求突出個性特徵，簡化造型，掌握意象的表現，於是開創出筆墨蒼勁、意象蕭疏，以寫意爲重的藝術境界。

老一輩的畫家林玉山在其「談雀與畫雀」文中就說，由當時的畫看來，可知畫家已體會出無須以形色華麗者爲美，反尚質樸淡雅者，並重視畫面筆墨之效果。在此繪畫思潮的影響下，麻雀「相貌」雖平凡，地位卻不遜於其他珍禽，成爲重要的鳥畫主角。

藝術是自由的，並無規範走向一定要如何。南宋之後，特別是文人畫家捨形從神，認爲只要能追求到物的精神，可以放棄形的眞實。無論重「形」或重「神」，後人只能說是中國人在藝術、文化上的一種選擇，是美學觀的轉變，無所謂對錯。但這樣的改變，仍然墊基在先有了仔細的觀察，再來轉化，也就是透過形的表現，但最終目的是要完成自我。

中國美術發展到後來，畫者皆相信只要思想豐富，不怕技巧拙樸、色彩簡單。人們認爲，人永遠達不到自然老師的境界，只有褪去華麗的表相，回到自然，也就是回到自己心中，靜下來聽自然與自己內在的對話。就像清朝八大山人的禽鳥作品，主題常是輕描淡寫幾筆而成，他的目的其實在藉鳥表現自己的心境，這才是中國文人心目中藝術的最高境界。

不經一番寒徹骨

可惜隨著時代發展，不論寫意或寫生，都被認爲不能跟上時代。少了形體，隨人自說自話，魚目混珠；另一方面，花鳥畫家漸失靜觀萬物的用心，如清朝畫評家方薰所說，宋人論花鳥是推崇能得生意者，「今人畫蔬果蟲魚，隨手點簇者謂之寫意，細筆鉤染者謂之寫生，以爲意乃隨意爲之，生乃像生而肖物者，不知古人立法命名之義焉。寫意寫生卽是寫物之生意也。」

除了少數畫家，花鳥畫在傳統中打轉，一般人遂誤解中國花鳥畫就是缺乏觀察、只知精雕細琢、臨摹古人的工筆畫。

此時在西方，經過一翻顚簸過程，自然觀察反而發展成爲一專門學問。

歐洲在十五世紀後的文藝復興、啓蒙運動，加上商業活動興起，個人生命由宗教的壓抑中解放出來，人自覺到本身的重要性，歌頌人存在與被人所征服的物質遂成爲繪畫題材。尤其是佔據過臺灣的荷蘭，最早向外拓展市場、尋找資源，新興的中產階級成爲畫

complexity to simplicity. By the Southern Sung, painting had developed to such extreme realism that brush work tended now to veer back towards calligraphic simplicity. In sketching flowers and birds, the Southern Sung painters gave up intricate lines of neat perfection in favor of bold and vigorous strokes and forsook rich colors in favor of more gentle pigments. Simplification of form could bring out the impressionistic meaning of subjects, opening the way to vigorous brush work, forlorn imagery and the world of impressionism.

In his writing *On Sparrows and Painting Sparrows*, Lin Yu-shan, a painter of the older generation, says that the paintings of that time already show an understanding that beauty can be achieved without florid shapes and colors but through a simplicity and plain elegance that emphasizes the effects of brush work. Under the influence of this tide of thinking, although the ''appearance'' of a sparrow might be plain, its status is not inferior to that of a rare species and it came to play an important role in bird painting.

Art is essentially free and there are no fixed rules to determine how it should be done. After the Southern Sung, painters forsook appearances for the sake of spirit, thinking it was enough to just pursue the spirit of things and that true appearances could be given up. Those who followed could only say that this was a choice made by Chinese artists, an aesthetic shift that could be judged to be neither right nor wrong. Yet even this kind of change was still built on the detailed observation by which it was preceded, passing through the intermediary of formal expression, although the ultimate aim was complete self fulfilment.

In the later development of Chinese art, artists on the whole came to believe that all that was needed was a richness of thought and no fear of crude

崔白的「雙喜圖」，形、神兼具，是宋朝寫生花鳥畫的代表作。

The unity of form and spirit evident in Tsui Pai's *Double Happiness* is representative of the Sung dynasty genre of flower and bird painting.

黃筌的「寫生珍禽卷」中，有九官、文鳥、白頰山雀等今天賞鳥者一一可叫出大名的鳥種。圖左中間則是今天仍每年過境臺灣的候鳥白鶺鴒。下圖爲郭智勇所拍的白鶺鴒。

Huang Chuan's scroll of rare species painted from life features birds the names of which today's bird watchers can call out with ease. To the left and center of the picture can be seen the migratory white wagtail which passes through Taiwan every year. Below is a photograph of the white wagtail taken by photographer Kuo Chih-yung.

市主要購買人，於是首先開展出人物肖像與平民化的題材，生活周遭相關的風景、動植物，如兔子、魚、鳥等動物常以靜物畫呈現。科學與醫學的發展，解剖學大盛，以被分屍的馬、被打傷的鳥爲主題的狩獵畫，也曾流行一時。但這種包含人類征服自然意味在內的畫，無法成爲主流。

中西接上頭？

十八世紀自然科學如大海澎湃，許多畫者追隨探險家，四處描繪來自各地的動植物。以科學記錄爲目的的繪畫，首要求眞，加上觀賞自然的工具發達、生態研究興起，保育風氣席捲，賞鳥人口大增，科學圖鑑大量出版，藝術市場需求也日多，遂發展出今天的生態繪畫。

西方在師法自然的過程，呼籲應尊重生態、描繪自然的畫家，還與傳統狩獵畫畫家有過唇槍舌劍的論戰。

如今塵埃落定，動植物的自然生態已成爲畫家喜愛的對象，尤其色彩鮮豔亮麗，種類、數量最多的鳥類，更成爲生態繪畫的主角。「至此中西方接續上了，」蔣勳認爲，雖然中國畫家在當時可能沒有西方生物學中動物吃些什麼，那一塊肌肉該凹、該凸等等的瑣碎觀察和記錄，「但科學作法上容有不同，尊重自然本性的精神層次，在藝術上已達極高的境界。」

由藝術長河看來，自然生態這一主題，中國曾經走在前頭；但美學發展的路向不同，無關乎誰快、誰慢，誰會、誰不會。今天西方講求所謂科學與藝術兼具的鳥畫，其繪畫理念、思想可能與中國古人不謀而合，只能說今人終於發現自然美妙、引人入勝之處，從而作爲繪畫體材。

寫實不難，寫生難

今天本土文化日受重視，人們關心本地特有的鳥類，加上科學講究品種，強調畫的是藍腹鷴、帝雉，對象必須很準確，繪鳥的意圖，除了藝術的欣賞，還需要讓觀者對鳥種一目了然，畫者的挑戰很大。

technique and simple colors. People felt that human beings would never be able to really attain the rigorous standards of such a teacher as nature and had to get rid of gaudy representations. The return to nature was thus rather seen as a returning to one's own heart, a quiet listening to the dialog of nature with the self. Such was the case of the animal and bird paintings of Pa Ta Shan Jen of the Ching dynasty in which the subject is portrayed with just a few minimal strokes of the brush. Here the aim of using birds as subjects is to express the territory of the artist's inner being. It is this that the Chinese literati came to think of as being the highest attainment in art.

If you do not suffer the winter . . .: Unfortunately, later developments, no matter whether of an impressionistic or realistic bent, have all been considered not in tune with the times. Lacking form and structure, following the whims of subjective opinion, ''fish eyes came to be passed off as pearls,'' while flower and bird painters gradually lost their interest in quiet contemplation of the myriad things of nature. The Ching dynasty artist Fang Hsun could thus say that while the flower and bird painting of the Sung was held in high esteem for its capturing of the vital significance of life, ''Today when people paint vegetables, insects and fish using blobs of ink it is called impressionist, while detailed outlines and filled in colors are called realistic; thinking that impressionism follows impressions and realism is like life and resembles objects, they know not the meaning of the laws and ordinances established by the ancients. In reality, impressionism and realism are merely the portraying of the vital significance of the being of objects.''

Apart from a small minority, most people arrived at the misunderstanding that Chinese flower and bird painting lacks the qualities of observation, only knowing the detailed craftsmanship and polish that harks after the refined brush work of the ancients.

At this time the West was passing through a turbulent stage in which observation of nature developed to become a special discipline.

In Europe after the Renaissance, the Enlightenment and the rise of commerce, the individual was released from the yoke of religion and people became conscious of the value of their own nature.

去年舉行「臺灣鳥木刻版畫展」的何華仁曾出走城市，遠至六龜進行鳥類觀察，他認爲賞鳥對作畫時，捕捉鳥類的神情與生態有所助益。圖爲他的木刻作品「褐鷹鴞」，圓而略帶童騃眼神的貓頭鷹，自然、樸實而不誇張。
(何華仁提供)

Ho Hua-jen, who held an exhibition of woodcut prints of Taiwan's birds, once left the city to travel to Liukuei and observe birds as an aid to capturing their spirit and ecology. This woodcut shows an innocent-looking wide-eyed owl in a style that is natural, simple and unexaggerated. (courtesy of Ho Hua-jen)

「畢竟寫意可以大膽揮灑，只怕沒有眞正的內涵，而要工筆到『工而能活』卻不容易，」林柏亭說。明、淸之後的工筆花鳥畫之所以落入窠臼，正因爲沿襲古人風格太久，已經工而不活。

事實上，只要經過正規繪畫訓練，要克服描繪上的求眞並不難，但要成就眞正的藝術，如宋畫大家的形、神兼具，畫家要克服的是藝術上的表現風格，不只是技巧的問題。

現代的畫鳥或許源自於西方，但就像曾經辭去工作，在中央山脈扇平從事鳥類生態觀察的何華仁，除了求眞的生態畫鳥，也嘗試以版畫表現臺灣鳥類，線條轉爲簡單、樸拙。這樣的轉變，說是西方的，不如說「更中國」。因爲就如古人一路行來，深切地體悟到，只描寫眞實、畫得惟妙惟肖，畢竟不是藝術的本質。若要回歸藝術，「眞不眞」就眞的只是其次的問題了。 S

(原載光華八十二年二月號)

Human existence and the materials conquered by people became the stuff of art. This was especially so for the Dutch, who colonized Taiwan and were the earliest to expand their markets and search for resources. When the ascending bourgeoisie became important buyers in the art market, this led to the rise of portraiture and the gentrification of subjects which came to include scenes common in everyday life, animals and plants, with rabbits, fish and birds often appearing in still lifes. The development of science and medicine and the flourishing of anatomy meant that gashed horses and wounded birds appeared in hunting scenes and became fashionable for a time. However, such art, revelling in the human conquest of nature, never became mainstream.

When China meets the West: With the eighteenth century explosion in the natural sciences, many artists followed explorers to all corners of the earth to portray the flora and fauna they found. Such painting for the scientific record was concerned primarily with seeking factual truth. When tools were developed to aid in the appreciation of nature and ecological research took off, conservationism swept the world and the number of bird watchers increased daily. Scientific illustrations were produced in great quantities and the demands of the art market gave rise to the ecological art of today.

While the West is in the process of learning from nature, Western artists appealing for respect for the environment and set on portraying nature still lock swords with the traditional painters of the hunt. Now that the dust has settled, the natural ecology of plants and animals has become a favorite subject for artists. Birds, with their resplendent colors and great variety, have especially come to play an important role. "In this way China and the West are connected," thinks Chiang Hsun. Although Chinese artists in the past did not have the Western biological knowledge of what animals eat and how their bodies are shaped, "There are different kinds of scientific method, and the level of respect for the fundamental essence of nature had already reached a very high point."

Looking at it from the long river of art, China once took the lead in ecology. Yet aesthetics developed to take roads leading in different directions, no matter who has been fastest or most able. The ideals and thinking given to painting by the Western meeting of art and science in bird painting might accidentally be the same as those of the classical Chinese, but it can only be said that people today have finally discovered the wonders of nature and drawn them to areas of great beauty. It is thus that it has become the subject matter of painting.

Realism easy — art from life hard: Today native culture is increasingly receiving more attention, and people are concerned about indigenous bird species. In science, great emphasis is placed on exact breeds and species, and painters strive to portray Mikado pheasants or other precise species of birds. What must be aspired to in painting birds, apart from aesthetic enjoyment, must still be to let spectators understand birds at a glance. The challenge for artists is very great.

"Ultimately, impressionism can paint with a bold freedom, although there is a fear that it lacks any real significance. But wanting to paint in a refined and detailed way that can bring your work to life is not at all easy," says Lin Po-ting. That the flower and bird painting of the Ming and Ching dynasties is said to have fallen into an ossified pattern was precisely because, having followed classical styles for too long, their work lacked vitality.

In fact, it is not that difficult to just go through a rigorous artistic training to overcome the difficulties of true representation. But to achieve the meeting of spirit and form, as did the artists of the Sung dynasty, artists must overcome the problem of style and not just technique.

Perhaps modern bird painting originated from the West. Yet an artist like Ho Hua-jen, who gave up his job to go to the central mountains to make observations, apart from seeking to paint birds realistically also wants to use his prints to represent the bird species of Taiwan with simple bold lines. Instead of being called Western, such an evolution might better be said to be even more Chinese in its travelling of the same road as the ancients. What must be fully realized is that just portraying a good likeness is not the essence of art. If you want to get back to real art, then questions of "real or unreal" are really of secondary importance.

(Chang Chin-ju/art courtesy of the National Palace Museum/tr. by Christopher Hughes/ first published in February 1993)

知天、知地、知物
——林玉山的畫鳥觀

Know the Sky, Know the Earth, Know the Creatures
— Lin Yu-Shan on Painting Birds

文·張靜茹

今年八十七歲的臺灣前輩畫家林玉山，喜愛畫鳥，在他的八十五歲回顧展中曾展出鷹隼、環頸雉、鴛鴦、鵲鴿等廿多種臺灣本土的野生鳥類。雖然過去受限於交通工具，無法更深入臺灣的山林觀察鳥類，因而所畫的鳥種不如新一代畫者豐富，但林先生的畫鳥精神卻不輸今人。

外師造化

他喜愛畫麻雀，為了了解麻雀的習性，常在田中以稻草蓋窩，藏於其中，觀賞鳥兒啄食穀粒、爭鬥、飛翔、鳴噪，常看得入神，彷彿自己也跟著雀羣跳躍。他也在野外寫過花鳥日記，內容是生態的觀察和研究，比如某種鳥類何時脫毛，冬、夏羽毛色澤如何區別等。

他曾為自己的寫生原則，整理出所謂的「三知論」。三知是指知天、知物、知地。知天，意指隨著時令與氣候變化，自然界的物體不免隨之產生不同現象，畫鳥人就應觀察到鳥類、花卉在四季不同的色澤與形態。

所謂知地，指須熟知大地地理環境的不同，高山、河川、海島等各種不同的地貌上，都有特殊的動植物存在；熱帶、溫帶、寒帶間則有不容並論的自然條件。而知物，也就是作畫時，須對動植物的生理、動態作入微的探究。

提倡自然寫生

林玉山有一套很中國的繪畫觀。他曾說，古人創作花鳥畫，無論勾勒色彩，或水墨寫意，都必須根據自然的觀察，不容率意描寫。雖主張意不在似的寫意畫，其所謂「不在似」，實係一種神似形離的超現實，根本就不是忽視自然的幻想畫。

他認為，寫生目的不在工整地寫實，而應寫其生態、生命，得其神韻。

「師造化必須深究自然，一方面由外體察，一方面從內孕育，所謂中得心源，須內外配合才可，」是他說過足以代表他寫生精神的一段話。

（原載光華八十二年二月號）

Lin Yu-shan, 87 years old this year, loves to paint birds. His retrospective exhibition at the age of 85 featured more than 20 different species of Taiwan's indigenous wild birds. Although travel difficulties which meant he was unable to go deep into the mountain forests to observe birds might have left his range of species somewhat less rich than that of the new generation of bird artists, the spirit of Lin's bird painting is by no means inferior to theirs.

Making nature your teacher: Lin is particularly fond of painting sparrows. So as to understand the habits of the sparrow, he would often build a hide in the fields out of rice stalks, where he would conceal himself so as to get a close view of birds pecking at discarded husks, squabbling, hovering and singing. He would watch so often to try to get into their spirit that it seemed he was hopping about with the flock. He also wrote a bird diary out in the wilds, containing ecological observations, such as at what times certain birds shed their feathers and how their plumage changed color between winter and summer.

For the principles of his own painting from life, Lin organized what he calls a tripartite epistemology. His three theories of knowledge include knowing the sky, knowing objects and knowing the ground. ''Knowing the sky'' involves following the changes of the seasons and weather: The objects in the sphere of nature cannot avoid producing different manifestations according to these changes and the bird artist should thus observe the different colorings and shapes that occur in birds and flowers through the four seasons.

What Lin calls ''knowing the earth'' involves becoming familiar with different geographical environments. High mountains, rivers and marine islands all have their different appearances and their own special types of plants and animals; tropical, temperate and frigid zones have their own natural conditions that cannot be confused. Finally, ''knowing objects'' means that when you are painting, you must do detailed in-depth research into the physiology and movements of plants and animals.

Advocating natural painting from life: Lin Yu-shan has a set of very Chinese ideas about painting. He once said that when the ancients did flower and bird painting, no matter whether it was outline filled in with colors or impressionistic splashes of ink, all had to be rooted in observations of nature and did not permit off-the-cuff depictions. Lin might well advocate impressionistic painting that does not pursue likeness, but what he means by ''no likeness'' is actually a kind of superseding of reality through a likeness to spirit that goes beyond form, and is not to be confused with fantasy painting that takes nature lightly.

Lin thinks that the aim of painting from life is not to be found in accurate realism. It should be to depict the relationships of things to nature, their life and to catch their poetic grace.

When making nature your teacher you must look deep into the natural world, on the one hand looking at the outside, and on the other getting nourishment from the inside. What is often called getting to the source of your inner self can only be possible through such a coordination of the inner and outer.

(Chang Chin-ju/tr. by Christopher Hughes/ first published in February 1993)

臺灣前輩畫家林玉山喜畫自然生命，尤其農村最常見的麻雀更深得他的喜愛。(林柏亭提供)
Senior Taiwan artist Lin Yu-shan loved to capture natural vitality in his work, and was particularly fond of the sparrows that are so common in farm villages.(courtesy of Lin Po-ting)

布農族民相傳上古時期，紅嘴黑鵯曾爲他們祖先渡海取火。(郭智勇攝)
It is legend among the Bunun people that in ancient times the black bulbul crossed the water to bring fire to their ancestors. (photo by Kuo Chih-yung)

飛越福爾摩沙的時空

Flights Over Formosa — Aboriginal People and the Birds

文・張靜茹　圖・張良綱

對於最早定居臺灣這塊土地的住民，「賞鳥」的歷史傳之久遠，只不過他們觀察鳥類行爲的目的，與今天的賞鳥者和鳥類學者都不同。

許多人都能說出一些與鳥兒相關的回憶，對於原住民，今天仍生存於臺灣野外的鳥類，卻是整個民族的共同記憶。

布農族的共同記憶

傳說遠古時期，臺灣發生大洪水，淹沒了陸地，全島只剩三千公尺以上的山頭露出水面。布農族民於是退到鄰近的卓社大山頂，天色昏暗、大雨不歇，忽然，與卓社大山遙遙相對的玉山頂上，茂盛的玉山圓柏灌木叢遭雷擊燃燒起來。

布農族人想，如果能到對山取回火種，烤食海魚與燃燒木材取暖，生命就有延續的希望。許多動物自告奮勇渡水前去取火，但路途遙遠，返程中火苗燃盡，動物忍受不了炙熱，火把紛紛掉入海中，只有鳥兒快速飛越玉山，啣著火種回到卓社山巔，但也因此被火灼傷。

這隻對布農族羣生命延續有恩的鳥兒，嘴巴被燙得火紅，身體燻成烏黑，至今仍優游於臺灣山林中，牠也就是廣泛分布臺灣森林中的特有亞種鳥類「紅嘴黑鵯」。

神話裡，鳥兒多

人是怎麼產生的？生命的延續過程，歷經過多少苦難？這樣的問題不時存在人們心中，因此每個民族都有一些含有創世紀意義的神話。根據人類學者的說法，布農族「紅嘴鳥兒渡海取火」的故事，始源於地質史上的第四紀冰河期，由於冰河溶解，當時也是地球的海漲期。同時期西方誕生了聖經中「諾亞方舟」與創世紀相關的故事；漢民族也有後來轉化爲「大禹治水、三過家門不入」的傳說。

對於長久居住自然山林中的原住民族，生活環境中蘊藏豐富鳥類，他們悠遠綿長的歷史，也可視爲一部與自然朝暮相處的生活史。而動物孕生人的傳說，與動物對人類族羣生存、繁衍有恩的故事，也就極其豐富。

帶領泰雅族射日

去年雲門舞集發表的新作「射日」，則是

For the aboriginal peoples, the earliest to settle on this island of Taiwan, "bird-watching" has a long history. The only thing is that their purpose behind watching our winged friends is completely different than that of contemporary bird-lovers and ornithologists.

Quite a few people can relate some memory about a bird. But for the aborigines, many of the wild birds that now exist in Taiwan are an ineradicable part of the collective memory.

The collective memory of the Bunun: It is said that in ancient times there was a great flood in Taiwan which inundated the land, with only the peaks above 3,000 meters still sticking out of the water. When the Bunun people retreated to the peak of Chuoshe Mountain, the sky was dark, heavy rains fell continually, and thunder and lightning ran jagged patterns across the sky. The densely shrubbed cypress forest of Yu Mountain, visible in the distance from Chuoshe Mountain, was struck by lightning and caught fire.

The Bunun people thought that if they could just get over to the peak there, they could bring back fire to cook fish and warm themselves, so that they would have some hope of keeping themselves alive.

Many animals bravely volunteered to swim across the floodwaters and fetch the flame, but the journey was long, and the kindling had already burned down by the time they could return. Unable to stand the pain, the animals had to drop the fiery wood into the sea. Only a bird, with its capacity to fly, could bring a live flame back to Chuoshe Mountain, though it too was injured in its endeavor.

This bird which did so much to extend the life of the Bunun people had its beak heated to red-hot and its body scorched to black, and it still sails over Taiwan's mountain timber belts — it is the red-billed black bulbul, a subspecies unique to the Formosan forests.

Mythical birds: How was man created? In the process of survival, how many trials have been overcome? These kinds of problems are more or less constants in the minds of human beings, so every culture has its own fables about the genesis of life. According to anthropological theories, the Bunun story of the "bulbul bringing fire across the water" has its origins in the Fourth Ice Age in geological

泰雅族生命繁衍過程中面臨苦難的故事。

在冰河期之前的暖溫期，也就是漢族后羿射日故事粗胚成形的同時，相傳天上多出了一個太陽，兩個太陽一上、一下，大地已失去白天、黑夜之分，天氣苦熱，傳染病孳生，萬物無法生息。爲了子孫生命的延續，在泰雅族族羣佔有重要地位的鳥（sisilika），就建議族民射下一個太陽。

在牠的引領下，一批勇士出發尋找太陽，一代接著一代，終於走到射日之巔大霸尖山。太陽被勇士們射中，噴出鮮血，四散成爲滿天星斗，血液流盡後化爲月亮。鳥兒的羽毛也被鮮血噴染，沉澱爲紫色斑點。雖然傳說中的鳥兒，不一定有具體形象；但雲門舞集的射日，仍選擇與臺灣原住民生活息息相關、身上有綻藍色澤的臺灣特有鳥種「藍腹鷴」，成爲 sisilika 的替身，出現舞臺上。

速度的象徵

在人類仍無法理解許多自然現象的洪荒時代，對鳥類的「超能力」，既敬又畏。鳥兒雖小，卻擁有人所沒有的飛行能力，能飛越人所不能及的空間。人們對鳥類又羨慕又忌妒的心情，也使鳥類在許多民族成爲重要地位的象徵。

在魯凱族與排灣族的成年禮中，必須舉行賽跑。因爲當有災難來臨時，腳力與體能最好的人可以儘快回到族內求救，或通知族人盡速避難，也能快速地追捕到獵物——強壯的身體是保護族羣的重要條件，也是成人必須具備的「美德」。

而族裏脚程最快的人，才有資格在服飾上繡鳥圖案。今天屏東霧台鄉新好茶村民，仍有在牆上雕刻紋飾的習俗，但全村只有兩個家庭可以在牆上雕刻燕子。鳥兒是速度的象徵，對於獸獵社會，這樣的象徵有重要的意義。

排灣族雕刻家薩古流，其弟由於跑步速度快捷，頭飾上可以裝揷五根臺灣特有鳥種「帝雉」的尾羽，而薩古流的頭飾卻只能有一根雉雞羽毛。

寄悲情於鳥

某些鳥兒對人們還具有警惕意義。傳說許多父母疏於照顧子女，等過一段時間想起小孩，孩子已經變成鳥兒飛到樹上了。蘭嶼雅美族有兩個兄弟遭雙親虐待，化爲鳥兒飛走的童話；泰雅族也有女兒變老鷹的故事。

霧臺新好茶村村民在自家牆上雕刻燕子，代表家中曾有過跑步健將。

A sparrow carved on the wall of a home in Haucha Village means that someone in the house has been a champion runner.

history. Because the glaciers melted, that period also witnessed the rise of the level of the ocean. From the same time period, Westerners have the story of Noah's Ark on the survival of man, as well as other tales of creation. The Han Chinese also later developed the legend of the official Yu, who was so occupied with and devoted to taming the floods that he passed his old home three times without even entering the door.

For the aborigines, who have long lived in mountain forests, their living environment has been rich with avian life. The lengthy history of their existence could be seen as a history of constant interaction and coexistence with nature. And legends about how animals gave birth to human infants, and of the kindness of animals toward mankind in its struggle for existence are also richly varied and widely passed down.

Leading the Taiya people to shoot the sun: Last year's Cloud Gate Dance Theater production, "Shooting the Sun," is actually a story of the repeated hardships faced by the Taiya aborigines as they tried to establish themselves and grow. During the warm period prior to the ice age, which is to say at the same time that the Han Chinese legend of Hou Yi (who shot down nine of the ten suns with his bow and arrow) was said to take place, the Taiya version has it that another sun appeared in the sky. There were two suns, one above and one below, so that the distinction between day and night was lost. The weather became oppressively hot, and illnesses spread rampantly; all living things faced extinction. To protect the future of their children and grandchildren, the sisilika bird, which has an important place among the Taiya people, suggested that the people bring down one of the suns with an arrow.

Under its guidance, a group of warriors set off to find the sun, moving generation after generation, until finally they had reached a peak close enough to shoot down their target. After being pierced by the warriors, the sun emitted fresh blood, which spattered across the heavens to become the stars. After it had been bled white, what was left became the moon. The bird's plumage was also colored by the blood, becoming spotted reddish-purple. Because today it is impossible to discover which bird the sisilika was, Cloud Gate selected as its bird the Swinhoe's pheasant, a species unique to Taiwan whose body is colored blue and also has purple markings.

Symbol of speed: In primitive times when people understood little of natural phnomena, they were both impressed by and fearful of the birds' "transcendent" abilities. Although small, birds could do something man could not — fly — and cross spaces where man could not go. People were both envious and bitter, so that many birds have become important symbols of one kind or another.

In the rites of adulthood of the Rukai and Paiwan people, it is essential to hold a footrace. Because those fittest and fleetest could in times of danger return fastest to the village for help, or warn others to flee the approaching danger, or could also quickly hunt down animals, a strong body became an important requirement for protecting the tribe. It was an integral part of being considered adult.

Only the swiftest individual in the tribe had the right to embroider a bird on his clothing. Today the villagers of New Haucha in Wutai Rural Township, Pingtung County, still have the custom of carving patterns into the walls. But only two houses in the village can carve swallows on their wall. Birds are a symbol of speed, which has great significance for a hunting society. Because the younger brother of the Paiwan sculptor Sakuliu finished first in the dash at the Taiwan Area Athletic Meet, he is allowed five Mikado pheasant feathers in his headdress; Sakuliu has only one pheasant feather.

Showing bad parents the bird: Birds also serve as warning signals to Taiwan's aboriginal

更有小孩因爲受到壓抑，夢見自己變成鳥飛了起來，「其實這是精神上的飛走，一種心理平衡作用，」人類學者洪田浚說。現代心理學者喜歡由心理學來解釋神話，這種人變鳥的故事，可以說是對不負責任父母的警剔，也是對整個族羣的啓示。

其實與鳥類相關的故事，各民族都不缺乏，今天鳥類在原住民族的一些象徵意義，在其他民族也都存在。

比如進一步對原住民命運有決定性功能的「鳥占」，中國早期也曾以鳥卜吉凶，做爲行事準則。美國作家伊雷克深入印尼婆羅州內陸叢林旅行，完成的「作客雨林」書中，也提到當地住民曾告知他預兆鳥的重要性。

洪田浚指出，對於原住民族，鳥占是包括夏威夷羣島、東南亞在內南島文化圈中的文化特色，臺灣九大原住民族中，除了蘭嶼島上的雅美族，包括平埔族在內，都有鳥占的習俗。

鳥占

原住民的占卜工具當然不只鳥類，但除了夢占，最重要的就是鳥類。舉凡是否築屋、打獵、播種、收刈，甚至出海，都要在特定的占卜地點舉行鳥占。若是半途中見到鳥兒象徵不吉祥的行爲，也得暫停前往，打道回府。一些小時跟大人打過獵的原住民，如今仍然淸楚記得鳥占的情形。

傳教士馬偕博士在「臺灣遊記」文中記載：「原住民以最深的尊敬，崇信小鳥的唧唧聲和動作，每回探險均事先考慮，尤其是狩獵。得先到部落外邊，投樹枝於樹上，驅走鳥類，如果鳥聲和飛走的方向一定，他們才會出發，否則頭目也鼓不起勇氣來行動。」

人類學者認爲，預兆與占卜是產自人們的好奇心與求知慾，原本都是人類自然的心理反應。但在原始民族之中，又會添加精靈的觀念，和對於不可思議現象的驚畏，再加上錯誤的推理，因而促進了對於預兆的信仰。基於這種信仰產生了占卜行爲，而這種原始禁忌，可以視同爲現代社會的法律。

不過上一代，鳥占仍是規範原住民行爲的重要依據。由於摻雜敬天畏神的精神在內，鳥占的影響力，遠大於今天的法令、制度。

捕鹿必聽鳥音

十七世紀，鳥占第一次被西方人記錄下來，由於臺灣東部產金、日本產銀，當時西方有所謂到遠東金銀島發財的熱潮。一六四〇年，當時的荷蘭統治長官布格(V. Burg)命令士兵自安平乘船前往臺東卑南採金，在勘查途中，卑南嚮導聽到鳥鳴，認爲不吉利，堅持回頭。荷人怎麼要求都不行，也不得不折回。

淸代出版的「皇淸職貢圖」書中也形容原住民：捕鹿必聽鳥音以占得失，以鳥聲之長短強弱、飛法等，做爲吉凶之判斷。淸朝記載當時臺灣住民生活的「番社采風圖考載」則認爲，鳥占就像漢人有風水之說，築屋、架橋喜歡請地理師勘輿，原住民雖不諳勘輿，但築舍也先行鳥占，確定良辰吉時再大興土木。

原住民的自然法則

與象徵速度一樣，占卜用的鳥兒地位通常都很高，每個民族用以占卜的鳥種也不同。資深鳥友劉克襄參考書上對鳥占的記載，認爲布農族常以小彎嘴畫眉、泰雅族以今天臺北植物園都能見到的繡眼畫眉爲鳥占主角。至於爲何選擇這些鳥種，除了是當地的土生鳥種，還有待進一步考證，而今天也僅剩泰雅族和曹族保存著鳥占起源的傳說。

嘉義縣誌記載，過去曹族尚未發明弓矢前，捕獸只能用陷阱。後來有一孤兒發明槍器，以其射擊鳥獸，鋒利如神，他晚年病弱，不能再入山狩獵，就告知族人，他死後將化爲華雀。族人出獵時，須注意牠的鳥鳴聲，聲音宏亮爲吉，細小代表凶兆。他死後，五體幻化爲華雀，飛入了福爾摩沙綠海般的森林。

曾經與原住民到山上打獵、觀看過鳥占的洪田浚認爲，人們不能以今天的邏輯來解釋鳥占，原住民解釋自然現象的方式本就與今人不同。

peoples. It is said that neglectful parents may find that their children have turned into birds and flown to the treetops by the time their parents remember to think about them. Two brothers in Lanyu who were ill-treated turned into birds and flew away; the Taiya also have a tale of a girl who changed into an eagle.

Then there are children who suffer too much pressure and dream of turning into birds and sailing away. "In fact this is a spiritual escape, a way to seek psychological balance," says Hung Tien-chun. Modern psychologists use their concepts to try to explain this. This kind of bird story could serve as a warning to irresponsible parents, and also as a lesson to all the people.

In fact, most nationalities have some feathered fables. The symbolic meanings expressed in birds among the aboriginal peoples are all present in other peoples as well.

For example, Chinese people long ago used bird divination as a basis for deciding the good or ill of a given circumstance; the same practice has had a critical function for the aboriginal peoples also.

Hung Tien-chun notes that bird divination is a special feature of cultures from the Austronesian linguistic family. Of Taiwan's nine aboriginal peoples, all, except for the Yami of Orchid Island, and even including the Pingpu, had the custom of bird divination.

A wing and a prayer: Of course, the tools for divination included more than just fowl. But, except for dream interpretation, birds remained the most important. For everything from building a home and hunting to planting and harvesting, and even going to sea, bird prognostications would invariably be held at the designated altar. If halfway through a bird was seen engaged in some activity thought to be ominous, it was necessary to call things off and go back home. Some aborigines who as small children went hunting with the elders still remember divination scenes.

In his book, *Journey Through Formosa*, a missionary named MacKay wrote, "The aboriginal people give the deepest respect to and sincerely trust in the calls and actions of the birds, and must consult them each time before striking out on some dangerous activity, especially hunting. First they go to the edge of the village and throw a branch up into a tree to scare the birds. They will only set off if the birds' calls and flight are a certain way; otherwise the leader won't be able to get up their courage to set off."

Anthropologists explains that prognostication and divination spring from man's curiosity and thirst for knowledge. Both are normal human psychological traits. But pre-modern peoples add the idea of the spirit world, and in their surprise at unfathomable phenomena, will further draw misleading connections, so that they come to have faith in fortune telling. Divination activities based on these beliefs and primitive taboos can be seen in much the same way we see law in modern society.

A little birdie told me: For the older generation, bird divination has still been an important basis for setting norms of behavior. Because they have completely internalized their sense of respect and/or dread of nature, the impact is far greater than modern laws or structures.

Bird divination was first recorded by Westerners in the 17th century. At that time because gold had been discovered in Taiwan and silver in Japan, there was a certain fever in the West to "seek one's fortune in the islands of the Orient." In March, 1640, the Dutch governor of Taiwan ordered his men to set sail from Anping to find gold in Peinan in eastern Taiwan. In the course of their search, their Peinan guide heard a bird call and found it to be inauspicious, and insisted on going back. He held to his view no matter how much the Dutch implored him, and though the latter were non-plussed, they had no choice but to call off their journey.

There are also descriptions of the aborigines in the *Huang Ching Chih Kung Tu* [*Descriptions of Ways of Life*]: When deer hunting, they would observe and listen to the birds, and, based on the length and strength of the warble, the way the bird was flying, and so on, determine whether or not it was a good idea to press on. The *Fan She Tsai Feng Tu Kaotsai* [*Descriptions of the Customs and Folklore of Barbarian Societies*] suggested that this was like the Han Chinese custom of geomancy when building houses or bridges, when a geomancer is asked to interpret the lay of the land, saying that although the aborigines knew nothing of geomancy, they had their own methods when building a house — they selected an auspicious day according to the calls of the birds.

A natural law: As we have seen in choosing

排灣族雕刻家薩古流跑步速度不如弟弟，頭飾上只能裝綴一根帝雉羽毛。

The Paiwan sculptor Sakuliu is no match for his younger brother in a footrace, so he gets only one pheasant feather in his cap.

鳥兒也有方言

事實上鳥占累積了原住民祖先觀察自然的經驗，譬如空氣潮濕、出現烏雲、充滿快降雨的氣氛時，鳥兒的叫聲就會有變化，聽到這樣的叫聲，他們就互相叮嚀最好不出門。今天的生態調查也證實，一般人聽來差不多的鳥鳴，可以因爲季節、環境不同而有上百種不同的叫聲。同一種鳥類除了共有的「國語」，還會因爲分布地區不同，而有不同「方言」。因此鳥占其實是人們觀察自然變化時，歸納自然界會出現那些相對應的異兆，再供後人遵循。

今天許多老人家仍愛依循黃曆行事，而人們也都看著月曆作息，每個人也都有自己的行事曆。決定原住民作息的「自然曆」，和所有人類老祖先一樣，他們看花開、葉落、鳥鳴等大自然的改變來決定「日出而作、日落而息」。於是春天忙著翻土、播種，夏天雨季一到，是幼小動物孕育、成長的時期，加上雨季深山多危機，就少往山裡走。秋冬一到，農地收成了，小動物也長大了，是打獵的時候了，再上山。

遵循自然曆

鳥占也是自然曆的一項。對於原住民，以鳥占卜，得凶兆的機會很多，他們只好等待另一個占卜時機，這段時間就悠閒地徜徉在山林中玩樂，讓自然、也讓自己休養生息。

鳥占中藏著人類敬畏造物主的精神，正是今人最缺乏的。從表面看來，今人比前人了解自然、也更會利用自然，但獨獨缺乏科技精進後更該有的敬畏自然的精神；只有覺得造物者的不可侵犯，人類才可能合理對待自然，否則只存人定勝天的自大想法，易造成對環境粗暴的開發與剝削。

只惜今天原住民傳統文化也在逐日瓦解，與鳥相關的習俗，僅留守山上的老一輩還能夠瞭解、信守。臺灣鳥類與住民在這塊土地上共同走過的千百年歷史，往後人們恐怕只能在故事裏追尋。 S

（原載光華八十二年三月號）

birds to symbolize speed, the status of birds was ordinarily quite high, though each aboriginal group used different types of birds for prognostication. In the section on divination birds in the reference book of Liu Ko-hsiang, a venerable friend to winged creatures, he says that the Bunun often looked to the lesser scimitar babbler, while the sibis, which can even today be seen flying about the Taipei Botanical Gardens, played the main divination role for the Taiya. It remains to be explained why certain types were selected, but unfortunately today only the Taiya and the Tsao peoples have retained legends explaining the origin of their birds of pray-tell.

In the local gazetteer from Chiayi County it is recorded that, before the Tsao had bows and arrows, they captured animals in traps. An orphan invented a projectile, and used it to shoot down birds and beasts, becoming as adept as a god. In later years, weakened and frail, no longer able to go up into the mountains, he told his fellow tribesmen: "When I die I will become a bird. When my call is clear and expansive, that is an auspicious sign; when it is thin and weak, that is an omen of evil. Before setting off hunting you must take note of this call." When he died, his limbs became birds, the type that are today known to Chinese as the "bamboo finch."

Hung Tien-chun, who has been hunting in the mountains with aboriginal people, says that one cannot explain this practice in contemporary terms; the way the aborigines explain natural phenomena follows a completely different logic.

Birds have dialects, too: In fact, bird divination is based on the cumulative wisdom of nature as observed by their ancestors. For example, when the air is humid, dark clouds appear, and rain seems imminent, birds' calls will change; when one hears that particular tone, it's best not to go out. Modern studies confirm that what most people take for identical bird calls are actually more than 100 variations which differ based on the season and the surroundings. Indeed, besides having their own "pidgin English" for communication with their fellows, a given species will also have "dialects" in different areas. Thus bird divination actually includes observation of natural transformations, which have been condensed into basic principles and then passed on to future generations.

Today many elderly people still like to do things according to the almanac, and Chinese society still has holidays on the basis of the lunar calendar; like the ancestors of all mankind, those of the first inhabitants of Taiwan followed the opening of the flowers, the falling of the leaves, and the calls of the birds to serve as reference in deciding when to be idle. Thus in the spring they turned the soil and planted seeds; when the summer rainy season came, that was the time to allow the little creatures of the wild to be born and mature, and anyway it being dangerous to go deep into the mountains in the rainy season, people held off on hunting. When autumn and winter arrived, it was time to harvest; it was also the sign that the animals had grown, so it was time to go back to the mountains and take up the chase.

Respect the natural timetable: Bird divination is a type of natural timetable. In fact, for these original settlers of Taiwan, there was a large chance that ominous ornithological omens would turn up, and when they did the people could only wait for the signs to change at some later time; in this period they could relax in the forests and have fun, giving both nature and themselves a little rest. Today, when people move at an ever faster pace, it seems like using bird divination to decide holidays might be worth a try.

Also part of bird divination is a sense of respect for the wonder of nature, what people today lack most. Moderns understand nature, and employ it more efficiently, than old-timers, but lack the respect for nature that should follow scientific and technical comprehension. Only if people see nature as something that cannot be simply invaded will people treat nature in a rational way; otherwise people will only retain their sense of self-importance and rudely develop and exploit nature.

Yet, the aboriginal culture of legends and myths has disintegrated, and only a few of the older generation ensconced in the mountains still understand and observe the avian-related customs.

Taiwan's birds and Taiwan's people have passed thousands of years together on this piece of earth, but in the future people might only know of it in stories. S

(Chang Chin-ju/photos by Vincent Chang/ tr. by Phil Newell/ first published in March 1993)

每年冬天，會有大批候鳥由北南來臺灣渡冬。圖爲停棲於泥濘灘地，數量足以遮天蔽「地」的濱鷸。(郭智勇攝)

Each year large numbers of birds fly south to Taiwan to pass the winter. Shown here covering the beach and sky are Pacific dunlins. (photo by Kuo Chih-yung)

有鳥自遠方來

Feathered Visitors from Afar

文 • 張靜茹　圖 • 鄭元慶

每年九月，大批紅尾伯勞鳥過境恆春；十月，被稱做「國慶鳥」的灰面鷲也準時向恆春半島報到；同時，西部沿岸的漁民開始在泥濘的沙灘上見到密密麻麻的水鳥覓食；寒流過境，臺北華江橋下忽然出現成千近萬體型圓胖的水鴨……，這樣的盛況，在春天降臨後，又陸續消失。

鳥從那兒來？去了那兒？爲何在某一時節忽然大批湧到，某些時候又芳踪全無？靠著「鳥類繫放」，這些問題逐漸都有了答案。

Every September large numbers of brown shrikes migrate to the Hengchun peninsula at the southern tip of Taiwan. In October gray-faced buzzards, which have been called the National Day bird, also punctually report in. At the same time, fishermen begin to notice flocks of waterfowl dotting the beaches and mud flats of the west coast. And when the first cold wave blows in, hundreds of thousands of pudgy wild ducks suddenly appear under Huachiang Bridge in Taipei. . . . With the arrival of spring, these grand sights gradually disappear one after another.

Where do the birds come from? And where do they go? Why do large numbers suddenly swarm in at a certain time of the year and vanish again at others? Through the technique of bird banding these questions are gradually being answered.

周末的臺北，有一羣人下了班、放了學，卽匆匆趕到淡水河口的關渡。只見他們扛著竹竿、拿著鳥網，其中幾個還掛著望遠鏡，一行人往堤防外沼澤地走去，然後在接近潮水處停住。有人舉起雙手，像在測風向，不久俐落地架起竹竿、掛上鳥網。結束工作後，回到堤防裡，等待夜晚來臨……。

同樣的周末，在彰化大肚溪口和臺南四草的鹽田上，各有一批人做著相同的工作。

這些人常被誤爲「捕鳥人」——他們的確在等鳥上網；但鳥兒到手後，接下來的任務却是上脚環，拿著尺、筆對著鳥兒比劃、記錄。直到東方旣白，大夥收起鳥網，在堤防外把鳥一一放走。

他們是追查鳥踪的繫放工作者。也許過個十天、半月，鳥兒們離開了臺灣，飛抵菲律賓、澳洲，當地的繫放人士也早架起鳥網「迎接」了。

燕子變青蛙

人類對鳥的好奇心開始得很早，但過去自然科學的研究不似今天有許多外力可以借助，鳥類的「忽隱忽現」往往惹人困惑。

被西方視爲最早研究鳥類的希臘哲學家亞里士多德，他在冬天見不到知更鳥，却見到紅尾鴝，因此認定知更鳥在秋天變成紅尾鴝，春天又變回知更鳥。這套動物會變的學說還被舉一反三。有很長一段的時間，古羅馬

After work and after school on Saturday afternoon in Taipei, a group of people rush off to Kuantu at the mouth of the Tamsui River. Shouldering bamboo poles and lugging bird nets, a few toting telescopes and binoculars, they file off to the marshes beyond the Kuantu embankment and come to a halt in the tidal estuary. One of them raises a hand as though checking the wind direction, and soon enough they agilely set up the poles and nets. When finished, they return to the embankment and wait for nightfall. . . .

The same day, others groups are engaged in similar activity at the mouth of the Tatu River in Changhua and at the Ssutsao salt ponds in Tainan.

These people are often mistaken for "bird catchers" — they are indeed trying to net birds, but once they have, the next thing they do is band them, measure them with a ruler and record the information with a pen. When the east lightens with dawn the next morning, they pick up their nets and let the birds go.

They are bird banders, gathering information on bird migration. In a week or two, the birds will leave Taiwan and fly to the Philippines or Australia, where local bird banders will similarly set up nets to "receive" them.

Swallows into frogs: Man's curiosity about birds dates back a long time, but naturalists of the past didn't have the tools and organizations to assist them that we have today, and the sudden appearance and disappearance of migratory birds often left

臺灣候鳥遷移概略路線

Avian Migration Routes in the Taiwan Area

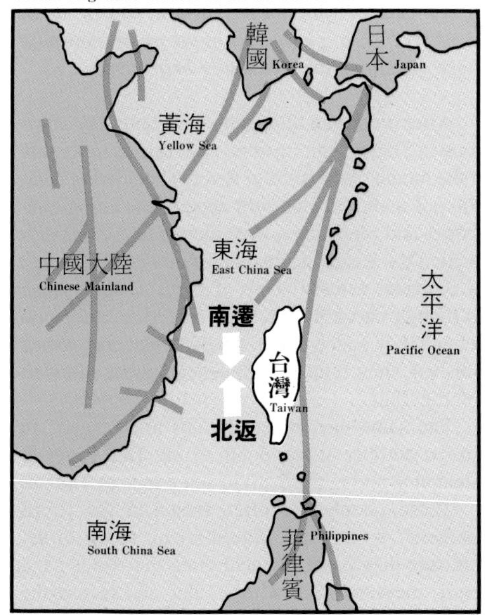

的博物學家將「燕子變青蛙」視爲不可辯駁的權威見解。

不過老祖先觀察日久，也逐漸明瞭某些鳥類的確有翻山越海、長途遷徙的習性。蘇武牧羊，若不是借鴻雁傳信，大概只能老死異域。

古人也常將自己捕獲的鳥加上標識。但不論迷途鳥兒知不知返，因古時交通不便，通訊困難，無法知道牠去過了那些地方，卽使靠著脚環被尋回，還不足以證明鳥類的移棲路徑。

眞正大量、有規模的對鳥類遷移作調查，開始於十九世紀末。通行無阻的交通，讓人類能在海上見到羣鳥飛翔，在其他國度發現和自己國家相同的鳥；更令人想知道鳥類如何遷徙，有無固定路線？

繫放是研究候鳥的基礎工作

近代鳥類學家以線圈套在鳥脚，想知道鳥的去向。但因線圈常被鳥啄掉，後來改用銀環，此後更不斷改進，今天使用的已是可經長年不朽的鎳、鉻合金製脚環。爲了更深入研究候鳥，除上脚環，也進行觀察記錄，使鳥類繫放工作能更科學、客觀。

鳥類繫放的分解動作是——

一、有效與安全地大量捕捉候鳥；二、繫上脚環；三、測量身體每個部位，如嘴長、全頭長、翼長、跗蹠長和重量；四、辨別鳥齡、性別和冬、夏羽；五、詳細記錄測量、觀察結果及發現地點；六、把鳥放走。

鳥兒留下資料、戴著脚環飛到另一地棲息，要是再被沿途繫放工作者捕獲，捕獲者除繫環外也做同樣的工作，並通知原繫放者，互做資料比較。一隻戴上脚環的鳥，被捕捉的次數愈多，我們對牠的資料累積也愈多。

由四面八方搜集來的資料，在綜合、分析後，除了可以了解牠的遷移方向，還能借著各種測量紀錄，看出牠身體的變化，由此推測、分析牠在我們見不到的遷移過程中的生態習性。因此，繫放不只是使我們知道鳥的來蹤去處，它也是研究候鳥最重要、客觀的一項基礎調查。

people bewildered.

Aristotle, considered the earliest person to study birds in the West, believed that robins turned into sparrows in the fall and back into robins in the spring because he couldn't see any robins in the winter or sparrows in the summer. This theory of metamorphosis was extrapolated to other species — that "swallows become frogs" was taken as incontrovertible by Roman naturalists centuries later.

After long observation, the ancient Chinese gradually came to realize that some birds had the habit of traveling long distances over lands and seas. The general Su Wu, held captive by the Hsiung-nu, used migrating wild geese to send a message back to the Chinese court and secure his release — otherwise he might have ended his days pasturing sheep among the barbarians.

The ancients sometimes attached markings to birds they caught. But transportation and communications weren't good enough in ancient times to determine where they went. Even if a bird or two was eventually sent back, there wasn't sufficient evidence to discover their migratory routes.

Truly large-scale studies of bird migration began in the late 19th century. With greater freedom of transportation, people saw flocks of birds on the wing at sea and found birds similar to their own in other countries, making them even more curious about where the birds went and whether they followed fixed routes.

Basic task: Ornithologists of an earlier day used to attach threads to birds' legs to trace their migration patterns. But because the birds would often peck it off, they later switched to metal bands. After constant improvements, what is used today is a durable band made of an alloy of nickel and chromium. At the same time, besides banding a bird, they also record various measurements and observations about it to make the work even more meaningful and scientific.

The process of banding a migratory bird involves six steps: 1) safely and effectively catching the birds in large quantities; 2) attaching a band with an address and serial number; 3) measuring the bird's weight and various parts of its body, such as its beak, head, wings and legs; 4) determining its age and type of plumage; 5) recording the measurements and detailed observations along with the location at which it was found; and 6) releasing it.

Leaving behind that information and sporting its new leg ornament, the bird flies to another stop along its migration route. If it is captured by another bander along the way, that researcher will perform the same tasks all over again, except adding a band, and will notify the original bander to compare notes. The more often a bird is caught, the more information can be built up on it.

Collected and compiled from all over, the data informs us not only about the bird's migratory pattern but also about its physiology, enabling scientists, after analysis, to deduce its living habits during the course of migration, which they might otherwise be unable to detect. So bird banding doesn't just tell us a lot about a bird's comings and goings; it is a critical, basic task in studying migratory bird species.

Although banding work may not seem difficult, each step and detail is important.

Comparing the weight of a bird with its weight at an earlier stage of the journey, for instance, indicates how much energy it consumed during its flight and what the food situation was like along the way.

Similarly, age records may show up a pattern such as "more adults are netted in October and more fledglings in December," leading to a hypothesis for further exploration, such as that adult birds set off on migration first.

International resource, international cooperation: This sort of "long-distance" banding work is aimed at migratory birds.

After years of observation and study, ornithologists have classified birds into two types according to their migratory behavior. Those that enjoy a sufficient food supply and do not migrate at fixed times over fixed routes are called resident birds. They live mostly in the tropics and semitropics. Resident birds are generally considered resources belonging to the areas in which they are found, and research on them is carried out independently by each country.

In colder areas with sharp seasonal climatic changes, where the food supply is unstable, many birds regularly head off for the tropics or semitropics before winter and wait for spring to return to the north and propagate. These are migratory birds.

Migratory birds are international resources, and

繫放工作看來雖然不難，每個步驟、細節却都很重要。

比如有了某一羣鳥的體重紀錄，就可以和牠們再度上網時的體重做比較，據以推測這段時間牠們飛行消耗的體力，和停宿地點的食物狀況。

同樣的，根據鳥隻年齡紀錄，可能就會出現某一種鳥「十月分成鳥上網較多，十二月上網多爲幼鳥」的資料，由此得出「成鳥比幼鳥先進行遷移」的假說，再進一步探討、研究。

這種「越洋式」的繫放是針對「候鳥」而來的。

尖尾鷸是臺灣普遍的過境鳥，有少部分爲冬候鳥。牠們在每年八月中旬至十月中旬過境台灣後，再繼續南下，翌年春天北返時，也取道台灣。(郭智勇攝)。

The sharp-tailed stint passes through Taiwan from August to October on its way south and returns on its way north the following spring. (photo by Kuo Chih-yung)

filling in the life history of a species depends on international cooperation. Bird banding work, showing the name of the country, the bander's address and a serial number, must be performed in each country along the way in order to complete a thorough survey. If one country fails to pull its weight, then a gap will appear in the story.

Major rest stop: Semitropical Taiwan is an important rest stop for many migrating birds in East Asia. From the maps of migratory routes drawn up by foreign scholars in the past and the studies of Chinese scholars in more recent years, it is evident that the hundreds of thousands of migratory wild geese, ducks and gulls that visit Taiwan on their way south in the winter start out from Japan, Korea, Manchuria, Siberia and Mongolia. Bird watchers here call them "winter birds."

Some of the birds only rest here for a time before continuing on their way to the Philippines or Indonesia, or even across the equator to Australia or New Zealand to enjoy the warm summer sunlight. Birds that treat Taiwan as a way station are called transit birds. But no matter whether the birds stay here for the winter or only pass through — and the numbers of each type are considerable — they all head back north in the spring, one after another, returning along the same routes.

There are also smaller numbers of "summer birds" that come to Taiwan in the summer to propagate and return south in the winter. More than 200 species of all three types are known at present. Considering Taiwan's size, that is quite a large figure, making the island comparable in importance to areas as vast as the United States or Australia.

The Republic of China embarked on banding migratory birds more than 20 years ago. At the time, scientists in the United States suspected that the spread of Japanese encephalitis was linked to animal migration and asked 13 countries in East Asia, including the R.O.C., to take part in a survey of migratory animals, the main focus of which was birds.

Setting out again: The study was completed seven years later. Although the R.O.C. was unable to continue bird banding because of limitations in manpower and resources, the experience did give us a much better understanding of the birds that visit our island and sparked a greater interest in bird watching.

The study of migratory birds continued unbroken internationally. Bird banding is especially active in Europe and North America, where it has a 100-year-old history. In the U.S. and Canada around 10 million birds belonging to 600 different species have been banded over the years, of which a million or so have been recaptured. Nearly a million are banded and 40,000 to 50,000 recaptured every year now. Fortunately, Africa and Latin America, which are on the same migration routes as Europe and North America, have picked up the practice so that completeness of information can be maintained.

In our part of the world, the history of bird watching and bird banding dates back the longest in Japan and Australia. During the past ten or so

國際資源靠國際合作

在鳥類學者長久對鳥類的觀察和研究後，把生活在氣候恒常、食物供應無缺，飛行沒有固定時間和路線，但只限在某一範圍遷移的鳥類稱爲留鳥，以熱帶和亞熱帶較多。

留鳥的活動範圍固定，大多被世界各國視爲自己的資產，獨力進行研究。

大多數棲息於寒溫帶的鳥類，因四季天候變化劇烈，尤其嚴多天寒地凍，食物供應不定。在冬季來臨前，他們固定往亞熱帶或熱帶移動，待翌年春暖再回北方繁殖。這就是我們口中的候鳥。

遠渡重洋的候鳥是國際資源，要完成牠生活史的記載，就必需靠「國際合作」——在候鳥的遷移線上，由每個停靠站所屬的國家爲鳥兒繫上印有國名、郵政信箱，和代號的統一脚環進行繫放調查。當然，如果其中有國家未做，該線上的候鳥生態就會出現一段空白。

旅鳥的重要驛站

地處亞熱帶的臺灣，正是東亞旅鳥南北遷徙的重要休息站。由過去外國學者發表的亞洲鳥類遷移路線圖和近年來國人的賞鳥紀錄得知，冬天，會有成羣結隊的雁、鴨和鷗科鳥類，由日本、韓國、大陸東北、西伯利亞、蒙古高原南來「過冬」，賞鳥人將之名爲「冬候鳥」。

有些鳥只在臺灣暫歇，還要繼續南下，飛往菲律賓、印尼，甚至穿越赤道，到南半球正是夏天的澳洲、紐西蘭享受暖洋洋的日光。這些把臺灣當成機場過境室的鳥羣，被稱做「過境鳥」。但是不管入境或過境，牠們的數量都很可觀，並在隔年春天陸續循原線北返。

另有極少數「夏候鳥」，夏季來到臺灣繁殖，冬天再轉往南方。目前已知的這三類鳥有二百多種。

以平均數來算，前來臺灣的候鳥羣，是美國、澳洲等一些幅員遼濶的國家、地區難以相比的，也因此地位相當重要。

我國在廿多年前就有過國際候鳥繫放的經驗。當時美國懷疑日本腦炎跨國流行，是與移棲動物有關，於是邀約包括我國在內的東亞十三個國家，展開「移棲性動物病理學調查」計畫，主要繫放的對象就是鳥類。

踏著前人脚步再出發

七年後，該計畫結束。限於當時的人力、經費，國人雖未能繼續從事候鳥的繫放，却因此對境內的鳥類有了進一步的認識。藉著繫放經驗，更帶動了野外賞鳥的風氣。

國際上的候鳥研究却一直未曾中斷。尤其有百年繫放歷史的歐、美更顯得積極。根據日本鳥類繫放手册記載，由目前留下的紀錄得知，至今美國、加拿大已大約給六百種、一千多萬隻的鳥戴上標識，接近一百萬左右有回收紀錄。

現在每年仍有近百萬的鳥被套上脚環，四、五萬隻被回收。有幸與歐、美在同一鳥類遷移線的非洲和中南美洲，也被帶動進行，因此能維持資料的豐富、完整。

在東亞到南半球的線上，則以日本、澳洲的賞鳥、繫放歷史較久。最近十幾年，蘇俄、香港、菲律賓、馬來西亞、新加坡……，也紛紛加入萬里尋鳥蹤的行列。

近年來我國學術界對留鳥的研究逐日增加，培養了不少鳥類研究人員；富裕的生活也使許多人有餘裕參加賞鳥活動。臺北、臺中、高雄鳥會陸續成立，並組成中華民國野鳥學會。有了這些人力做後盾，七十五年十一月在日本鳥類學者市田則孝的建議和農委會支持下，三個鳥會拾起前人留下的棒子，開始投身東亞候鳥繫放的工作。不過，繫放可不像賞鳥那樣輕鬆。

繫放樂趣不下賞鳥

在寒、溫帶，冬天河川湖泊結冰，南遷的旅鳥以棲於各種水域的水禽爲多；至於大部分生活在樹林、草叢中的陸鳥，因爲較易覓得避寒處所與食物，尤其以水果和穀糧爲主食的鳥，他們所受的環境壓力不如水鳥大，故移棲種類不如水鳥多。也因此繫放工作大

years, the Soviet Union, Hong Kong, the Philippines, Malaysia and Singapore have all joined the ranks of countries tracing birds on their distant paths though the skies.

In the R.O.C. the study of resident birds and the training of ornithologists have increased in recent years, and greater prosperity has given more people the leisure to engage in bird watching. Bird watching clubs have sprung up in Taipei, Taichung and Kaohsiung and have joined together to form the R.O.C. Bird Watching Society. With the backing of these clubs and support from the Council of Agriculture, local enthusiasts began to take up bird banding work once again in November 1986.

As much fun as bird watching: In frigid and temperate zones, where lakes and streams freeze over in winter, most of the birds that travel south every year are waterfowl. Not as many species of land birds migrate. They face less pressure from the environment: Most of them live in forests and other places where they can find food and shelter from the cold, especially those that subsist on fruits and grains. Since the majority of migrating birds are waterfowl, banding is usually carried out near rivers, lakes, wetlands or other watery areas.

Most birds that migrate long distances have extremely sharp vision, meaning that nets must be set up at night if they are to be effectively banded in large numbers. Staying up all night isn't easy, and there aren't all that many people willing to engage in bird banding for a long time. Banding in Taiwan, so far, is limited to places with large numbers of birds near bird watching clubs, such as Kuantu in Taipei, the mouth of Tatu Creek in Taichung and Ssutsao in Tainan.

Over the past two years, nearly 10,000 birds have been banded around the island. The work is arduous but it has its rewards, and the information accumulated has shed some interesting sidelights.

Chuang Yung-hung, an assistant researcher with the Taipei Bird Watching Society, says that several bird watching clubs often put up nets at the same area. He recalls that a yellow-footed snipe that was banded at Kuantu was netted and banded again two weeks later at the same spot. The bird was fine. In fact, Laurel had become Hardy: Its weight had more than doubled, from 82 grams to 174.

''In view of the fact that it stored up so many calories in such a short period of time, it was doubtless staying here on Taiwan to eat its fill before taking off to its next stop, which must be a good distance away,'' Chuang says, concluding that Kuantu must be an important supply stop for it along its route.

Although much can be learned from a single set of records, the biggest hope of banders is that the birds they have banded will be found by someone overseas or will be recovered by themselves.

After being banded, where do the birds go? What detailed routes do they follow? What places do they stop at? How fast do they fly? Are there any variations from year to year? These and other questions depend for an answer on our being notified of findings made in other countries. If we manage to recover the birds during their return back north or during another migration south, besides experiencing the thrill of meeting up with an old friend we can also observe the changes they have undergone during their absence.

A migratory bird was once recovered in the United States that had been banded 20 years previously. It was only then that scientists realized that a migratory bird could actually survive the perils of life in the wild for so long.

Recovery depends on everybody: Recovery records in the R.O.C., which has only two years of experience in banding, are naturally not that impressive. The prime example so far is a domestic swallow, banded by the Taichung Bird Watching Association, that flew to Japan and was recovered 26 days later. This April the association caught a sandpiper that had been banded two years before in Japan, and in May another kind of sandpiper that had been banded 38 days earlier in Australia and had flown more than 2,000 kilometers to get here.

''It's quite normal not to have any recoveries at all during the first few years,'' Chuang Yung-hung says. With so many birds, the likelihood of one being renetted in another country is slim indeed, not to mention being recaptured at the place it was originally banded. The recovery ratio is less than one in a thousand in East Asia and still only five in a hundred in Europe and North America, where banding has gone on for decades.

''You can complete a research report on a resident bird species in a year or two,'' says Chuang Yung-hung. ''But for migratory birds you may not

鷹斑鷸(上)、蒙古鴴(下)均爲臺灣春秋二季有規律的過境鳥。(陳永福攝)

Photos show a wood sandpiper (above), a Stegmann's Mongolian plover (below). They all visit Taiwan regularly in the spring and fall. (photo by Ch'en Yung-fu)

多在河岸、湖泊或涉入沼澤、鹽田等水域進行。

長途飛行的鳥類，多半擁有異乎尋常的視力，想要有效、大量繫放，就得在晚上架網作業。熬夜的滋味不好受，願意長時間投入的賞鳥人仍然有限。因此臺灣目前的繫放工作只能局限在離鳥會較近、鳥況還不錯的地點進行。像臺北關渡、臺中大肚溪口和臺南四草都是重要的據點。

兩年多來，北、中、南三地繫放的鳥已接近萬隻，繫放工作雖然辛苦，也有不少樂趣，累積的資料中也可看出一些「端倪」。

臺北鳥會資深鳥友莊永泓表示，許多鳥會在同一棲地重複上網，他印象最深的是在關渡繫放的一隻黃足鷸，牠在上網繫放後兩個禮拜，又在原地入網。別來無恙，只是勞萊變成了哈台，小兄弟的體重竟然由原來的八十二克直升爲一七四克，足足兩倍重。

「由牠短時間大量儲存熱量的情形看來，牠無意在此久留，吃飽，拍拍屁股就要走，而且下一個停靠站大概不會離臺灣太近，」莊永泓猜測，關渡是牠糧食的重要補給站。

雖然借助自己的紀錄可以從事一些判斷；但繫放者最期盼的仍是鳥兒多多被國外或我們自己「回收」。

繫放後，鳥又去了那裡？詳細的路徑，停靠那些驛站，飛行的速度，每一年有無不同？……這些都需要靠其他國外人士發現，再告知我們；要是鳥兒在回程、或來年再進行遷徙時被我們自己回收，除了一股久別重逢的興奮，也可觀察牠在這段時間的變化。

美國就曾回收自己在廿年前繫放的候鳥，大家才知道，要是能安全度過路程中隨時可能發生的暴風、急雨和被捕捉吃掉等等意外，候鳥的生命原來可以這麼長。

「回收」靠全民

至於只有兩年經驗的我國，回收成績，自然還未能進入狀況。目前只有一隻臺中鳥會繫放的陸鳥「家燕」，在廿六天後由日本回收。此外，今年四月鳥會進行繫放時，曾捕獲日本在兩年前繫放的反嘴鷸，五月，又有

be able to reach a conclusion in five or ten.''

Chen Chao-jen, a member of the wildlife protection division in the Council of Agriculture, says that if banding work is to get on track and research made more scientific, more people will have to join in. That includes specialists from related fields, such as statisticians to perform data analysis and avian veterinarians to do in-depth exploration of bird diseases and lower the fatality rate in banding. Equally indispensable are banders and recorders to work the front line.

In Europe and North America, Chen says, bird banding has long been performed mainly by volunteers, who look on it as a form of recreation. That is something we should emulate and promote.

''Recovery, in particular, should become everyone's job,'' Chuang Yung-hung maintains. It's not enough to rely on the efforts of a limited number of banding workers. If each of us possessed a little common knowledge about banding and recorded the basic information of any banded bird we saw and informed the banding agency after letting it go, records would be much more complete.

Manmade factors: Another problem that confronts bird banding work in Taiwan is the environment. More than 180 species of birds were once counted in the Kuantu tidal area, which is rich in food, but the number has declined in recent years because of waste land accumulation, water pollution and other factors. In other words, even though Taiwan is located in a migration belt, the deterioration of its environment would adversely affect the willingness of birds to come here and force them to choose another route.

The species and numbers of birds banded by the Taipei Bird Watching Society at Kuantu were not as numerous this October as they were last year or the year before. ''Later, maybe the only birds we'll be able to see will be ones with strong ability to adapt, like sandpipers,'' Chuang says with some discouragement.

At the Tatu estuary, in central Taiwan, the rich bird population is the result of a combination of favorable natural factors and not-so-favorable manmade ones, which many of us may not be aware of. Chen Chao-jen says that the whole of the west coast of Taiwan was originally a bird paradise before the large-scale development of fish hatcheries, the establishment of industrial districts and the expan-

一隻卅八天前由澳洲繫放，飛行二千多公里才抵達臺灣的黃足鷸。

「剛開始的幾年沒有回收算很正常，」莊永泓表示，鳥那麼多，在此上網，不見得在另一國度就會入甕。目前東亞繫放的鳥，能在別處再度上網的機率不過千分之一，更何況舊地重遊，再度光臨。除非我們繫放的候鳥大量增加和長久持續，像歐美累積幾十年，回收率才能高達今天的百分之五。

「留鳥的研究可以一、二年有一篇報告，候鳥可能五年、十年也不一定能得出一個結論，」莊永泓說。

農委會保育科陳超仁則指出，要使繫放工作更上軌道、學術研究更專業化，必須要有更多的工作者願意投入。所謂的工作者，包括各相關學門的專業人士，比如能做資料分析的統計專家；能對鳥疾病作深入探討，並減少繫放過程中鳥類死亡率的禽病醫學學者等；更不可缺少的是堅守第一線的繫放工作記錄者。

陳超仁表示，在歐美，候鳥繫放早就以業餘人士爲主，他們將之視爲一種休閒活動來參與，值得我們借鏡、提倡。

「『回收』更應該發展成全民工作，」莊永泓進一步強調，候鳥翔翔，光靠有限的繫放工作者是不够的，如果每個人都能略具繫放常識，在看到脚上繫環的候鳥時，記下基本資料、通知繫放單位，再予以「放行」，那麼，紀錄就會更完整了。

天時、地利，只欠「人和」

臺灣鳥類繫放工作面對的另一課題是環境的變遷。以關渡鳥食豐富的潮間帶爲例，過去此地曾有過一百八十幾種鳥類棲息的紀錄；近幾年來因爲廢土堆積、河水汚染等因素使鳥類逐漸減少。換句話說，臺灣雖然在候鳥的遷移帶上，然而，所能提供的環境一旦改變，就會影響鳥兒自遠方來的意願，轉而借道他處。

臺北鳥會今年到十月爲止，在關渡繫放的鳥種、鳥數都沒有前年、去年多。「以後搞不好只能看看適應力較強的濱鷸了，」莊永泓有點無奈地說。

再說中部，目前看來鳥況豐富的大肚溪口，除了天時、地利使然，還有一般人可能沒有注意到的「人和」。陳超仁以爲，其實這是因爲過去整個臺灣西部沿岸都是鳥的樂園，在大量開闢魚塭、建工業區，如梧棲海岸成爲臺中港後，鳥類只好轉向大肚溪口去安身立命。而這也已是候鳥在中部停棲的最後樂園了。

令人擔心的是，工業局將以大肚溪口爲火力發電廠堆灰場的計畫雖已擱淺，並轉手給省府管理，但在省府十月初的公告使用計畫中，並未打算爲候鳥保存一塊淨土。可以想見，若此地也不再適合鳥類時，某些候鳥無枝可棲，無法在「中途站」補給，就是牠們生命受威脅的時候了。

至於高雄鳥會在原本鳥況頗盛的高屏溪口和林邊海岸過度開發後，只好選擇臺南作繫放。

別做崖底羔羊

人類開發與自然生態是否相衝突，已被視爲老掉牙的話題。

有一幅漫畫倒是一針見血，省去不少口舌之辯——一羣羊在懸崖上吃草，牠們的脚都有鎖鍊相連，其中一隻已經墜落崖邊，哀哀懸空；第二隻眼看著也岌岌不保，就將被拖下山谷去；而離懸崖最遠的那隻肥羊却意態從容地安心吃草。

道理很簡單：皇天后土，萬物並生，如果瀕臨滅絕的鳥類，是那隻再吃不着嫩草的崖底羔羊，我們萬物之靈，不就是那隻傻乎乎兀自吃草的肥羊嗎？一個「千山鳥飛絕」的孤寒之境，能不是「萬徑人蹤滅」的絕地？

爲鳥兒留些淨土，顯然不是什麼捨人爲鳥，犧牲開發利益、發揚民胞物與的情操。

有鳥自遠方來，不亦樂乎！何不善盡地主之誼，爲牠們保存幾處安適宜人又宜鳥的地方？否則，當候鳥失去臺灣這個避風港，我們所將失去的，又豈僅是分享這種國際資源——候鳥的權利！ S

（原載光華七十七年十一月號）

sion of Taichung harbor made Tatu the last refuge for migratory birds in central Taiwan.

What is particularly worrying is that even though the Industrial Department Bureau shelved its plan to turn the estuary into a waste dump for coal-fired power plants and handed it over to the provincial government for management, the plan that the government announced in early October didn't mention preserving the estuary as a pristine bird sanctuary. If this area is no longer suited for them, migratory birds will lose another way station to stoke up at, further threatening their survival.

As for Kaohsiung, now that the Kaoping estuary and the Linpien coast, which used to have quite abundant bird resources, have been overdeveloped, the Kaohsiung Bird Watching Society has had to head off to Tainan for banding.

Sheep by a cliff: The question of whether environmental protection and industrial development conflict is a hoary one. There's a cartoon that hits the nail on the head. A flock of sheep, their feet linked together by a long chain, are grazing near a cliff. One of them has already fallen over the edge and is plummeting down in mid air. The next is being pulled over and about to tug the rest with it. But the plump sheep grazing farthest from the edge is still blithe and oblivious.

The principle is simple: All living things are interdependent. If endangered birds are the lamb toppling over the cliff, then aren't we, "the beauty of the world, the paragon of animals," that fat, stupid sheep still munching calmly away? Will a time come when "the sedge has withered from the lake/ and no birds sing," leaving us "alone and palely loitering"?

Preserving a pristine piece of earth for birds clearly isn't a question of favoring animals over people or sacrificing development for sentimental ideal.

As Confucius once said, "A visitor from afar is a joy indeed!" Why not play the good hosts to our feathered visitors from afar as well? Otherwise, should migratory birds lose Taiwan as place of refuge along the way, we will have lost much more than the privilege of sharing in this splendid international resource.

(Chang Chin-ju/photos by Cheng Yuan-ching/ tr. by Peter Eberly/ first published in November 1988)

上網的鳥帶回室內，要先上脚環，再測量鳥的嘴長、翼長和全頭長等。

In bird handling work, the band must be put on first before measuring the bird's mouth, wings, head, and so forth.

都是鳥仔惹的禍？

Chiku Industrial Park — A Victim of Fowl Play?

文・張靜茹　圖・郭東輝

據說，臺南縣七股鄉民近來受盡「鳥」氣。

兩百多隻在世界鳥類圖像中瀕臨絕種的黑面琵鷺，偏就選在七股鄉工業區預定地的沙洲上棲息。

行政院農業委員會希望縣政府調整工業區的規劃，以免影響鳥類生存；縣政府則打算在工業區附近另覓樹林，請黑面琵鷺「搬」過去……。

這莫非又是一個開發與保育衝突的故事？

據說，堤防才是衝突的起點。

民國七十六年，臺南縣政府爲開發海埔新生地，在曾文溪河口圍起馬蹄形的堤防，形成一片漲潮時淹沒、退潮時露出水面六百多公頃的沙洲。

海水上下淘洗，豐富的營養塩孕育了無數沼地生物，堤防與沙洲間的溝渠又成爲隔開人爲干擾的屏障。人類無心插柳，沙洲竟成候鳥的最愛；每年五十多種、上萬隻的鳥族，羣聚這片「世外桃源」，使原本鳥況就極驚人的曾文溪口，更加熱鬧了。

大自然的賞賜

早在堤防興建的前兩年，一個來自南部鄉下的年輕人郭忠誠，在喧囂的臺南市不知如何消磨年輕的精力。徬徨少年於是拿起了望遠鏡、一本簡單的鳥類圖鑑；往來於南臺灣西部河口。

民國七十四年冬天，他來到曾文溪河口，在來自太平洋的刺骨寒風中架起望遠鏡，鏡頭裡竟然出現了一百多隻在臺灣已近六十年沒有成羣紀錄、鳥友只能在圖鑑上看到的黑面琵鷺。

黑面琵鷺與大陸聞名的朱鷺同屬於朱鷺科鳥類。在廿八種朱鷺科鳥類中，過去只有朱鷺、黑頭白䴉與琵鷺，在臺灣有過驚鴻一瞥的個位數紀錄。黑面琵鷺體型比鷺科鳥中的大白鷺稍大，身體雪白、頭部有黃色飾羽、脚與臉爲黑色，覓食時，長而扁平的嘴巴伸進水中左右掃動，掃中魚蝦後順勢拋起，張口接住，成爲河口有趣的一景。

郭忠誠只將此次收穫，當成是賞鳥歷程中一次大自然的賞賜。歷經幾個寒暑，鳥兒去了又來，這羣罕見的訪客，也僅限少數鳥友知曉。

「世外桃源」原是工業區預定地

此後，賞鳥人口不斷增加，黑面琵鷺又在蘭陽溪口、大肚溪口等地被零星發現過一、兩隻，但只有曾文溪河口這片具有天險般的沙洲留住了黑面琵鷺。牠們每年依時前來，且數量還呈穩定增長……。

同時，堤防內有一些事也在默默發生。

海埔新生地開始露出海面。縣政府已計劃在此設立工業區，希望未來能以土地售價和工業區稅收，增加地方收入。工業區所在的七股鄉，地方人士也希望在農、漁業日漸沒落之際，工業能帶來人潮和提供當地就業機會，給只有兩萬多人口的七股鄉注入繁榮生機。

經濟部工業局也曾經表示，希望這片沙洲能提供給覓地不易的國內企業開發、使用。縣政府受到鼓勵，圍堤、環境影響評估遂一一展開……。

魚與熊掌的局面，在四年前香港觀鳥會資深鳥友彼得·甘乃利寄給臺北市野鳥學會的一封信後，逐漸棘手起來。

全世界只剩兩百八十八隻！

香港在一九七〇年代設立的米埔自然保護區，近來每年冬天也都有黑面琵鷺前往避寒，甘乃利希望臺灣能提供觀察紀錄，以更進一步了解牠們的分佈與遷移。

全世界有九千多種鳥類。雖然地球鳥類圖像，在鳥類學家與觀察者努力下，拼圖已愈具體，但對擁有飛行天賦的鳥族，未知的領域仍然很廣，面臨絕種壓力的黑面琵鷺卽是其中的漏網之魚。

根據過去的鳥類紀錄，人們只知黑面琵鷺繁殖於中國東北與朝鮮半島，冬季時移棲至

漲潮時，七股工業區預定地盡數淹沒在海水中，成了釣魚場。(卜華志攝)
When the tide is in, the proposed area for the Chiku Industrial Park is inundated, making a handy fishing spot. (photo by Pu Hua-chih)

It is said that the people of Chiku Rural Township in Tainan County have recently become overbirdened.

More than two hundred black-faced spoonbills, which are endangered, have chosen to rest on the sandbank that has been selected as the Chiku Industrial Park.

The Council on Agriculture hopes that the county government will adjust the plan for the industrial area to avoid having an impact on the lives of the birds; the county government, on the other hand, is looking for some forest area near the industrial zone to which they will ask the black-faced spoonbill to "move house."

Is this another case of preservation vs. development?

It is said that it all started with a dike.

In 1987, in order to develop new land by the seaside, the Tainan County government built a horseshoe-shaped dike at the mouth of the river. This created an area of more than 600 hectares which, while at this point still inundated at high tide, comes above water level at low tide.

With the sea sweeping in and out and cleansing the area, rich nutrients are able to support countless forms of marsh life. Also, the drainage ditch between the dike and sandbank has become a moat keeping out human interference. Quite unintentionally for its human creators, the sandbank has become a favorite stop for migratory birds. Every year tens of thousands of birds of fifty different varieties come to this Shangri-la, making the already thriving situation at the mouth of the Tsengwen River even more exuberant.

A gift from nature: As early as two years before the construction of the dike, a young man from the countryside in south Taiwan, Kuo Chung-cheng, was unable to find an outlet for his youthful energy in noisy Tainan City. So this leisure youth picked up a telescope and a basic ornithology guide, and headed off to river mouths in southwestern Taiwan.

In the winter of 1985, he came to the mouth of the Tsengwen River. He set up his telescope in the bone-chilling breeze off the Pacific, and in his lens

大陸華南、臺灣、越南等地的開濶潮間帶活動。

如今東亞各國鳥會互通信息，比對紀錄，結果發現黑面琵鷺族羣數，僅維持在兩百八十隻左右，比起由一百多個國家簽訂的「華盛頓公約」所公告的許多瀕臨絕種動物，數量還要稀少。

過去史料記載，大陸東南沿海不時可見黑面琵鷺，但大陸近來進行大規模的濕地調查，牠們出現最多的一次，只有十五隻。臺灣曾文溪口卻擁有一百多隻黑面琵鷺，佔了總數的三分之二，遂成為國內外鳥會關注的焦點。

天外飛來「橫禍」？

包括世界野生動物基金會、國際鳥類保護總會，與亞洲濕地保護學會在內的卅幾個國外機構，紛紛致函主管野生動物保育的農委會，希望設法保護黑面琵鷺，以免滅絕。

原來只希望鳥兒能安靜在曾文溪河口渡冬的鳥會，今年五月也發起一人一信拯救「黑面舞者」的活動，希望勸阻臺南縣政府開發工業區，至今已有兩萬多人寫信支持。

在擔心開發工業區影響鳥兒生存的情況下，農委會希望縣政府能調整七股工業區的規劃，並在七月一日正式將黑面琵鷺公告為保育類動物，任意獵捕或破壞其棲息地，將受刑法處分。

這時候，期待工業區振興地方的七股鄉紳，則對於家鄉忽地跑出一種佔世界半數以上的稀有動物而惶惑不已，「幾隻憑空飛出來的鳥仔，為什麼能影響工業區的開發?!」他們說。

七股鄉長王龍雄(左)說：「七股鄉民生活窮困，需要開發工業區以繁榮地方。」(卜華志攝)
Chiku Rural Township mayor Wang Lung-hsiung (left) says: "Life is hard for people in Chiku. We need to develop the industrial park to make this area prosper." (photo by Pu Hua-chih)

人類共有的財產

兩百多隻黑面琵鷺的生死存亡，顯然不是少數人有無候鳥可賞的問題。

一個物種的滅絕，是全世界的問題，因為它暗示了人類生存的環境已出現危機。今年六月在巴西舉行的地球高峯會議，就將「物種保護」列為主要議題之一。

物種的滅絕原本是自然的事，但過去自然界只有遇到巨大的天災，才可能有大規模的變化，否則環境變化極其緩慢。以鳥類滅絕的速率而言，過去平均五十萬年才有一種鳥類消失。

到了現代，人類改造自然環境的能力和速度都異常駭人，受威脅的動物種類大量增加，滅絕速率也快馬加鞭。而自然界各種生物之間，彼此相互依賴，關係複雜，當一種生物絕種時，也許有其他物種可代替其功能，但生物不斷消失，就只可能惡性循環，造成一個個更快速的折損……。等到造成重大影響，已不及彌補。

就譬如人們平時不覺得砍伐高山上的森林和自己有何關係，等到旱季缺水連連、雨季洪澇不斷，才會想到「種二千萬棵樹救水源」的道理，但此時森林已非一朝一夕可以重現。

國外爭相設立保護區

再由基因保存的角度來看，任何一個物種

appeared more than 100 black-faced spoonbills — of whom there had been no record of any flock in Taiwan for over sixty years and which bird lovers could see only in their books.

The black-faced spoonbill, like the famous Crested Ibis of mainland China, belongs to the Ibis family. Among 28 types of Ibis, in the past Taiwan has only seen the Crested Ibis, the Oriental Ibis, and the white spoonbill in anything more than single-digit numbers. The black-faced spoonbill is somewhat larger in size than the white spoonbill; its body is snow white, with some yellow feathers on the head, with the feet and the face being black. When it feeds, it sweeps its long, flat bill through the water scooping up fish and shrimp, making one of the more intriguing sights at the river mouth.

Kuo Chung-cheng only saw his experience as a great moment to savor in his bird-watching career. Even after several summers and winters of coming and going, these rare visitors were still known to only a limited number of bird lovers.

The Shangri-la Industrial Park: The bird watching public has grown rapidly, and one or two black-faced spoonbills have been spotted at the mouths of the Lanyang River and the Tadu River. But only in this naturally defended area at the mouth of the Tsengwen River does the black-faced spoonbill come regularly every year, and moreover in steadily increasing numbers.

Meanwhile, other things were quietly happening inside the dike.

Originally the reclaimed land area by the seaside would surface at low tide. The county government had already planned to set up an industrial park here, hoping that in the future they could raise local revenues with land sales and taxes from the park. Local people in Chiku, site of the industrial area, also hoped that — with farming and agriculture on the decline — industry would bring a tide of people and job opportunities, turning the small township of only 20,000 people into a boom town.

The Industrial Development Bureau of the Ministry of Economic Affairs has expressed hope that this piece of land might be given over to the land-starved business community. Thus encouraged, the county government then built the dike, undertook an environmental impact study, and step by step....

Amidst this dilemma, things really got sensitive after the receipt of a letter four years ago from Peter Kennerly, a senior member at the Hongkong Bird-watching Association.

Only 288 left in the world: In the 1970's, Hongkong set up the Mipu Natural Protected Area; in recent years every winter black-faced spoonbills come here to escape the cold. Kennerly hoped that Taiwan would provide observation records to better understand their distribution and movements.

There are more than 9,000 types of birds in the world. Although with the hard work of birdlovers and ornithologists, the picture is increasingly complete, there is much we do not know about these creatures that have the whole sky to sail about in. The endangered black-faced spoonbill is precisely one of those who has fallen through the cracks.

According to past records, people only knew that the black-faced spoonbill flourished in northeast China and Korea. In the winter it would move its activities to wide open areas like south China, Taiwan, and Vietnam.

By comparing records of all the birdwatching associations in east Asia, it was discovered that there are only about 280 black-faced spoonbills left, making it even rarer than some species of birds defined as endangered by the Washington Conference Treaty, signed by more than 100 member nations. Historical records show that the black-faced spoonbill appeared on the southeast coast of China from time to time, but in a recent large scale survey of wetlands in the mainland, the largest single group in which they appeared numbered only 15. The mouth of the Tsengwen, on the other hand, has more than 100, accounting for two-thirds the number, thus becoming a focus of concern for foreign bird-watching associations.

Disaster from out of the blue? More than 30 international organizations, including the International Council for Bird Preservation and the Asian Wetland Bureau, have sent letters to the Council on Agriculture — which is the governing body for wildlife — hoping the latter will protect the black-faced spoonbills and prevent their extinction.

The Wild Bird Society, which had originally simply hoped the birds could quietly pass the winter at the mouth of the Tsengwen River, began a one-man, one-letter campaign to save the "black-faced dancers." They hope to convince the Tainan County

的產生，都是經過億萬年時間長河、無數生物不停演化，才有它的出現。卽使路邊的一根小草，它的基因以再高的科技也無法發明，因此任何一個物種都極爲寶貴。

「人類追求的是世代能永遠生活在這一塊土地上，不是只爲我們短暫的這一代，自然沒有人希望寶貴的物種斷送在我們手裏，」中華民國野鳥學會秘書長陳明發說，稀有生物也因此特別受人關照。

以棲息中國大陸山西省僅存數十隻的朱鷺爲例，由七〇年代以來，在保育人士奔走下，不僅成立保護區，更有生態學者進行追踪、復育，歐洲還有一份保育雜誌以之爲名，時時提醒人們，不可忘記物種的生存危機。

國外對黑面琵鷺的保護也積極展開。除了大陸、日本等數量出現極少的地區，香港早設有保護區；北韓有五十隻左右，在春天固定返回北方四小島繁殖、育雛，當地已被列爲人跡禁地；移棲越南紅河三角洲過多的六十隻黑面琵鷺，也因河口設置六千多公頃的紅樹林保護區，而有了安頓之所。

臺南鳥會的鳥友只希望所有的候鳥都有枝可棲。（卜華志攝）

The Tainan Wild Bird Society hopes that all the migratory birds have a place to rest. (photo by Pu Hua-chih)

一百年前就來過了

除了橫向聯繫，鳥友也找出臺灣鳥類史，發現有如下紀錄：一百年前，英國探險家拉圖許最早記下黑面琵鷺翔翔臺灣的情景。六十年前，日本鳥類學者風野鐵吉連續十四年，在臺南安平港附近發現約五十隻。此後，因缺乏調查紀錄，鳥踪再度如風箏斷線。

又一個六十年，今天，黑面琵鷺羣再度「出現」，卻面臨了是否還有下一個六十年的考驗。

五月，七股工業區環境影響評估未能通過環保署審核，黑面琵鷺遂成爲衆矢之的。鄉長怒氣沖沖地說，工業區若蓋不成，要到河口下毒，「不想再見到這些『黑面』的鳥仔！」

保育與開發的問題再度被搬上枱面。農委會也認爲，雖然鳥類保育是爲全人類共同的未來，但在未能說服所有人之前，也不能忽視地方迫在眉睫的開發心願。但撇開鳥事不談，僅以工業區開發來分析，地方眞的能如願以償，得到繁榮？

我們需要那麽多工業區嗎？

臺南縣有七個工業區，七股工業區屬於都市計劃中的麻豆生活圈；依內政部營建署估計，到民國九十年，麻豆生活圈工業用地只需七百三十公頃，現行計劃卻已超過一千二百多公頃。在土地、勞動資本愈來愈高，工廠紛紛外遷的今天，超額的工業用地眞能「地盡其利」嗎？

針對環境影響評估，環保署綜計處環境影響評估科長劉宗勇也指出：「卽使沒有黑面琵鷺，這份評估報告也絕對過不了關。」

包括多位經濟學者在內組成的審核小組，針對開發可能對當地環境造成的影響，提出了五十幾項質疑。比如目前曾文溪河口，因爲中、上游工廠林立，工業廢水排放逐年增多，已佔河口水污染的百分之五十二，未來工業區工廠廢水若仍排放河口，對沿岸養殖業與漁場均有影響。而臨近海邊的工業區，

government to stop development of the industrial park. Already more than 20,000 people have sent letters in support.

With concern about whether the industrial park will affect the existence of the birds, the COA hopes that Tainan will adjust its plan for the zone. Moreover, on July 1, it formally declared the black-faced spoonbill to be a protected species, with criminal punishments for hunting them or disturbing their living environment.

At this time, the leaders of Chiku Rural Township, who expected the industrial zone to bring prosperity to the area, were more than a little panicked by having more than half of all that remained of a rare species suddenly show up in their back yard: ''Why is it that a few birds that appear out of thin air can affect the whole development of an industrial zone?''

Common asset for mankind: The survival of the black-faced spoonbill is obviously not an issue that just involves the pleasure birdwatchers derive from watching migratory birds. ''The disappearance of a single variety is a question for the whole world,'' because it warns that there is a crisis in the environment in which mankind lives. The ''Earth Summit'' held in Brazil this June thus listed ''biodiversity'' as one of the main issues.

The disappearance of a species was originally a matter of nature, but there were only dramatic changes in nature if some cataclysm occurred; otherwise ecological change would be gradual. In terms of the rate at which bird varieties disappeared, on average, only one type disappeared every 500,000 years.

At the present, man's ability to transform nature, and the pace at which it is done, are startling. The number of endangered species has risen greatly, and the rate of extinction has accelerated. Moreover, the relations and mutual dependence of all types of animals in nature is complex. When only one type disappears, perhaps some other type can fulfill its function. But when species continuously are eradicated, this could create a vicious cycle so that they die off increasingly quickly . . . until there is a huge impact which can never be compensated for.

It is just as people cannot fully understand the relationship between themselves and deforestation — until there is no water in the dry season and constant flooding in the rainy season. Only then do they understand the logic of ''planting 20 million trees to save the water sources.'' But by then it's impossible to restore the forests in a short time.

A series of protected areas abroad: From the point of view of gene preservation, the creation of any given type of animal has taken hundreds of millions of years of evolution through countless variations before it can appear. Even if we are just talking about a blade of grass by the roadside, no amount of higher technology can recreate its genes, so that every type of living thing is precious. ''What people want is that successive generations can live forever on this planet, and not just this one generation stay here temporarily, so naturally no one hopes that precious creatures will die in our hands,'' says Chen Ming-fa, secretary-general of the Wild Bird Society of the ROC.

Take for example the Crested Ibis, which settles in mainland China and of which there only several tens left, in Shansi Province: Since the 1970's, with activities by preservationists, not only has a protected area been established, ecologists have gone in to investigate and care for the birds. One European conservation magazine even uses it as its name, constantly reminding people not to forget the crisis of existence of many living things.

Activities to protect the black-faced spoonbill are being actively pursued in other countries. Besides mainland China and Japan, where an extremely small number have appeared, Hongkong early on set up a protected area. There are about fifty in North Korea. Four small islands where the birds fly annually to reproduce and nurture their young have been declared forbidden areas. And the sixty or so black-faced spoonbill who winter in the Red River Delta in Vietnam have a safe place to rest because of the establishment of a 6,000 hectare protected forest at the mouth of the river.

They came 100 years ago: Besides communications from other present day allies, birdlovers have also gone through ornithological records and discovered the following: One hundred years ago, the British explorer Swinhoe recorded the activities of the black-faced spoonbill in Taiwan. Sixty years ago, a Japanese natural history scholar discovered about 50 in the Anping area of Tainan for fifteen consecutive years. Thereafter, because no records were maintained, the trail of the birds gets cut off ''like the cutting of a kite string.''

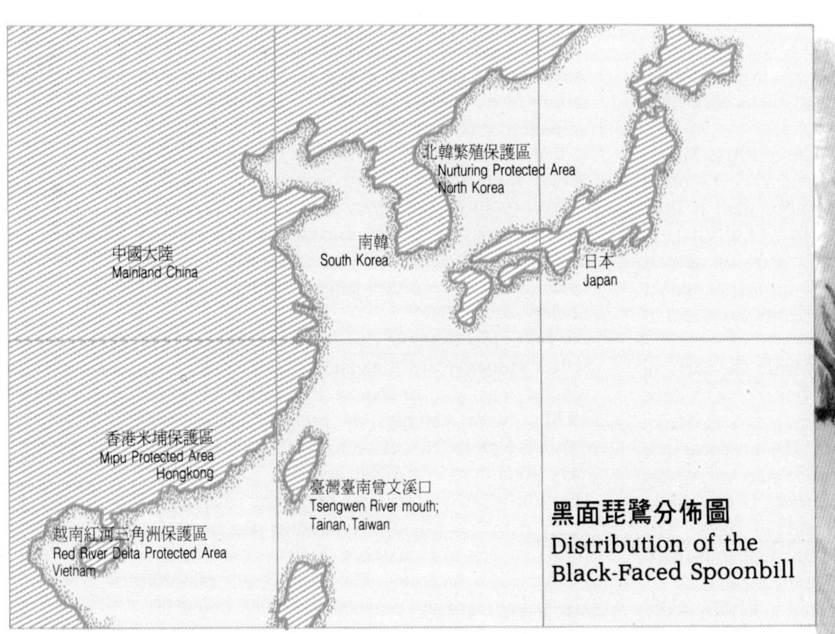

(中華民國野鳥學會提供) (map courtesy of the Wild Bird Society of the ROC)

全世界黑面琵鷺遷移季節估計數量
Estimated Number of Black-Faced Spoonbills in the World (Migratory Season)

(倪淑雲繪圖) (drawing by Ni Shu-yun)

用水如何取得？未來空氣品質的監測、引進產業廢水特性如何，報告中均付之闕如。

能否開發，與鳥無關

六百多公頃的地若眞能標出去，縣政府預估將有六十億的收入。

「不過佔臺南縣半年的預算，就要把祖產賣了？」從小在曾文溪口長大的臺南縣民李老師指出，如今漲潮仍淹沒水中的「工業區」，未來計劃在曾文溪抽砂塡土，此舉可能造成曾文溪河床崩塌、海水倒灌，危及當地居民生命安全，「到時候用六十億都不一定能解決！」他強調。

但這些問題在極力主張開發者眼中，卻不比「眞實」存在的黑面琵鷺嚴重。在他們看來，評估報告「反正是做資料，多做幾次就會過！」七股鄉民代表陳朝來就說。

在一次環保署勘察後的會場上，臺南縣長李雅樵坦承，七股工業區是臺灣第一個縣政府自行規劃的工業區，經驗比較不足，希望中央與學者們能提供建議。

他並提議，曾文溪口腹地遼濶，不如將沙洲東北方的防風林，劃爲黑面琵鷺保護區。

但以黑面琵鷺目前在各國棲息的情況看來，牠們並不居住樹林中，而且「牠們需要的警戒區比其他鷺科鳥遼濶，這正是七股沙洲吸引牠們的原因，」臺南縣崑山工專共同科講師翁義聰說。

現地保育最經濟

由一九八五年在溪口發現黑面琵鷺至今，曾文溪口最大的改變是什麼？「一句話，沙愈來愈少，」每年仍維持到河口賞鳥上百次的郭忠誠說。西海岸原本是上升地形，卻由於曾文溪上游的兩條支流，各有一水庫攔截了沖刷而下的泥砂，加上西海岸大量非法濫墾魚塭，如今海岸新生地不增反減，要爲鳥兒另覓一片相似的沙洲並不容易。

鳥是憑自由意志飛翔，人們再努力經營，鳥兒若不來，就是不來；人們認爲毫無是處的荒地，大自然反而使之生命繁盛。因此，自然保護區往往是「現地保育」，因爲它是最經濟、最有效的保育方法。

「要請鳥兒『搬家』也並非不可能，」高雄市野鳥學會前任會長曾瀧永說，過去人們圍堤，誤打誤撞，使堤內成爲黑面琵鷺的最愛，如今也可以試著在外海再築堤，產生相同的沙洲，盡量避開人爲干擾，說不定鳥兒會往外遷移。只是這樣的做法，需要長時間嘗試，人們要有耐心等待。

只要鳥兒有枝可棲，工業區的環境影響評估能達到要求，並能確實遵行，「沒有人反對地方開發工業區，」一位希望地方政府能心平氣和溝通的農委會官員說，與鳥距離最近的是當地人，黑面琵鷺的保護，首要取得當地人的支持。

魚與熊掌兼得

至於是否在當地設鳥類保護區，和東亞各國爭相設立的情況比起來，我們沒有理由不加緊脚步。根據聯合國一九七一年在法國藍薩簽訂的濕地合約，一地佔有某一稀有候鳥數量的百分之一，就應將該地列爲保護區。

農委會表示，目前人們對黑面琵鷺的了解太少，牠們的食性如何、覓食區多大……，均無人能提出確切答案。農委會已請師大生物系進行相關研究，只希望在結果與建議出來之前，能保持沙洲現狀。

「其實一個魚與熊掌得兼、開發與保育不衝突的規劃案並不難，」一位參與環境影響評估審查的學者指出。

如果工業區能開發成功，吸引的不過是工業人口，以現有臺灣開發的工業區而言，許多工業區附近的城鎮，不但沒有繁榮，反而環境糾紛不斷，人人避之不及。

若能將有無數候鳥停留的沙洲，規劃成生物龐雜度高的自然公園，除了具有長遠的環境教育意義，也可以吸引人們來此休憩。然後在鄰近發展由當地人經營、具地方特色的工業，生產可以配合觀光消費的產品；當地漁民也可藉此發展觀光漁業，讓養殖業起死回生。

而工業區附近有自然公園，不僅工業型態被限制，重污染工業進不來；縣政府還可要

Sixty years on, today, the black-faced spoonbill has been "rediscovered," only to face the challenge of whether it can have another 60 years.

In May, the environmental impact study done by Chiku was not approved by the Environmental Protection Administration, with the black-faced spoonbill the bone of contention. The mayor of the township angrily declared that if the industrial park were not built, he would put poison in the river and "not have to think anymore about seeing those black-faced birds!"

With the conflict between development and conservation again at center stage, the COA believes that, although the preservation of species is for the future of all mankind, you can't ignore the urgent wishes of the locality to develop. But by not talking about the birds and simply assessing the development of the industrial park, will the locality really be able to get what it wants and achieve prosperity?

Do we need so many industrial parks? There are seven industrial zones in Tainan County. The Chiku Industrial Park is part of the Matou Living Area in the urban plan. According to estimates of the Construction and Planning Administration of the Ministry of the Interior, by the year 2001 Matou will require only 730 hectares of industrial-use land. The current plan already allows for more than 1200. With land and labor costs increasingly high, and industries moving abroad, will surplus industrial-use land really be profitable?

As for the environmental impact assessment, Liu Tsung-yung, head of the Office of Environmental Impact Assessment in the Bureau of Comprehensive Planning at the EPA, says that "even if there were no black-faced spoonbill, this assessment report would absolutely never pass."

The review committee, which includes several economists, raised more than 50 doubts about the impact the industrial zone might have on the local environment. For example, at present because of the large number of factories upstream and midstream of the Tsengwen River, the amount of waste water has been constantly increasing, already accounting for 52% of the pollution at the river mouth. If in the future the factories in the new zone also expel their waste water into the river mouth, this will affect coastal fisheries and breeding. Also, where will an industrial park near the sea get its water? There was nothing in the original study about air quality monitoring, about what special conditions would apply for bringing in industrial waste water, and so on.

To develop or not does not depend on a bird: Even if all 600 hectares could be sold off, the income of NT$600 million estimated by the county government would amount to no more than half the annual budget of Tainan County, "and for that you want to sell out our heritage?" wonders Mr. Li, a teacher and resident of Tainan County who grew up by the mouth of the Tsengwen River. Today the "industrial park" is still inundated at high tide, so it is planned in the future to take rock and soil from the river to fill in, which might result in the collapse of the riverbed. If sea water flows back up the river, this will endanger the safety of the local residents, "and you wouldn't be able to solve that problem even with NT$600 million!"

But, in the eyes of the strongest advocates of development, these arguments are not more serious than the very real existence of the black-faced spoonbill. As they see it, the environmental impact report "is just a matter of compiling the data; if we do it a few more times it will pass," in the words of Chiku township assemblyman Chen Ming-lai.

In a meeting following an inspection by the EPA, Tainan County magistrate Li Ya-chiao stated frankly that the Chiku Industrial Park was the first one in Taiwan to be directly planned by the county government. The county government is relatively inexperienced, so he hoped that the central government and scholars would offer their suggestions.

He also stated that the hinterland of the Tsengwen River is very broad, and the best thing was to set aside the wind-blocking forest at the northeast of the sandbank as a preserve for the black-faced spoonbill.

But, judging from the situation of the black-faced spoonbill in various other countries, they do not live in wooded areas, and anyway "they need a broader undisturbed area than other Ibises, which is why they were attracted to the Chiku sandbank in the first place," says Ong Yi-tsung, a lecturer at Kunshan Industrial College in Tainan County.

It's most economical to preserve the current land: What has been the biggest change in the Tsengwen river mouth area since the discovery of the black-faced spoonbill in 1985? "In a sentence, there's less and less sand," says

嘴巴扁平如琵琶，是黑面琵鷺最主要的特徵。
A flat bill like the shape of a Chinese guitar *(pi-pa)* is the most distinquishing feature of the black-faced spoonbill.

求業者，若鳥類因污染去而不返，將會受罰或必需關廠。如此業者不僅不敢違反環保法規，還會試著更進一步去保護鳥類。結果不論在經濟發展或生活品質的維護上，最大的受益者都將是當地人。

永續利用才是長遠之計

農委會官員認爲，若能把這其中的利害，分析給地方上了解，相信臺南縣民都會願意選擇一個資源能永續利用的規劃。

工業區到處有，黑面琵鷺卻是臺南縣特有的資源。有一天，人們專程到臺南看黑面琵鷺，就像人們到四川看貓熊一樣。這不是不可能的。

（原載光華八十一年八月號）

Kuo Chung-cheng, who still goes birdwatching over 100 times a year. The western seacoast was originally rising land, but because each of the two tributaries to the upstream portion of the Tsengwen River had dams built on them, plus a great amount of illegal pisciculture on the western seacoast, today reclaimed land on the seacoast is declining, not increasing. It would be very difficult to find a similar sandbank for the birds.

Birds fly as they will, and no matter how carefully people manage, if the birds don't come then they don't come. What people might see as empty wasteland is for nature a place for life to flourish. Thus naturally protected areas are generally "protection of the existing area," because that is the most economical and most effective method of conservation.

"But it's not impossible to get the birds to 'move house,'" says Tseng Lung-yung, an anesthesiologist by trade and former president of the Kaohsiung Wild Bird Society. In the past people built a dike, and without thinking made the area inside the dike into something the black-faced spoonbill loves. One could try building another dike farther out to sea to produce a similar sandbank, in order to avoid human interference as much as possible. And who knows? The birds might move there. But this method would require a long-term trial, so people would have to have the patience to wait.

If the birds only have a place to go, and the environmental impact assessment for the industrial zone can be brought up to requirements and implemented, "no one would oppose the locality developing an industrial park," says one official at the COA who hopes the local government can enter into reasonable, calm dialogue. Since those closest to the black-faced spoonbill are the local residents, to protect it first requires getting local support.

A dilemma: As for whether or not to set up a protected area, the COA says that at present people don't understand enough about the black-faced spoonbill, about their feeding habits, about how large a feeding area they need . . . no one can give precise answers. The COA has already asked the Biology Department at National Taiwan Normal University to do the relevant research. They only hope that they can maintain the status quo in the sandbank until the results and suggestions come out.

"In fact, you can have your cake and eat it too — it's not hard to lay out a plan where development and preservation won't conflict," says one scholar who participated in the environmental impact assessment review.

If the industrial park can be successfully developed, nothing but workers will be attracted. From the development of the industrial parks currently in Taiwan, for many surrounding cities and towns, not only is there no prosperity, on the contrary there are endless environmental disputes and outmigration.

If there can be a sandbank where countless migratory birds can stay, which can be made into a natural park teeming with life, besides having long-term importance for environmental education, it could attract people to come for recreation. Thereafter industries run by local people, with local flavor, would develop in the neighborhood, producing things for tourist consumption. Local fishermen could develop recreational fishing, bringing the pisciculture industry back to life.

If there is a nature park in the vicinity of the industrial park, not only will the types of industry be limited (heavy polluting industries would not be permitted), the county government could even demand that operators pay fines or close factories if the birds fail to come back because of pollution. In this way, not only will operators not dare to violate environmental regulations, they will try to figure out new ways to protect the birds. Whether in terms of economic development or maintaining the quality of life, the greatest beneficiaries would be the local people.

Sustainable development is the best long-term solution: One COA official concludes: "If the costs and benefits can be analyzed and provided to the local people, I think that the people of Tainan County would choose a plan that provided for sustainable use of natural resources."

There are industrial parks everywhere, but the black-faced spoonbill is a resource unique to Tainan County. One day will people come to Tainan to see the black-faced spoonbill just like people go to Szechuan today to see the Panda? It's not impossible.

(Chang Chin-ju/photos by Kuo Tung-hui/
tr. by Phil Newell/
first published in August 1992)

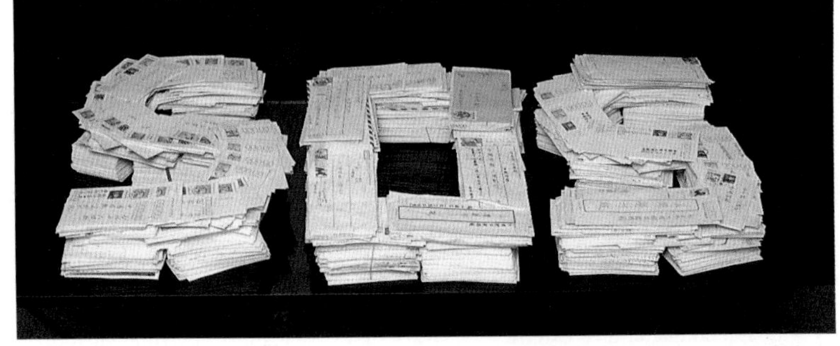

(卜華志攝) (photo by Pu Hua-chih)

中華民國野鳥學會舉辦的一人一信拯救「黑面舞者」的活動，已收到兩萬多封信，本刊摘錄其中幾則小朋友所寫的內容。看完他們的想法，大人又怎麼想？(信件由中國時報人間副刊提供)

The Wild Bird Society of the ROC is having a "one person, one letter" campaign to save the "black-faced dancer." They have already received 20,000 letters. Sinorama has selected excerpts from a few written by small children to see how they think. What about the adults? (letters courtesy of the China Times)

洪瑩眞：黑面琵鷺沒有落腳處，就像人沒有家一樣，臺灣能佔有琵鷺數目的一半，我們在世界上沒能和任何國家比生物的多寡，只有黑面琵鷺爲我們爭光，爲什麼我們不保護牠們，要讓牠們對人類產生仇恨和恐懼呢？

Hung Ying-chen: The black-faced spoonbill having no place to land is like a person having no home. Taiwan has half the world's black faced spoonbills. We can't compete with any country in the world in terms of the number of animals; only the black spoonbill brings any repute. Why do we not protect them, but produce in them a hatred and fear of mankind?

陳弘傑：我希望人類不要把這些黑面琵鷺趕走，能夠讓牠們住在曾文溪裡過冬，假如我有一筆龐大的錢，我會建造一個開闊的草原和森林，讓快要覺（絕）種的動物住在這一個寧靜的地方。

Chen Hung-chieh: I hope that people will not chase away the black-faced spoonbill, and can let them winter in the Tsengwen River. If I had a vast amount of money, I would build a huge open space and forest, to let endangered animals have a quiet place to live.

于佩君: 難道人們眞的狠得下心腸，讓黑面琵鷺不只來過冬，而在黑色的河水和吵鬧的噪音中永遠睡著，永遠消失在這世界上，每個人都說要保護動物，却沒有人以行動來說明，從前，就因爲人們的貪心濫捕濫殺而造成許多動物瀕臨絕種，到最後才說要把牠們當作「國寶」。

Yu Pei-chun: Is it possible that people are so hard-hearted that they won't let the black-faced spoonbill merely pass the winter, but would see it sleep forever amidst blackened rivers and cacophonous noise, to let it disappear forever from this world? Everyone says they want to protect animals, but no one does anything about it. In the past, just because of people's greed and lust for killing, many animals approached extinction, and only after were they called "national treasures."

徐仁國: 讓我爲黑面琵鷺說句話，我們的機器壞了，可以利用我們的智慧來修補，但大自然的生物死了，難道可使牠們復活嗎？雖然科技使我們很方便，但我們更要重是(視)大自然。

Hsu Jen-kuo: Let me say something on behalf of the black-faced spoonbill. If our machines break, we can repair them using our intelligence, but can we bring these birds back to life? Though technology makes life easier, nature is even more important.

莊業偉: 從小，我就很嚮往住在大自然的生活，天天可以在草地上跑步，細聽鳥兒的歌聲。如今，這夢想似乎已成了再也看不到，做不到，更聽不到的幻想，柔軟的草地變成硬梆梆的柏油路，可愛輕脆的鳥語聲，竟然變成了一捲捲的錄音帶，這，並不是我們所希望的，不是嗎？

今年，是天安門的三週年記(紀)念，人要自由，鳥也要自由，每一個人出一份心，出一份力，救救牠們吧!!

Chuang Yeh-wei: I've always longed to live in a natural environment ever since I was small, running in the grass, listening to the birds. Today, this dream has become a mirage that you can't see, do or hear. Soft grass has become hard asphalt, and the adorable, crisp and clear call of the birds has become the sound of cassette tapes. This is not what we wanted, is it? This year is the third anniversary of the Tienanmen Square incident. People want freedom, and birds also want freedom. Let everybody devote a little heart and a little effort to save them!

劉拓驛: 我想過，人可以自由，爲什麼鳥不可以呢？如果我們竟然還要眼睁睁看著牠們從自己的土地消失，那還有什麼事値得我們迫切去關心呢？

Liu To-yi: I have thought that people can be free, so why can't birds? If in the end we are willing to watch them disappear from our own soil, then what kinds of things can be considered important?

農民與鴨子的戰爭

A Seesaw Battle Between Farmers and Wild Ducks

文・張靜茹　圖・張良綱

旗海飄飄，玉里鎮農民在田間插滿旗幟以嚇阻雁鴨。

Yuli farmers bedeck their fields with waving banners to frighten off wild geese and ducks.

如何在利用資源與保護資源間找到平衡點？自生態保育觀念興起後，這樣的話題一直被討論著。若是生態保育者認爲應該保護的動物，侵犯了人類的利益，人類又該採取什麼樣的態度？

近來花蓮秀姑巒溪畔發生候鳥——雁鴨——侵入農田覓食，影響農民收益的事。農業人員追根究柢，却發現了和表相完全不同的答案。

在中央山脈與東部海岸山脈狹窄的縱谷間，秀姑巒溪流域由南而北，綿延一百多公里。豐沛的水域，孕育了兩岸村落；中游由花蓮縣玉里到富里間，引秀姑巒溪水灌溉的連綿稻田，成爲臺灣東部的魚米之鄉。

同樣在秀姑巒溪，寬廣的溪床，則是每年冬天由西伯利亞、中國東北前來避寒的候鳥——雁鴨——的最愛。石塊堆疊成的小水窪深藏魚、蝦、水生昆蟲，佈滿水域內的綠色水草、各種水生植物，供養千里迢迢南來的雁鴨，年年牠們得以補足體力，安然北返。

許多年來，候鳥以溪床安身，農民以兩岸農田立命，秀姑巒溪滿意地餵養著兩羣不同的生命……。

水鴨失常，冒犯農田

日子卻因原本各守分際，和平相處的農民與雁鴨起了衝突，不再平靜。

最近兩、三年來，玉里鎭附近農民發現，稻田常受不速客騷擾。尤其從九月到隔年春天，當一、二期稻作播種後，秧苗屢遭連根拔起，根部種子被吃掉，然後歪歪倒倒地浮在水面上，有如頑童惡作劇。

老鼠不會肆虐水田，平時就愛落脚稻田的鷺鷥也只吃魚蝦；因此，每年九月後翩然造訪秀姑巒溪畔，且葷素不忌的雁鴨，成了最大的嫌疑犯。

損失一穀一粒皆感心痛的農民，日夜追踪，結果發現了過去少見的鏡頭：日盡黃昏與天將破曉，常有野鴨幾十隻成羣由河流方向飛來，從空而降，落在稻田。等農民上前探究竟，鴨子早機靈地飛走，或迅捷地躲入田邊雜草叢生的灌漑渠道中。

稻田曾受野鴨光顧的玉里鎭長良里農民李敏雄，由秧田受損情況推測：「鴨子用嘴巴伸到泥土中，找尋根部尙未腐爛的稻穀種子吃，因此受損的都是播種不到兩星期的秧苗田。」

由於受損的秧苗，東一撮、西一撮，凌亂四散，很難再以機器補植，必須人工代勞，但工人不好請、工資又高昂，農民頻呼受不了。李敏雄就說，今年寒假他在臺中、臺北讀書的孩子回家過年，結果還得下田補植秧苗。

農民的防禦戰

農民不堪其擾，反制行動也一一展開。除了各種極盡眞人打扮的稻草人不斷粉墨登場，如今沿著與秀姑巒溪平行的玉里產業道路迤邐而行，但見溪畔綠汪汪的稻田間，花、白各色旗海飄揚，尼龍繩、塑膠袋做成的萬國旗於空中飛舞。

然而，這些傳統的防鳥方法，對嚇阻雁鴨的效果不大。許多農民反應，雁鴨膽大又聰明，稻草人、旗海頂多阻遏幾天，過後照樣來攪攪一番。因此農民又在黃昏與破曉時佐以放沖天炮、燒廢輪胎、點油燈……，甚至在田中架起捕鳥網。「但鴨子很少中計，」行政院退輔會在長良里的農田管理員余石廷搖搖頭，以對水鴨莫可奈何的表情說。

雖然各種嚇鳥招式盡出，但近年每逢秋末，人鳥拉鋸戰仍然準時在秀姑巒溪畔上演。去年九月，這場人鳥之戰，引起花蓮區農業改良場植物保護股注意，並派出工作人員到實地對「鳥害」展開調查。

花蓮農改場植保股助理研究員徐保雄表示，花東縱谷由北而南的三條溪流，花蓮溪、秀姑巒溪、卑南溪沿岸，農民偶爾都會見到雁鴨羣，但侵入農田、造成損害主要集中在

農民說雁鴨專吃秧苗下尚未腐爛的稻穀。

Farmers say the wild geese and ducks go for intact seeds at the foot of rice shoots.

How does one strike the balance between utilizing natural resources and protecting them? Such questions have been discussed ever since the idea of ecological conservation caught on. If animals conservationists think should be protected happen to infringe upon man's own interests, what attitude should we take?

Recently, migratory birds — wild geese and ducks — have invaded farmland in search of food beside Hualien county's Hsiukuluan River, affecting the local rice farmers. Investigation by agriculturists, however, has come up with answers radically different from appearances.

Flowing northwards for over 100 kilometres along the narrow valley between the Central Range and Taiwan's east coast, the Hsiukuluan River's mighty waters nourish villages on both banks. Along its middle reaches between Yuli and Fuli, rice fields irrigated by the Hsiukuluan River constitute the breadbasket of East Taiwan.

Every year the Hsiukuluan River's broad bed is also home to migratory wild geese and ducks escaping the Siberian and Manchurian winters. The river's green water weeds and water plants, together with the fish, shrimp and aquatic insects that shelter in its rock pools, provide nourishment for these distant southern migrants who thus build up their strength each year for the long journey back north.

For many years these migrants settled on the Hsiukuluan River while farmers made their living along its banks, the river happily sustaining both.

But now conflict has arisen between these two groups, disturbing their former peaceful co-existence.

During the past two or three years, farmers near Yuli have found their rice paddies being damaged by unwelcome guests. From September to the following spring, when the first and second rice crops are sown, the rice shoots are being torn up by the roots and the seedlings eaten, leaving the paddies strewn with fallen shoots as if by some mischievous child.

The farmers knew that rats don't infest paddy fields, and that the little egret which habitually stalks rice paddies is only after insects; so the omnivorous wild ducks which visit the Hsiukuluan River's banks from September onwards became the prime suspects.

Jealous of every grain of their rice, the farmers kept watch day and night and soon discovered a previously rare sight: at dusk and dawn, dozens of wild duck were flying over from the river and settling in flocks on the rice paddies. By the time the farmers approached, the ducks had either flown off or quickly taken refuge in weed-grown irrigation ditches.

From the damage to his fields, affected farmer Li Min-hsiung of Yuli's Ch'ang-liang village concludes: "The ducks probe the mud in search of intact rice seeds, so they always go for the roots of seedlings planted within the past fortnight."

The damaged rice seedlings, scattered right and left, cannot be recovered mechanically and have to be picked up by hand, and the farmers began complaining. Li Min-hsiung's children, home from college in Taichung and Taipei, had to spend their Chinese New Year holiday salvaging rice seedlings

被破壞的秧苗田中，只留下一個有蹼的脚印，其他是鷺鷥與秧雞的足跡。

In this damaged rice field there's only one webbed footprint, the others are egret and rail tracks.

秀姑巒溪中游玉里一段。在玉里二百多公頃的稻作面積中，據當地農民表示，約有三分之一會於插秧期間遭到雁鴨「叨擾」；嚴重時，一片田中有三成以上的秧苗受損。近來甚至在夏天收成季節，也可見到鴨子在田間徘徊、覓食稻穀，索性不回北方了。

流連在秀姑巒溪中游

令人好奇的是，爲何雁鴨不再安於河床而侵犯農田？爲了找出原因以對症下藥，徐保雄帶著同仁對雁鴨進行追踪。

臺灣由於瀕臨太平洋岸和位在南北半球中間地帶，成爲候鳥南北遷移的一個重要中點站。由西伯利亞、東北、韓國、日本一路南下的候鳥，進入臺灣後一分爲二，有的借道西岸，有的選擇東岸而行。

臺灣西部海岸有許多開濶的河口，潮汐造成的肥沃沼澤、泥灘，吸引了大多數的候鳥；花東沿岸候鳥的鳥數、鳥種因此大不如西岸。但往來南北半球的候鳥中，被農民一律

crippled by wild ducks.

Irritated farmers took steps to control the menace. In addition to lifelike scarecrows springing up on all sides, the green rice paddies beside the Hsiukuluan River were decked with a motley show of white and colored flags, nylon raffia, and plastic bags fluttering in the breeze.

But these traditional methods weren't very effective against wild ducks. As many farmers discovered, the ducks were so bold and intelligent they just ignored the scarecrows and flags after the first couple of days. So around dusk and dawn the farmers set off rockets, burned old tires and lit oil lamps, and even set up bird nets in the fields. "But the ducks seldom fall for it," says Yu Shih-ting of Ch'ang-liang's VACRS farm management committee, shaking his head and at the end of his tether over these waterfowl.

They've been through every bird-scaring trick in the book, but still this seesaw battle between man and bird goes on beside the Hsiukuluan River as regular as clockwork every autumn. Last September it came to the notice of the plant conservation department at the Hualien District Agricultural Improvement Station, who sent personnel to carry out an on-the-spot investigation into this bird pest.

According to Hsu Pao-hsiung, a research assistant at the department, farmers along all the Taitung-Hualien rift valley's three north-south flowing rivers, the Hualien River, Hsiukuluan River and Peinan River, occasionally see flocks of wild geese and ducks, but they only invade rice paddies in the area concentrated on the middle reaches of the Hsiukuluan River, in the Yuli area. Up to a third of the over 200 hectares of rice paddy in Yuli are damaged by waterfowl during the planting season, with at worst over 30 percent of the seedlings in each field being laid waste. Wild ducks have even been seen lingering in the rice fields during the summer harvesting season, never bothering to return north at all.

Curious as to why these waterfowl should leave their river bed home and invade farm fields, Hsu Pao-hsiung and his colleagues launched an enquiry to find the reasons and so pinpoint solutions to the problem.

On the Pacific rim and midway between the northern and southern hemispheres, Taiwan is a major way-station for north-south migrating birds. Migrants from Siberia, Manchuria, Korea and Japan reaching Taiwan follow along either the west or the east coast.

The majority are attracted by the numerous broad river estuaries and rich tidal marshes and mud-flats along Taiwan's west coast, so there are far fewer migrant birds, and of fewer species, on the steeper eastern coast. Many of the wild geese and duck species which migrate between the hemispheres, all of which Taiwan's farmers lump together as waterfowl, do not just transit through Taiwan but stay for the winter. They stay for quite

視爲水鴨的許多種雁鴨科鳥類，不只是過境臺灣，而是留下渡冬，由於停留時間較長，牠們不一定棲息在海岸、河口，反而常常往河口上溯，或直接翻山越嶺，飛到內陸流域，尋找更安全的打尖地點。開發程度較低的秀姑巒溪，因此「招攬」了許多呼朋引伴而來的雁鴨。

根據中華民國野鳥學會調查，秀姑巒溪下游是泛舟地點，人煙較多，因此雁鴨集中在中游，這是玉里農民較易見到雁鴨的理由。至於鴨子冒犯農田，農改場在對環境做客觀探究後，發現答案出在秀姑巒溪身上。

溪流的嗚咽

「近年來秀姑巒溪改變太大了，」徐保雄對溪流面貌變遷感到驚訝。他表示，過去河域寬濶、潮濕；如今築了堤防，河川取直，河面窄縮，且河水下滲、溪床乾涸，已不適合雁鴨棲息。

秀姑巒溪上游的清水溪因爲開採蛇紋石礦，牛樟、烏心石等原生濶葉林又被大量盜伐，影響水土保持，砂石沿著河水滔滔而下，堆積河床，中游一帶河床如今幾乎都與橋面同高。

爲了充分利用土地，縣政府把河川地租給農民開墾，兩岸新闢的水田，要用大量溪水灌溉，河水更加枯竭。農田又帶來農藥污染，影響了溪中生物的繁衍。

目前秀姑巒溪玉里段新築的堤防，平均又比舊堤往內移了六十公尺以上。過去河川彎彎曲曲，有泛濫平原、曲流，及自然生長的溪邊植物，如今兩岸植物被清除，河道雖然變得整齊，但却貧瘠。還有不少非法濫墾者在乾涸的河床開墾高經濟價值、生長期短的西瓜和玉米等雜糧作物，河床成爲單調的旱田。

鳥兒豈懂人類的界限

雁鴨和許多水鳥一樣，仰賴岸邊植物掩護棲息，需要水中生物維生；但人爲的開發，許多原來適合魚、蝦、水生昆蟲、植物生存的環境已消失。「雁鴨爲了生存，只好往兩岸農田覓食，」徐保雄在調查報告的「雁鴨族羣爲害原因探討」一章中，寫下這樣無可奈何的詞句。

他的看法和東海大學環境科學系教授張萬福不謀而合。在秀姑巒溪沿岸小村莊春日里長大的張萬福，特別關注秀姑巒溪環境的變化。他回憶，二、三十年前，附近小孩最愛在溪邊追逐成千上百的水鳥；但最近十幾年來，農耕地不斷擴展，河川地被墾殖，如今開發的觸角更延伸入河床，溪流生態體系已完全兩樣。「再也難見衆多野鴨在溪床嬉耍的場面，」他說。

類似秀姑巒溪畔農民和雁鴨的戰事，其實時常發生。麻雀流連曬穀場，趕都趕不走；西海岸則常有鷺鷥偸吃養殖池中魚蝦的事件。事實上，恐怕早在人類懂得耕種，鳥類發現農田有充裕的食物，而牠們又不懂得人爲界限，人與鳥的糾紛就已開始。

但人鳥之爭，追根究柢，「更多時候是因爲牠們原有的覓食環境被破壞，才會退而求其次，侵入人類的疆界，」野鳥學會資深鳥友許建忠指出。

水鴨數量只減不增

由於人類對環境需求膨脹，其他生物生存的範圍因此日愈縮小，甚至危及生命的延續。以鳥類爲例，除了麻雀等少數種類能在人類生存環境求得一席之地外，鳥類數量、種類一直在大量減少。

十幾年來，野鳥學會每年對臺灣候鳥鳥口做調查，西岸由淡水關渡河口、大肚溪口到臺南曾文溪河口，因爲魚塭、工業區不斷開發，加上油污、垃圾等污染，鳥口一直是負成長。

近三年來秀姑巒溪同一時間出現的候鳥數量，也都不超過一千隻。野鳥學會東部鳥口調查負責人、臺東豐里國小總務主任廖聖福認爲，這可能是爲什麼秀姑巒溪鳥類只集中在幾個地方的原因。由整個大環境來看，鳥害比起病蟲害、鼠害對農作物的影響還是微

a lengthy period, and don't necessarily settle on the coast or on estuaries, often following rivers upstream or flying directly over the mountains to inland watercourses in search of somewhere safe to feed and rest on their journey. The relatively undeveloped and pollution-free Hsiukuluan River draws multitudes of wild geese and ducks, who gather in droves.

An ROC Ornithological Society survey has shown that wild geese and ducks congregate along the middle reaches of the Hsiukuluan River because the more densely inhabited lower reaches are frequented by boaters, which explains why they are most often sighted around Yuli. As for their invasion of the paddy fields, an objective environmental study by the Agricultural Improvement Station has established that the answer lies in the Hsiukuluan River itself.

"The river has changed out of recognition in recent years," says Hsu Pao-hsiung, expressing his amazement at the transformation that has overtaken it. The flood plain used to be wide and marshy, but today its banks have been narrowed by levees, its course straightened, and the bed of the river raised, so that the water has soaked away leaving dry river bed inhospitable to waterfowl.

Mining for ophiolite and unlicensed logging for hardwood timber such as shoghu has affected the Chingshui River, an upstream tributary of the Hsiukuluan, creating an alluvium of gravel that has collected along the river bed, raising it almost level with the bridges across its middle reaches.

Land on the flood plain has been leased to farmers by the county government, and irrigation for these new paddy fields has further depleted the river. Pollution from pesticides has in turn interrupted biological life cycles in the river.

From bank to bank, the new section of levee at Yuli is on average 50 meters narrower than the old one. Where the river used to meander along and overflow into the flood plain, watering the vegetation growing along its banks, all this vegetation has now been cleared away leaving the flood plain tidier but barren. Illegal cultivation of high-value, quick-growing crops such as water melon and sweetcorn has turned the dry river bed into monotonous market gardens.

Like many water birds, wild geese depend on the riparian vegetation for shelter and live on food from the water, but development is causing much of the habitat of fish, shrimps, aquatic insects and plants to disappear. "To survive, geese and ducks must feed in the paddy fields," concludes Hsu Pao-hsiung's report on the problem.

His view is shared, coincidentally, by Tunghai University professor Chang Wan-fu, who grew up in the village of Ch'un-jih-li on the banks of the Hsiukuluan River and has a special interest in environmental change affecting the river. He recalls that twenty or thirty years ago local children would chase the myriad water birds on the river, but in the past decade farming has come to the flood plain and development is extending even to the river bed, transforming the river's ecology out of all recognition. "Now you no longer see large flocks of wild ducks sporting on the river bed," he says.

Battles like that between the Hsiukuluan farmers and the wild ducks are commonplace. Sparrows haunt threshing floors and cannot be driven off; on the west coast, egrets often steal fish and shrimps from breeding ponds. In fact, birds probably discovered farmers' fields were a rich source of food at the very dawn of human history, and it may be that conflict between men and birds has its roots in remote antiquity.

But such conflicts tend to arise "generally because their original feeding habitat has been destroyed, so they go for the next best thing and invade human territory," indicates Hsu Chien-chung of the ROC Ornithological Society.

Man's demands on the environment are reducing the scope for other living things, and even threatening the continuation of life. Bird populations and species numbers are declining drastically, for example, apart from those few, like sparrows, which can cope with the human environment.

The ROC Ornithological Society has been studying Taiwan's annual migrant bird population for the past ten years or so, and has found steady negative growth along the west coast from the Kuantu estuary at Tamsui southward via the Chu-an and Tatu estuaries to the Tsengwen estuary at Tainan, due to aquaculture and industrial development, along with oil and garbage pollution.

In the past three years there have never been more than 1,000 migratory birds at any one time on the Hsiukuluan River. According to Liao Sheng-fu, general affairs supervisor at Taitung's Fengli

人們對自然資源的過度開發，影響了候鳥的棲息地，要拍到這樣的鏡頭愈來愈難了。圖爲淡水河華江橋下的小水鴨。

Overdevelopment of nature affects the rest areas of migratory birds. Photos like this will be more and more difficult to find. The photo shows ducks under the Huachiang Bridge over the Tamsui River.

小得多。徐保雄就說，花蓮農改場還是第一次做「鳥害」調查。

但鳥兒既已對農田造成損害，也不能要農民犧牲收益，成全水鴨。在農民與鴨子各自的利益中，又要如何取得平衡？

補償農民、承租耕地

廖聖福表示，對於日漸稀少的鳥類侵犯農地，有些先進國家政府會要求農民不要捕殺，再透過客觀的調查數據，由政府加以補償。「就像對農作物風災損害的補償一樣，」他比喻。

日本九州鹿耳島為保護每年冬天由西伯利亞來的鶴羣，當地政府就承租牠們休憩的農地，做為鶴類保護區。在鶴羣離去的季節，農民仍可照常耕種。

然而，一個無可奈何的事實是：在許多工業先進國家裡，因為強烈的開發壓力，候鳥的棲息地往往也只剩下被設為保留區的地方了。

為了保留一塊地方給候鳥，臺北市政府也於民國七十二年在關渡設立全臺第一個候鳥保護區。但和開發脚步相較，保護區設立往往困難重重。為了徹底保護候鳥棲息地，市政府欲把保護區擴大，成立關渡自然公園，關渡地區幾年來地價漲了二、三十倍，市府籌不足經費收購關渡提防內的土地，自然公園因此一直無法成立。

在美國，由政府編預算收購土地、設保留區，同樣緩不濟急。因此有保育團體將平時募得的款項，拿來做為搶救自然環境的基金，等政府預算下來，再拿回基金，用在下一個目標上。

麻雀也能立大功

但，是否保護包括鳥類在內的自然資源，就和人類開發資源的利益衝突呢？事實上，自然界的各種生物彼此相依，關係錯綜複雜，每種生物都有它存在的價值，甚至表面看來影響人類利益的鳥類，對人類的生存也有其意義。

譬如鳥類中被農民視為最大敵人的痲雀，除了穀粒，還吃很多昆蟲。過去大陸上曾有段時期，各鄉鎮發動滅絕痲雀運動，結果那幾年蟲害大增，五穀雜糧、蔬菜全部欠收。

「我們不能簡單地以牠對人類的利害關係，來評定某種鳥類是益鳥或害鳥，害鳥就加以捕殺，」承辦秀姑巒溪「鳥害」案的農委會保育科技士陳超仁說，若侵害稻田的雁鴨並不稀有，且對稻作眞的造成劇烈損害，以人為方法有限度的控制鳥類數量，並非不可行，但前提是必須對水鴨族羣有相當了解。

牠是否眞的已超過這一地區所能容納的量？牠對農作物的影響程度到底如何？是那些種類的雁鴨侵入農田？牠們的習性如何？……有了這些資料，才能擬出一套合理的管理辦法。否則僅憑主觀的說殺就殺，萬一又發現牠們是大自然環節上一個重要的齒輪，再想挽救，恐怕為時已晚。

秀姑巒溪的反撲？

若能進一步了解水鴨，說不定很容易就可找出嚇阻水鴨侵犯農田的良方。但這些資料都必須靠平時累積，碰到問題才能立即反應。我們近來對鳥類的研究已逐漸增加，但多集中在留鳥身上，對屬於國際性資產的候鳥，只有數量調查，生態習性的了解仍十分有限。

秀姑巒溪畔的人鳥之爭，已提醒了主管單位。農林廳技術室保育組已決定，除了多提供農民一些驅鳥的方法，也商請野鳥學會對秀姑巒溪雁鴨習性、數量做更深入調查。

但要徹底平息這場人鳥之爭，恐怕還是要減少秀姑巒溪河床的開發與污染，鳥兒吃住有了著落，人鳥才得以相安無事。花蓮農改場就呼籲縣政府應積極取締非法濫墾河川地的行為。

除了人與鳥，我們也該還給默默孕育許多生命的秀姑巒溪一個自然、清新的面貌。水鴨的「失態」，說不定只是秀姑巒溪給人們的一個「暗示」。

（原載光華八十年四月號）

Elementary School and the man responsible for the society's East Taiwan bird population survey, this may be why the Hsiukuluan River bird population is only concentrated at certain points. Within the environment as a whole, birds are an insignificant crop pest compared to the damage from disease, insects and rodents. Hsu Pao-hsiung adds that this is the first time the Hualien Agricultural Improvement Station has ever carried out a "bird pest" survey.

But damage is being done, and farmers cannot be expected to sacrifice their income for the sake of waterfowl. How does one balance the separate interests of farmers and wild ducks?

As Liao Sheng-fu points out, governments concerned for the environment ask farmers not to kill birds that encroach upon their fields, and pay them compensation calculated on the basis of objective data. "It's rather like compensation for typhoon or flood damage," is his analogy.

To protect the flocks of cranes that migrate from Siberia every winter, the local government in Kagoshima, on the Japanese island of Kyushu, rents the farmland where they roost as a bird sanctuary. Once the birds have left, the farmers may till the land as usual.

Pressure of development in many advanced industrial nations means that bird sanctuaries are the only place left where migratory birds can roost.

In 1983 the Taipei City Government decided to set up Taiwan's first migratory bird sanctuary at Kuantu, but their plans have been overtaken by the pace of development. Land prices in the area have risen 30-fold, and the city government has still not officially established the sanctuary due to difficulties over funding purchase of the land.

In the United States, provision for buying land for conservation areas is included in the federal budget, but this is slow in coming. So conservation groups usually raise funds to rescue the natural environment, and once the federal budget is approved the funds are withdrawn for use elsewhere.

Is there a conflict of interest between protecting natural resources (including birds) and human development of resources? In fact, all living species in nature are bound by complex relations of interdependency. Each species has its own value, and even superficially harmful birds have their value for human survival.

"We can't simply decide whether a species of bird is beneficial or harmful on the basis of its affect on man, and then just kill off the harmful ones." Chen Chao-jen, a specialist with the Council of Agriculture's conservation division who is handling the case, explains that if large flocks of wild ducks are seriously damaging farmers' crops, then conservationists have no objection to artificial means being used to control their numbers, provided they have a real understanding of the species.

Is the bird population really too large for the area? How much effect do they really have on crops? Which species of wild ducks are invading the fields? What are their habits? . . . . This data is necessary for drawing up a proper control strategy. If people just kill them because they're a nuisance, once the birds are found to be an important part of the natural cycle it'll be too late to do anything about it.

If more is understood about water birds it should be easy to find a way of stopping them invading paddy fields. But this information needs to be gathered routinely, for use as and when required. More ornithological studies are being carried out now, but they tend to concentrate on resident birds, and apart from the size of population we still have only limited knowledge about the migrant birds that are part of our global heritage.

The Hsiukuluan River conflict between man and bird has alerted the authorities. In addition to providing farmers with methods of scaring birds, the Department of Agriculture and Forestry's conservation section has decided to commission a deeper study from the ROC Ornithological Society of the habits and numbers of wild ducks on the Hsiukuluan River.

But resolving the conflict may depend on reducing development and pollution of the flood plain, since peace will only be restored once the birds have somewhere to feed and live. The Hualien Agricultural Improvement Station is calling upon the county government to actively put a stop to illegal farming on the flood plain.

Birds and humans apart, we ought to restore the Hsiukuluan River, host to numerous life forms, to its natural clean state. The wild ducks' abnormal behavior may just be the river's way of giving man a hint as to what should be done. S

(Chang Chin-ju/photos by Vincent Chang/ tr. by Andrew Morton/ first published in April 1991)

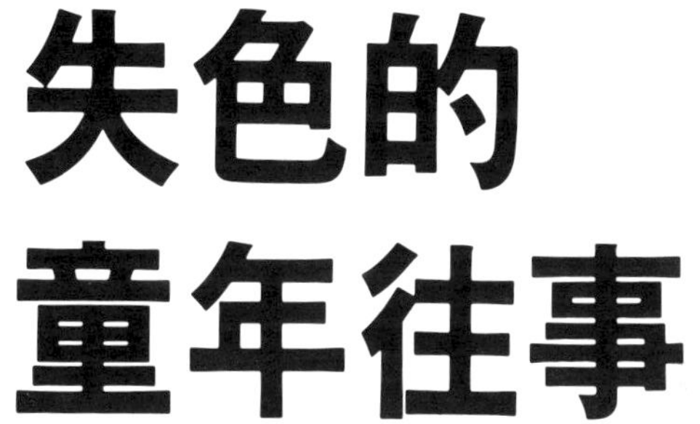

失色的童年往事

——野生動物回憶錄

Faded Childhood Memories of Wildlife

文・張靜茹　圖・本刊攝影小組／郭智勇

家燕常選擇屋簷與樑柱築巢，是少數可以在都市見到的動物之一。(郭智勇攝)

The swallow often makes its nest in eaves or a beam, and is one of the few animals you can see in the city. (photo by Kuo Chih-yung)

鄉村的動物也在銳減，小朋友忙碌半天，只撈到幾粒「蛤仔」。(卜華志攝)

Animals are sharply declining in rural areas as well; these kids worked at it a long time and only scooped up several clams. (photo by Pu Hua-chih)

當研究人員在三千公尺高的玉山進行黑熊生態調查，在墾丁復育已絕種於野外的梅花鹿，在梨山大甲溪放養人工繁殖的臺灣鱒……，許多瀕臨絕種的本土動物陸續受到關注；而一向飛舞在人們身邊的蝴蝶、螢火蟲，池塘與溝渠中的蝌蚪、大肚魚，盤旋空中的老鷹……，卻也在我們生活中陸續變成「稀有動物」。

如何讓牠們回到都市之中、人們身邊，是世界各地都市化後面臨的一致課題。

暑假日記，民國六十年七月五日，晴，五年乙班王小明。

今天下午和弟弟妹妹到家後面的溪邊玩，路經竹林時，妹妹跑得好快，因爲昨天我們在這裏看見一條青竹絲。我們赤脚在水中撈大肚魚和蛤仔，成對的蜻蜓由草叢間飛來，停在水面。弟弟忙著用蒼蠅紙黏樹上的蟬。

回家後我急忙跑到屋簷下，看樑上的燕巢是否有蛋孵出來。弟弟卻對著院子的芒果樹大喊：「芒菓被鳥吃了一個大洞！」我想一定是常逗留在我家樹梢的白頭翁幹的好事。

吃過晚飯，我最喜歡的事是看壁虎捕捉飛蛾，弟弟在紗門上抓到兩隻金龜子，還用線綁住，讓牠們在空中飛舞。忽然一隻蝙蝠不知由何處飛進客廳，妹妹嚇得大叫起來。

× × ×

暑假日記，民國八十年七月五日，晴，五年十七班李小梅。

昨天媽媽幫我們買了一套動物百科全書，裏面有很多圖片。這是我第一次看清楚許多動物的長相，哥哥一直在算有多少種是市立動物園有的。

我最喜歡的動物是梅花鹿，可惜媽媽說梅花鹿已在野外消失。去年臺北市選擇區運代表動物時，我還投了牠一票。

我最喜歡的電視節目「臺灣的生物世界」今天介紹藍鵲，牠是一種生活在原始濶葉林的鳥類，有好長的尾巴。我拜託媽媽趁暑假帶我到郊外看藍鵲，媽媽卻說她也不知道那裏有藍鵲，還是在家看動物圖片好了。

影片結束前，一羣藍鵲由樹林往山谷中飛去，正當我看得目瞪口呆，忽然有隻蟑螂由電視機後面飛出來，我嚇得大叫起來。

With researchers 3,000 meters up Yushan undertaking a study of the ecology of the Formosan Black Bear, bringing back the Formosan Sika — long extinct in the wild — at Kenting, and stocking the Tachia Stream with Taiwan Trout raised artificially, many local species on the brink of extinction are getting greater attention. Meanwhile, the butterflies and fireflies that once fluttered around us, the tadpoles and mosquito fish which occupied ponds and ditches, and the eagle searing through the air, have all become "rare animals" in our daily lives.

Getting these creatures back into our cities and by our sides is a problem faced by cities worldwide.

From the summer diary of fifth-grader Wang Hsiao-ming: July 5, 1971, clear and sunny.

Today I went to the stream behind the house with my little brother and sister to play. When we passed by the bamboo grove, my little sister started to run really fast, because yesterday we saw a poisonous snake there. We scooped up the mosquito fish and clams in the stream with our bare feet. Dragonflies came flying out of the grass in pairs, and stopped on the top of the water. My little brother was running around with flypaper catching the cicadas on the trees.

When I got home I ran over to look under the eaves, to see if any of the eggs in the swallow's nest on the beam had hatched yet. But my little brother was looking at the mango tree in the courtyard and yelling, "Hey! Some bird ate a big hole in this mango!" I thought for sure that it must be the work of that bulbul who often sits on the ends of the branches of our trees.

After eating dinner, my favorite thing to do was to watch the geckos catch moths. My little brother caught two tumble bugs on the screen door, and tied them down with some thread, so they did a flying dance in the air. All of a sudden, I don't know from where, a bat flew into the living room, and my sister screamed.

From the summer diary of fifth grader Li Hsiao-mei: July 5, 1991, clear and sunny.

Yesterday mommy bought us an animal encyclopedia, with lots of pictures. This is the first time I ever really got to see the faces of these animals, and my older brother kept count of how many are in the city zoo.

I like the sika best of all the animals. It makes me sad because mother said they have disappeared in the wild. Last year when they chose the official mascot for the Taipei athletic meet, I gave this animal my vote anyway.

My favorite television program, "Taiwan's Animal World," introduced the "Swinhoe's blue pheasant." It's a kind of bird that lives in primitive deciduous forests, and it has a really long tail. I asked mommy to take me to the suburbs during the vacation to see a blue pheasant, but she said she didn't know where any were, so it was just as well to stay home and look at animal pictures.

Right at the end of the show, a bunch of pheasants took off from the trees and flew into a valley. I was sitting there watching the screen with all my attention, when a cockroach flew out from behind the TV, and it startled me so much I screamed.

From Sika to Rats: "In twenty years, the sea can turn into a mulberry plantation." This proverb is usually used to describe the vicissitudes of human affairs, but when this becomes a fact of life in the natural world, the ones left bemoaning it are the animals.

If you look back at historical records and all kinds of biological evidence, you can discover the following:

In the Taiwan of the beginning of this century, people were surrounded by the Formosan Sika, the Chinese Otter, the Formosan Rock Monkey, and other large mammals. There were more than 100 different types of birds, more than 20 types of amphibians, more than 100 kinds of reptiles, and a large number of insects (the number is still not clear today).

By the 1950's, the sika had disappeared, and, except for rats and bats, almost all remaining mammals were confined deep in the densely forested mountains. But even then the animals still had "air supremacy," and you could lift your head and see eagles, pheasants, and all kinds of birds; insects like butterflies and fireflies, and amphibians like toads and lizards were active, accompanying postwar babies as they were born and went through childhood to become the generation which is today in the "prime of life."

And the next generation? Today's kids grow up with carefully nurtured and pampered pets which

農地大面積單一化的栽種稻作，麻雀成爲最強勢的平地鳥種，鳥相也「單一化」了。（鄭元慶攝）

As farmland becomes increasingly standardized into single-crop rice growing areas, the sparrow has become the unchallenged ruler of the plains. It seems the bird life has also become "standardized." (photo by Cheng Yuan-ching)

仁愛路上的行道樹，招來了一群白鶺鴒。（郭智勇攝）
The trees along Jen-ai Rd. have attracted some bulbuls. (photo by Kuo Chih-yung)

梅花鹿到老鼠

悠悠廿年，滄海成桑田，通常用來形容人事變遷，而當它成爲自然界的事實，要大加慨嘆的可就是野生動物了。

如果由史料和各種生態記錄循線追踪，可以發現——

本世紀初的臺灣，人們身邊常打轉的有梅花鹿、水獺、臺灣獼猴等較大型的哺乳動物；還有一百多種鳥類、廿多種兩棲類、一百多種爬蟲類與目前仍不淸楚數目的昆蟲。

到了一九五〇年代，梅花鹿消失，哺乳動物除老鼠、蝙蝠之外，幾乎都已遠避深山密林。不過，動物的「制空權」還在，人們抬頭可見到老鷹盤旋，畫眉、藍鵲、雲雀……各種鳥兒飛翔，螢火蟲、蝴蝶等昆蟲與盤古蟾蜍、攀木蜥蜴等部分兩棲類也仍活躍，伴著戰後嬰兒潮出生、現在正値壯年的一代走過童年。

我們的下一代呢？今天相伴的是細心養護、抱著摟著，不時要上醫院打預防針、修毛理容的寵物，野生動物常見的似乎就只剩下老鼠、蟑螂等「害蟲」。

老鷹展翅、水獺抓魚？

「老鷹捉小鷄」的遊戲一代傳一代，但它並非無中生有的發明。俗稱老鷹的「鳶」在老一輩口中就常出現。根據一九六〇年美籍鳥類觀察者謝孝同的報告，當時老鷹在臺灣仍是非常普遍的留鳥。在今天臺北市區內的三張犁、大直及北縣五股等沼澤區，都可以見到老鷹居高臨下，展翅盤桓，甚至俯衝攫食的場景。

民國六十三年中華民國鳥類學會的鳥口調查顯示，臺北地區只有廿一隻老鷹，往後繼續逐減。民國七十年至今，臺北的「鷹口」數字都掛零。

身體五彩斑斕，常被錯認是異國熱帶魚的臺灣鬥魚（俗稱三斑），原本廣游於臺灣溝渠。「八、九年前溫州街臺大宿舍尚未改建，我們常在那兒抓三斑，」研究魚類的臺灣大學動物系博士班研究生曾晴賢說，如今研究人員翻遍郊區池塘、灌溉渠道、水田，已遍尋不著三斑游踪。

三、四十年前，平地乾淨的溪流、溝渠還可以看見水獺抓魚；走在大臺北的路上都能

as often as not are down at the vet's getting an innoculation or off at the "hairdresser" for a trim. The only "wild animals" left are harmful ones like rats and cockroaches.

Where Eagles Fear to Tread: The game "The Eagle Catching the Chick" has been passed from generation to generation, but it wasn't just invented from nothing. The "eagle" is the term people use to refer to the "Formosan Black-Eared Kite," and it was often uttered by the older generation. According to a 1960 report by the American bird observer Hsieh Hsiao-tung, at that time the eagle was very common in Taiwan. In Sanchangli and Tachih, now part of the Taipei metropolitan area, and Wuku in Taipei County you could see eagles spreading their wings and swooping down from on high, even catching prey.

As for animals that only eat one type of food, that only live on one kind of tree, or whose mobility is low or who require a high degree of concealment, these are often the "first sacrifices" to the development of the environment. The Orchid Island Monarch Butterfly, which consigns its young only to "birthwort," or the guinea fowl, which likes to hide in the dense thickets on the edge of a field so that often people can only hear its squawk, are both examples.

Reduction of Species: The restructuring of nature undertaken by man invisibly is a kind of "unnatural selection" for animals. A small number of animals, which have a broad diet, can reproduce rapidly, and don't have any preferences about heat and humidity are able to "keep in step" with human development. For instance the Hartert's Chinese Bulbul, which naturally prefers open ground in deciduous forests, has taken over where man has opened up new land, replacing the animals which desire dark and damp places, becoming one of the few bird species to "take off." With mankind making way for it, you can even hear the call of the bulbul in trees on city streets.

Moves toward uniform planting on agricultural land have provided abundant feed for sparrows and certain kinds of insects. Just as the dragonfly, firefly, tadpoles, and a number of shallow water fish have been wiped out in droves beneath an assault of pesticides, a few types of insects which are harmful to agriculture — like moths — have built up their resistance, testifying to the old adage that, "a centipede dies hard."

Densely populated urban centers provide even tougher challenges for animals. Besides the cats and dogs roaming the streets, the eternal piles of garbage attract and become the survival laboratories for more daring animals. The wastemeisters — rodents — are going in precisely the opposite direction from other mammals; while others are declining in number, they are increasing daily, running rampant, contaminating food, and spreading disease. Similarly the pigeon, not exactly a connoisseur of fine dining, grows fat being fed by urbanites. In European cities, their excrement corrodes brick and stone buildings and centuries-old statues.

The smell of the Fu-teh-kang garbage dump is indeed repellent, but it also offers a beautiful sight, white egrets flocking in to hang out over the resources tossed away after human consumption. The openings in city air conditioners have been known to replace tree branches, with nests of sparrows or bulbuls. Perpendicular skyscrapers provide a resting environment not all that different from sheer cliffs. There are stories from overseas of kites landing on the top of a building and gliding down along its face. But these scenes which occasionally startle and delight the urbanite are, after all, only very rare exceptions.

Nearer Is Better: Cities are, in the end, places made by people for themselves, and are really too different from the natural environment, and the number of animals which can coexist with mankind is extremely limited. Ants, pigeons, rats, and cockroaches are some of the important urban migrants, and are a universal result of urbanization.

Thus today various nations are rushing to create national parks and preserves, with one of the goals being to preserve a pristine environment for animals in the wild. Recently, studies of "urban animal life" have appeared, with the purpose being to enable other life forms to return to the human environment.

"For a person living in the concrete jungle, a Muntjac on the top of Yushan is not nearly as meaningful as a bulbul that you can see everyday," repeatedly stresses Lin Yao-sung, director of the Department of Zoology at NTU. There should be undomesticated animals in the living environments of people; only then can we remind people that man lives in close interdependence with nature.

Lin Jun-yi of Tunghai University adds that

(鄭元慶攝)(photo by Cheng Yuan-ching)

(本刊資料)(*Sinorama* file photo)

(張良綱攝)(photo by Vincent Chang)

「撿」到分佈在臺灣北部的臺北赤蛙，而水獺與赤蛙今天都已芳踪難尋。

螢火蟲臣服於路燈

「我們小時候到處可見的動物，大部分數量都在萎縮，」今年四十歲的臺灣省林業試驗所森林保護系主任趙榮台肯定地說。獵捕壓力、農藥過度使用、河水污染等都是野生動物減少的原因，而人類大規模開發，更是生物學界認定的主因。

例子俯拾皆是。民國六十六年夏天，臺大植物病蟲害系主任楊平世和賞蝶學會在新店五峯山，一天可以見到二、三十種蝴蝶。三年後，草木被伐除、山被剷平，樓房代之而起，蝴蝶幼蟲缺乏寄生植物，跟著消失。

都市路燈則是螢火蟲的剋星。因爲夜晚燈火通明，使牠們無法在光亮中覓得另一半，

because in the process of growing up today's young people have no contact with nature, have never really felt nature, and have not learned from nature, they don't understand how moving life is. ''To touch a tree leaf every day, to see a butterfly, to grow up with nature — that's the secret to building a healthy character,'' he says. As for today's kids who answer ''7-Eleven'' to the question ''where does milk come from?'' or wonder ''how is it they have long hair?!'' when they see a genuine chicken, Lin — who feels that concern for nature is really concern for humanity — is distraught.

But given the limits of man-made environments, how can the overall number and number of species of animals be increased? Protection and management of green space in urban areas, and retention of as yet undeveloped natural settings in the suburbs, have become extraordinarily important.

Frog Tunnels: For example, in public recreation areas, land could be reserved for large ponds

(郭智勇攝) (photo by Kuo Chih-yung)

以繁衍子孫。

生物歧異度最高之處

生物學者喜歡用地形縱剖面來說明野生動物的生態變遷。由平地到海拔五百公尺是臺灣植被最豐富、生態最多樣、氣候最溫暖的地方，餵養的動物種類、數量也最多。「這是生物學上所謂『生物歧異度』最高之處，」林試所育林系研究人員劉一新說，但它同時也最適合人居，不可避免地成爲開發程度最高，都市化、人工化最劇烈的地方。

不久前臺北流行的一首「捉狂歌」，有句歌詞說起過往：「忠孝東路夠卡（再）過去是蒙仔甫（墳場）」。過去都市規模和今天眞是不可同日而語。以臺北爲例，四十年前除大稻埕、萬華等少數地區是熱鬧的聚落，其他多是大片菜園、稻田。今天臺灣居住都市的人口已接近百分之八十，僅臺北、高雄與臺中三個都會區就佔總人口的三分之一。

都市向外擴張，原始濶葉林被開墾爲農地，農地又劃爲建地，山坡地則廣闢別墅區、遊樂場、高爾夫球場，加上公路大量開鑿，工廠進入鄉間……。「今天連鄉村的動物也被人嚇跑了，」家住臺南縣的李剛說。

然而，芸芸衆生適應環境的能力畢竟個個不同，生死興衰也就漸漸步上殊途。最明顯的例子是，當許多臺灣本土動物在人類的開發脚步下節節敗退時，有些外來動物，像吳郭魚、福壽螺，在引進者無心揷柳的情況下，自行覓食存活，反而子孫綿延。（另詳「不是猛龍不過江——外來種野生動物」文）

物競天擇，適者生存。誰是人類開發巨掌下的「不適者」？

大型動物先受害

「需要活動空間大、生殖數量少的動物，首先禁受不起，」臺大動物系副教授李玲玲舉例說，在食物鏈上端、體型較大的哺乳動物，如雲豹、石虎等貓科動物就都是第一波受害者。

東海大學生物系教授林俊義曾整理四十年來臺灣哺乳動物分佈的高度表，發現牠們棲息地平均升高了約一千公尺。在「鄭成功開臺史」中被形容爲「麋鹿處處」的水鹿，如今就只有在二千公尺以上的深山，才可能驚鴻一瞥。

對於只吃某一種食物、住一種樹上，或移動能力差、要求禦蔽程度高的動物，往往也是環境開發的先烈。比如只以馬兜鈴爲幼蟲寄主植物的珠光黃裳鳳蝶；喜歡躲在田邊密叢中、人們往往只能聽到它「鷄狗乖」叫聲的竹鷄都是。

動物種類單一化

人們對自然環境進行的改造，無形中對動物做了大篩選。少數食性廣、繁殖力強、對溫濕度不挑剔的動物就與人們的開發「齊步走」。譬如原本就喜歡開濶林地的白頭翁在人們開墾過的地區，據地爲王，取代了其他喜好陰暗潮濕的飛禽，成爲鳥類中少數的「先趨鳥種」。有了人類爲牠作嫁，甚至都市行道樹都聽到白頭翁喧鬧的叫聲。

農地走向單一化的栽種，也爲麻雀與特定昆蟲提供了豐富的糧食。當蛞蝓、蜻蜓、螢火蟲、蝌蚪和許多淡水魚類紛紛葬身在農藥的威力下，一些危害農作物的飛蛾、螟蟲，却愈挫愈勇，抵抗力愈磨愈強，爲「百足之蟲，死而不殭」做了見證。

人口密集的市中心，對動物的考驗更多。除了遊走街頭的流浪貓狗，都市死角終年不斷的垃圾，則成了動物的求生實驗場。消耗人類廢棄物的高手——鼠輩，就與同屬哺乳類的其他動物呈兩個極端，其他動物數量日減，牠們則蒸蒸日上，四處橫行，汚染食物，傳染疾病。同樣不挑嘴、外型又討好的鴿子，也往往被都市人餵養得又肥又大。在歐洲都市，牠們的排泄物甚至腐蝕了石塊、磚瓦、上百年的建築與藝術雕刻。

臺北福德坑垃圾場臭味令人退避三尺，卻也有美麗的一景——成羣白鷺鷥翩然飛臨，在人們消耗過後的剩餘資源上留戀不去。都市冷氣機口則代替了樹枝，偶爾會有麻雀、白頭翁等鳥築巢；垂直的大厦提供和懸崖面差不多的棲息環境，在國外就有鳶、隼停佇

or lakes to serve as "ports in a storm" for migratory birds like ducks and geese. In Europe, the US, and Japan, it is common for people to build bird feeders in their yards or on rooftops; there are feeding stations resembling bats' nests; in London, England there is a man-made passage beneath the road used as a "frog tunnel," so that they are not flattened by automobiles as they try to cross the road. Drivers going by there will see a sign warning: "Beware of the Frogs."

A survey of the bird population by the Wild Bird Society of R.O.C. in 1974 showed only 21 eagles in the Taipei area, and the number declined thereafter. The Taipei eagle population has racked up nothing but zeros since 1981.

The Taiwan Rumble Fish, whose body is so colorful that it is often mistakenly assumed to be a tropical fish from abroad (also known as the "Three Spotted" fish), was originally widespread in ditches and drains throughout Taiwan. Tseng Ching-hsien, a doctoral candidate in Zoology at National Taiwan University (NTU) who specializes in the study of fish, recalls that "eight or nine years ago, before the university dormitories were built on Wenchow Street, we often went down there to catch Three Spotted fish." Today, researchers scour the ponds, drains, ditches, and rice paddy fields in the suburbs but can't find a trace of the this type of fish.

Thirty or forty years ago, when the water in streams and irrigation ditches was clean, you could still see otter catching fish. You could even spot Taipei Red Frogs, which were spread across north Taiwan walking the streets in Taipei. Both of these animals are facing extinction today.

Lights Out for the Firefly: "A lot of the creatures we could see everywhere when we were young are gone, and for those that remain, the numbers have shrunk," affirms 40-year-old Chao Jung-tai, director of the Department of Forestry Preservation at the Taiwan Forestry Research Institute. Hunting and trapping, excessive use of pesticides, and water pollution are all reasons for the decline of wild animals, but the main reason according to biologists and zoologists is large-scale development by the human race.

There are examples everywhere. In the summer of 1977, Yang Ping-shih, chairman of the Department of Plant Pathology and Entomology at NTU

哺乳類動物是人類開發的第一波受害者。圖爲已成稀有種的臺灣獼猴。(鄭元慶攝)

Mammals are the first victims of man-made development. The photo shows the Formosan Rock Monkey, already an endangered species. (photo by Arthur Cheng)

and the Butterfly Watcher's Association could see 20 or 30 species of butterflies in a single day on Wufeng Mountain in Hsintien. Three years later, the vegetation had been cleared and the mountain levelled, and highrises and houses had taken over. The immature butterflies had no plants on which to feed, and the butterflies consequently disappeared.

Urban streetlights have been the curse of fireflies. Because it is still bright at night, they cannot find partners in the light in order to produce offspring.

Highest Biological Diversity: Biologists prefer to use altitude to explain the changing ecology of animals in the wild. The areas, from sea level to 500 feet are the places in Taiwan with the most profuse plant life, the most diverse ecologies, and the warmest temperatures, and thus the places which can support the most varieties and largest numbers of animals. "In biology this is referred to as an area with the highest 'degree of biological diversity,'" says Liu Yi-hsin, a research fellow at the Department of Reforestation of the Forestry Research Institute. However, these also happen to be the areas

樓頂，沿大樓立面往下滑翔的情形……。但這些令都市人大開眼界的情景，仍只是少數例外。

近鄰更重要

畢竟都市是人類爲自己創造的地方，和自然環境相去太遠，能和人類共同生存其間的野生動物十分有限。壁虎、螞蟻、鴿子、家燕、老鼠、蟑螂是幾個重要的都市移民，也是全世界都市化的一致結果。

如今各國紛紛成立國家公園、自然保護區，目的之一就是爲野生動物保存一塊淨土。近來更出現「都市動物」研究，爲的也是使其他生物回到人類的環境。

「畢竟對生活在水泥森林的都市人來說，玉山頂上的山羌，意義遠不如身邊一隻天天可見到的白頭翁，」臺大動物系主任林曜松不斷強調，人的生活環境中要有野生動物，

most suitable to human habitation, so inevitably they have become the most developed, urbanized, and man-ipulated.

Not long ago, a song that was popular in Taipei had lyrics that went: "Past Chunghsiao East Rd. there used to be a graveyard." You really can't even mention the scale of today's cities with those of the past in the same breath. Taking Taipei for example, forty years ago, besides a few bustling areas like Wanhwa, most everything else was rice paddies. Today nearly 80% of Taiwan's population lives in urban areas, with one third of the population in the three cities of Taipei, Kaohsiung, and Taichung alone.

Cities expand outward, and primitive deciduous forests are cut away to make agricultural land; agricultural land turned into construction land, and the mountain slopes are opened up for villas, recreation and amusement parks, and golf courses; and then you add on massive highway construction, factories moving into the countryside.... "Today the animals have even been frightened away from rural villages," says Li Kang, a resident of rural Tainan County.

Nevertheless, every living thing has a different capability to adapt to the environment, and takes a different path on the road to survival or extinction. The clearest example is that, while many species native to Taiwan have begun the path to extinction, some imported species, like the Tilapia or the *Ampullarius canaliculatus* having been brought here unwittingly, have spread in the search for food and survival. (See the article on foreign species in Taiwan.)

Survival of the fittest. And who are the best "adapters" under the powerful blows of mankind?

Large Mammals Suffer First: "The animals that reproduce in low numbers and require a lot of space are the first ones that can't stand up to it," says Li Ling-ling, an associate professor of Zoology at NTU. The bigger mammals, at the end of the food chain, like the big cats — the clouded leopard or the Chinese tiger cat — are the first wave to be hit.

許多野生動物因「懷璧其罪」，數量銳減。圖爲農委會焚毀走私進口的象牙。(卜華志攝)

A lot of animals sharply declined in numbers because they had something desirable about them. The photo shows the Council on Agricultural Planning and Development destroying illegally smuggled ivory. (photo by Pu Hua-chih)

Lin Chun-yi, a professor of biology at Tunghai University, has compiled a chart of the altitudes of various mammals in Taiwan over the last forty years. He discovered that the areas in which they stay have increased about 1000 meters. For example, the Formosan sambar, which was described as "ubiquitous" in the *History of Koxinga's Exploration of Taiwan*, can only be occasionally glimpsed in deep forests above 2000 meters.

The ancients said, "Take pity on the moth, and do not light the lantern." The light attracts insects far from the places they stay, and causes them to spawn in places without food; a number of species of large moths in Taiwan have disappeared in this way. Students and professors of the Department of Forestry at NTU discovered in experiments at Hsitou that if the lights are simply changed from white to a dusky yellow then there is a sharp decrease in the number of insects found dead within 20 meters of the lamps the next day. The Hsitou Forest Recreational Area has already begun the work of replacing its white path lights.

Recently, domestic ecology scholars have begun extending the theory of an "island ecology" from biology, and are pointing out ways to increase the number and types of animals in a man-made environment.

The Island Ecology Theory: It is suggested in "island theory" that if an island had once been connected to a continental land mass or is not far from such a land mass, then migration of animals back and forth should be relatively easy, so that the number of animal species will be more abundant. Moreover, the larger an island, the longer the food chain it can contain, and the more visible will be the facets of life.

By the same logic, the scattered parks and gardens in cities are like islands in the sea. If the area of green space is adequately large, it can sustain and contain a large number of animals. Or, if a sufficient "land bridge" (like a series of trees that animals can cross to get from one place to another) can be maintained between "island" and "island" or between green space and green space, then there will be a linkage between green spaces, which is

才能提醒人們與大自然間有密切的聯繫。

林俊義教授也指出，現在的年輕人成長過程中，由於沒有接觸自然、感受自然，由自然得到啓發，也因此不懂對生命心存感激。「每天摸摸樹葉，看看蝴蝶，在自然中成長，才是創造健全人格的秘密，」他說。對今天有些小孩被問及牛奶何處來，回答「7－eleven」；看到眞正的鷄卻大駭道：鷄怎麼長毛？認爲關懷生態也就是關懷人性的林俊義憂心忡忡。

但在人爲環境的局限下，如何增加動物數量、種類？都市中綠地的保留與經營，和保存市郊尚未被開發的自然環境，就變得異常重要了。

青蛙地道

例如，在公共遊憩場所保留大片的池塘、湖泊，做爲雁鴨等候鳥羣聚的避風港。歐、美、日本許多人家則流行在庭院或陽臺設置鳥食臺與仿蝙蝠巢的食臺；英國倫敦有一條人工穿鑿、橫跨公路的「青蛙地道」，以免牠們穿越馬路時，不幸被車輛輾過。經過該地的司機則會看到「小心青蛙」的告示牌。

古人說「憐蛾不點燈」，燈光往往引誘昆蟲遠離棲地，使卵孵化在沒有食物之處，臺灣許多大型的蛾類卽因此消滅。臺大森林系師生就在溪頭實驗發現，只要將白色日光燈改成暈黃色，翌日直徑廿公尺內死掉的昆蟲就少了很多。溪頭森林遊樂區已開始進行換掉白色路燈的工作。

國內生態學者曾將生物學上的「島嶼生態」理論衍伸，指出在人爲環境中增加動物類別的方法。

島嶼理論

島嶼理論中指出，大海中的島嶼若曾和陸塊相連或離陸塊較近，彼此間動物遷徙交往容易，動物種類也就比較豐富。此外島嶼愈大，可容納的食物鏈環節愈長，生命相也就愈可觀。

同理，都市中零星的公園、植物園就像海洋中的島嶼，綠地面積若够大，能餵養容納的動物就愈多；或在「島」與「島」，卽綠地與綠地間保留一條够寬的「陸橋」（如行道樹），使綠地彼此間有連繫，等於棲息環境加大，對動物種類、數量的成長都有所幫助。

當然，都市有其限制，不能期望看到需要幾十公頃活動面積的熊，或生活在高海拔的帝雉、山羌。「但設計良好的公園、河濱綠地，仍然可以保有多種動物，」師範大學生物系教授王穎表示。

民國七十三年中華民國野鳥學會的一項調查中也發現，僅植物園一地，一年中就出現過五十幾種鳥類，雖然其中有許多是候鳥，但在闤闠塵世中已屬難得。由於植物園的目的就是展示多樣的植被，且經營已近百年，樹高葉茂，林地上潮濕、蔭蔽，又限制人們進出踐踏，加上一座水漾漾的荷花池，等於製造了一個生物最好的棲息場所。

「植物園已自成一個能進行能量、養分循環的生態體系，」辦公室就在植物園邊的林試所研究員趙榮台觀察到，園裏的松鼠最多時達廿隻。林試所甚至能在此做「野外調查」——以無線電追踪記錄松鼠，了解其生態習性。

關渡多鳥，公館有蛙

師大生物系的「兩棲類資源調查紀錄」也顯示，以羅斯福路公館爲中心，方圓四公里間竟然還有十幾種青蛙。研究青蛙的楊懿如就對臺大校園中的蛙口分佈如數家珍：「農場有澤蛙、醉月湖有虎皮蛙、黑框蟾蜍……。」畢竟還是有些小動物對環境並不苛求，只要一個小水池，少許植被就心滿意足，這也是都市郊區仍蛙鳴處處的原因。

貫穿大臺北的淡水河雖不甚清澈，但由河口竹圍、關渡到中興橋與華中橋一帶，是許多遷徙性鳥類必經與棲息的重要區域。每年冬季仍可見成千的雁鴨載沉載浮於河面。

都市周邊發展程度不如市區，創造動物生存的空間更容易。五年前，木柵動物園搬新家，就在園內找到一塊十幾公頃的山坡地栽種蝴蝶所需的寄主與蜜源植物，例如過去在

爲了防範燕子的排泄物毀損建築，美國耶魯大學在雕飾上裝了「避鳥針」。(卜華志攝)
In order to avoid damage to structures by bird excrement, Yale University has attached "bird needles" to statues. (photo by Pu Hua-chih)

equivalent to expanding the environment in which animals can reside, which in turn is helpful to raising the number of both animals and species.

Of course, cities have their limits, and one can never hope to see bears which require tens of hectares of land or pheasants or Chinese Muntjac which only live at high altitudes. "Nevertheless, a well-designed park or riverbank preserve can sustain many types of animals," states Professor Wang Ying of the National Taiwan Normal University (NTNU) Department of Biology.

In 1984, the ROC Wild Bird Association revealed in one survey that in only the Botanical Garden, more than 50 types of birds appeared in a single year. Although many of them were migratory birds, it was an amazing thing amidst all the bustle and dust. Because the purpose of the garden is to display all kinds of plant life, and it has been operated for almost 100 years, the trees are tall and vigorous, the floor beneath the trees is damp and dark, and people are prohibited from walking in certain areas. Added to all this, there is a rippling pond. This is the equivalent to manufacturing an optimal environment for living things.

"Botanical gardens have already become ecological systems capable of undertaking energy and nutrition cycles," observes Chao Jung-tai, whose office is on the edge of the Botanical Garden. The number of squirrels peaks at 20, and the Forestry Research Institute can even conduct "wildlife surveys" here — tracking and recording the squirrels with wireless radio to understand their behavior.

How Many Frogs in Kung-Kuan?: The *Amphibian Resources Survey* of NTNU further reveals that there are more than 10 types of frogs within a four-kilometer diameter circle with its center at Kung-Kuan. Yang Yi-ru, who studies frogs, lists the amphibians to be found on the NTU campus like precious family heirlooms: "On the experimental agricultural land we have the rice frog, at Moon-Drunken Lake we have the speckled toad." In the end having a few small animals is not all that exacting on the environment, you just need a small pond, and a few trees to keep them happy — that's why you can still hear frogs croaking in the suburb.

Although the Tamsui River which threads through Taipei, is not very clean, from Chuwei at the mouth of the river and Kuantu to the area around the Chunghsing Bridge and the Huachung Bridge, there are several important locations where migratory birds have to stop and rest. Every winter you can still see tens of thousands of geese and ducks on the river.

The extent of development on the edge of a city is never as great as in the city proper, and it is even easier to create space for animals to exist in. Five years ago, when the Mucha Zoo moved to its new home, they set aside land of more than ten hectares in the park to plant the types of trees butterflies need, such as orange trees beloved by the one ubiquitous Monarch Butterfly. Today the zoo is filled with butterflies, so that "often people go into the butterfly exhibit but can't see all that many butterfiles, but as soon as they go out the door, and they see the flourishing condition of the butterflies, they know they haven't wasted a trip," relates Chen Chien-chih, head of the Expansion Committee of the zoo.

Unable to Bridge the Gap?: There are still many places that need work. The ecology has always

平地相當普遍的馬兜鈴、柑橘類植物。如今動物園滿山遍園的蝴蝶，「常常人們在蝴蝶館看不到多少蝴蝶，一出門口，見到蝴蝶迎面的盛況才驚嘆，不虛此行，」動物園推廣組長陳建志說。

「陸橋」效果不彰

要努力的地方仍多。生態在與開發的競爭中一向居於弱勢，何況要在寸土寸金的都市，或隨都市膨脹、行情飆漲的城郊佔一席之地。已被指定爲自然保護區的關渡沼澤地，臺北市政府原計劃購買堤防內的土地，規劃爲「關渡自然公園」，卻因地價節節上漲，遲遲未能定案；堤防外五十五公頃沼澤濕地則成爲廢土、垃圾傾倒處。七十二年在當地的鳥口調查有一百卅九種，去年鳥口銳減爲八十四種。

而臺北市的公園內蓋滿建築物，綠地稀疏，招徠動物的效果有限。除了仁愛路、敦化南路等林蔭大道，大部分道路的安全島卻如車海中的渺小孤島；許多行道樹也如路燈排隊，間隔過寬，不能發揮使公園綠地連成一氣的「陸橋」效果。

自然課本或童話故事？

有一回曾晴賢帶一羣居住在臺北市，和一羣住陽明山上的小朋友做陽明山國家公園之旅。當他問小朋友爲何大屯山頂不長草時，山上的小朋友以親身體驗發揮想像力，不斷發表意見；帶著錄音機、照相機的都市兒童則忙著拍照和錄音。「如果沒有實際的生活經驗，叫他們從何想像起？」與小朋友接觸頻繁的陳建志也說。

當我們兒時，身邊有許多動物，卻因爲相關的自然書籍缺乏，無法更深入了解牠們，直到民國六十年代才出現了第一本和本土有關的自然雜誌；如今本土生物的研究不斷增加，相關的書籍、雜誌也陸續出現，但生活周遭已沒有多少野生動物了。這一代的自然課本，會成爲下一代遙不可及的「童話故事」嗎？ S

（原載光華八十年十二月號）

been at a disadvantage in competition with development, all the more so in a city which measures inches of land in ounces of gold. In the Kuantu wetlands area, already designated as a natural preserve, the Taipei City government had originally planned to buy the land inside the dikes and had mapped it out as the "Kuantu Nature Park," but because land prices lurched upward, they have not been able to finalize the plan; the 55 hectares of wetlands outside the dikes are already wasteland, a place where people dump their garbage. A survey of the bird population there in 1983 had 139 types; last year the number had dropped sharply to 47 kinds.

Also, all Taipei city parks have manmade structures in them, and green space is very sparse, and they have limited impact in attracting animals. Besides Jen-ai Rd. and Tun-hwa S. Rd., the safety islands on most roads are like isolated little desert islands; a lot of the trees on sidewalks are lined up like streetlights, and the distance between them is too great, so they cannot have the effect of being a "land bridge" between green spaces in the city.

Nature Textbook or Fairy Tale?: Once Tseng Ching-hsien took a group of children from Taipei City and a group of children from Yangming Mountain up to the Yamging Mountain National Park. When he asked them why there is no grass atop of Tatun Mountain, the children who lived on the mountain, drawing from their own experiences, used their imaginations to continually come up with opinions; the camera-carrying kids from the city were just running around snapping photos and recording sound. "If they don't have this kind of life experience, how are they going to come up with any answers?" says Chen Chien-chih of the zoo.

When we were children, there were many animals by our sides, but because we lacked relevant nature textbooks, we couldn't understand them more deeply; the first nature magazine about Taiwan only appeared in 1971. Today research about local animals is increasing constantly, and related books and periodicals have appeared one after another. But there is already very little "wildlife" in our lives. Will the nature textbooks of this generation become the unreachable "fairy tales" of the next? S

(Chang Chin-ju/tr. by Phil Newell/ first published in December 1991)

人類消耗所剩的垃圾成了鳥兒的資源。圖爲福德坑垃圾場吸引了一群雪白的鷺鷥。（張良綱攝）

Man's unwanted waste has become a resource for birds. The photo show the Fu-teh-kang dump attracting some egrets. (photo by Vincent Chang)

外來動物具有觀賞、食用等價值，但也潛伏許多問題。
While exotic species may be pleasant to the eye and tasty to the tongue, they bring with them no small number of problems.

不是「猛龍」不過江
——外來種強壓地頭蛇

"A Fierce Dragon" Can Cross the River After All—
Exotics Are Driving Out Taiwan's Native Species

文•張靜茹　圖•卜華志

當許多本土動物隨人們脚步節節退卻，一些「外來客」卻乘隙據地稱雄，反客爲主，在人們周遭安身立命。

誰說「強龍不壓地頭蛇」？

根據日據時代的紀錄，世代長住淡水河的淡水魚有石鱝、鯽魚、鯝魚……等林林總總五十多種，還有五十多種洄游性及偶爾由河口進入的海水魚類。四十年後，淡水河的天下已更換旗幟。

民國七十二年經濟部水資源規劃委員會針對淡水河魚類所做的調查發現，來自非洲的吳郭魚，才是現任淡水河盟主。在新店秀朗橋一帶，牠佔了總量的百分之六十；愈往下游，汚染愈嚴重，存活的魚類愈少，吳郭魚所佔比例也愈高。在景美橋、中正橋一帶，均超過百分之八十。

外來種進駐寶島

翻開「吳郭魚在臺史」：民國四十年代由非洲引進，五十年代農政單位大力推廣。之

According to the old Chinese maxim, "Only the fiercest of dragons can cross the river." Yet as numerous native animals have withdrawn in step with man's encroachment, some "foreign guests" have seized the opportunity to take over, colonizing a foreign land at man's side.

Who said, "A strong dragon doesn't push out the local snakes"?

According to records kept during the era of the Japanese occupation, more than fifty kinds of fish, including the golden carp and *Xenocypris argentea*, had inhabited the Tamsui River for generations. There were an additional fifty species that returned with the currents or occasionally entered the mouth of the river. Today the situation has vastly changed.

In 1983 a study on fish in the Tamsui River by the Department of Water Resources of the Ministry of Economic Affairs found that the Tilapia, native to Africa, is now lord of the river. In the area by the Hsiulang Bridge in Hsintien, it accounts for 60 percent of all fish. The pollution gets worse the further one goes downstream, where the proportion of Tilapia in the river rises, exceeding 80 percent

福壽螺進駐寶島水域，給農民帶來不少困擾。
The *Ampullarius canalicultus* has been the nemesis of Taiwan's farmers.

後，牠從魚塭流入河川，六十年代佔國內水域魚量的百分之五，七十年代劇增至百分之四十。如今連高雄澄清湖，都可以隨手撈到吳郭魚苗。

福壽螺是另一個數量緊追吳郭魚之後的外來水族。最初商人引進是爲了推廣食用，可惜牠不對中國人口味，命運遠不及吳郭魚。慘遭惡意遺棄於荒郊野外之後，靠著旺盛的生命力，以農作物「裹腹」，卻因此入主寶島成功。如今地無分南北、時不分四季，只要有農田，均可見到牠粉紅色的卵，附著在離水面一尺的作物上。

「宜蘭平原已被巴西龜佔領了，」喜做野外活動的楊景星在一項研討會中指出。水族館引進的觀賞水族，除巴西龜在野外漸成氣候，近來釣魚客在溪裏發現食人魚、銀帶魚、老鼠魚、螯蝦的事也屢有所聞。

臺灣的天空也在起變化。許多具異國風情的鳥種，藉著國人養鳥風氣興盛，或由籠中「逃亡」，或被放生，不斷奪取領空權。國內的生態攝影者和鳥類學會的賞鳥族，常在野外見到巴西雀、南美產的鸚鵡、大陸畫眉等非候鳥的「舶來品」。

其中數量拔頭籌的是善歌的畫眉。爲了贏取畫眉鳥鳴唱比賽，鳥販除了在野外捕捉臺灣畫眉，也大量進口大陸畫眉。雄畫眉買賣熱絡，不會鳴唱的雌鳥因缺乏市場而被放生，如今連植物園都可以見到「唐山過臺灣」的大陸畫眉，牠們更在野外求偶，與臺灣畫眉配對、繁衍，產生一支新鳥族。

「強龍」顯威

「強龍」輩出的情形不只在寶島。

美洲大陸有個享譽全球生態界的例子。早期美國移民因思鄉由歐洲引進椋鳥，如今椋鳥除在北美洲大肆掠奪其他鳥種的巢穴外，白天聚集在市郊農田吃穀物，黃昏則成千上萬飛入鄉鎮，影響住家安寧。

鄰國日本也有「血淋淋」的教訓。琵琶湖是日本最大的內陸湖，也是魚源豐富的漁業湖泊。自從該湖引進美洲鱸魚（大口鱸），湖中原生魚種逐漸消失——後來人們在鱸魚肚子發現各種原產湖中的蝦兵蟹將魚族。美洲鱸魚至今困擾著日本漁業界。

事實上，動物不論是被人有意或無意地帶到一個陌生國度，因氣候、棲息環境、食物來源等生存條件與「故國山河」不同，除非人們細心豢養，大部分無法在野外自求多福，找到生存位置。

但外來種在原生地長久衍化，往往另有天敵會控制其數量；初到異地，如果能活得下去，常又因爲不被當地生物了解、沒有天敵，或「尅」牠的動物沒碰到一塊……，因此，只要有存活的機會，常常能表現傑出。

in the area by the Chingmei and Chungcheng bridges.

Foreign Occupiers of a Treasure Island: Lets leaf through *The History of the Tilapia in Taiwan*: Introduced in the fifties from Africa, in the sixties it was widely promoted for breeding by agricultural and fishing agencies. Afterwards, it entered the rivers from the fish ponds, and in the seventies it constituted 5 percent of all fish in Taiwan, rising to 40 percent in the eighties. Today, one can catch Tilipia fry easily — even in such bodies of water as the Chengching Lake of Kaohsiung.

The *Ampullarius canaliculatus* is another aquatic outsider, second only to the Tilapia in numbers. At first businessmen introduced this snail for human consumption. Unfortunately, its flavor didn't go over well with Chinese eaters, and its eventual fate was far from that of the Tilapia. After being discarded in the wild, this tenacious snail has thrived by helping itself to crops. Today, in the north or south, during any of the four seasons, you can see its pink eggs on crops within a yard of water.

"The Ilan plain has already been invaded by the Brazilian tortoise," stated outdoorsman Yang Ching-hsing at a research symposium. Sport fishermen have recently been catching such aquatic exotica as piranha and *Osteo glossum bicirrhosum* which — like the flourishing Brazilian tortoise — were originally introduced in Taiwan by pet fish stores.

The air above Taiwan is also being transformed. Whether by escaping from their cages or being set free, many foreign species of bird have managed to get up there and fight for their place in the sky. Ecological photographers and bird watchers from the Wild Bird Society have frequently spotted such nonmigratory "imported birds" as the Brazilian Sparrow, parrots native to South America, and the Mainland *hwa-mei* (*Garrulax canorus canorus*).

Among them, first in numbers is the mainland *hwa-mei*, a bird noted for its vocal abilities. In order to win *hwa-mei* singing competitions, bird mongers have, in addition to capturing the Taiwanese *hwa-mei* in the wild, imported large numbers of its mainland cousins. Business for males is very good, whereas females, with neither songs nor buyers, are set free. Today you can see them, far from their mainland homeland, in such places as the Taipei Botanical Garden. They have also begun to seek mates in the wild. Crossbreeding with the Taiwanese *hwa-mei*, they have created a new species.

A Fierce Dragon Blows Its Smoke: Formosa is not unique in its problem with foreign invaders.

On the American continent, there is an ecological case known around the world. Out of homesickness, early immigrants to America introduced starlings from Europe. Today, in addition to brutally taking over the nests of native birds, starlings congregate by day in farmers' fields outside of cities, where they devour the crops, and at dusk fly in hordes into villages, driving the residents to their wits' ends.

The nearby country of Japan also has learned a painful lesson. Lake Biwa is the largest inland lake in Japan, one that used to teem with life. Since the American large mouth bass was introduced there, native species have gradually disappeared. Later people discovered every kind of native shrimp, crab and fish in the stomachs of the perch. Up to the present, the large mouth bass has been a scourge of the Japanese fishing industry.

In truth, whether man introduces a foreign species intentionally or accidentally — because the climate, habitat and sources of food in its adopted land are all different from its native environment — the newcomer will have little chance to survive unless man feeds and cares for it.

Having lived for generations in its native habitat, however, a species will usually have natural enemies there to control its numbers. When first entering a foreign land, it may be without natural enemies — because other animals have no understanding of it, because it hasn't yet met the animal that can overcome it, or for other reasons — and it will have a chance of surviving. "Even if a species has only a one in a hundred chance, there will always be some that will beat the odds," says Li Ling-ling, associate professor of zoology at National Taiwan University (NTU).

When the Natives Meet the Newcomers: This chance exists because the native animals have already lived with each other for tens of thousands of years, resulting in an ecological balance wherein things go along happily and pretty much unchanged. Hence, the natives are no match for tough recent arrivals eager to establish a place

當本土動物遇上外來種

由於本土動物彼此已相處成千上萬年，早已互相適應，生態維持在一個衡定狀態，日子過得樂天知命，頗為自在，因此往往個性馴良，不是初來乍到、急於安身立命、表現強勢的外來種對手。

「尤其是島嶼生物，一向在與外界隔離狀態下生活，較缺乏侵略性，遇到外來的天敵多半不知如何逃避或因應，」臺灣大學植物病蟲害系主任楊平世解釋，島嶼生態比陸塊生態脆弱，臺灣也因此給外來者較多可乘之機。楊平世舉例，土產龜悠遊於臺灣河川不知幾代，生活安逸、性情溫和，就不是巴西龜的對手。

萬一不慎「引狼入室」，招來過江猛龍，本土動物就更難招架。譬如被利用來清除水族箱垃圾的老鼠魚（琵琶鼠），原本就是臭水溝都能活的魚種。牠的皮極厚韌，是原產南美洲土著做挫刀的材料，臺灣河川又沒有鱷魚等堪與為敵。牠初到臺灣時身價高昂，後來數量一多，又不容易「夭折」，人們養得不耐煩，牠也就隨著下水道而下。「如今淡水河、高屏溪、桃園的池塘到處可見其踪跡，」研究魚類的曾晴賢說。

一個蘿蔔一個坑

「島嶼生態可以說是『一個蘿蔔一個坑』，多跑進一個蘿蔔，就有一個要找不到坑了，」臺大動物系博士班研究生楊懿如比喻。

種種理由，使得外來客的來臨，不但未使本地動物更多樣化，反而進行了一場場的「反淘汰賽」。

遊過日月潭的人都知道，當地原產一種因為老總統蔣公喜愛，而被稱為總統魚的翹嘴紅鮊（又名曲腰）。由於總統魚沒有護衛稚魚的習性，魚卵產在水草上，成為吳郭魚口到擒來的美食。如今總統魚數量每下愈況，一斤價格高達一千元。

國內許多淡水魚種如溪哥、香魚根本無法與什麼都吃、有耐污能力，甚至因生命力強而被美國太空實驗室選為「太空用魚」的吳郭魚一較長短，遂族羣漸失。

螯蝦落井下石

主客易位的例子同樣也發生在鳥類身上。「國際鳥類保護總會」過去卅年多次全球性的鳥類調查都證實：一地原生鳥類絕種的原因中，「無法與外來種鳥類競爭」佔了百分之廿強，比環境污染的影響還高。

「賞鳥時看到大陸畫眉與臺灣畫眉的交配種並不稀奇，」鳥類學會的資深鳥友陳葉旺擔心這可能使臺灣特有亞種畫眉絕跡。

外形華麗、容易飼養的環頸雉也有相同危機。進口的大陸環頸雉在人為或半自然狀態下與臺灣環頸雉雜交，也使臺灣環頸雉逐漸失去基因中的特點。

缺乏天敵，適應力與繁衍力驚人的外來種，殺傷力也可能波及人類。歷史早有教訓，十四世紀十字軍東征，無意中由東方帶回的戰利品——黑死病——老鼠和牠身上的跳蚤，將腺鼠疫傳到人類身上，曾造成歐洲人口四分之一死亡。

當臺灣南部因為今夏雨量不足，水資源缺乏，有些農田不得已休耕之際，又傳來美國螯蝦落井下石的消息。

由水族館引進的螯蝦，在蝦輩中屬體型碩大之流，長十幾公分，攻擊性強，不僅水族箱內的大小水族皆對之豎白旗，即使人們抓住其後胸，還是會被牠靈活後仰的螯夾傷。牠因此步上福壽螺與老鼠魚的後塵，在野外四處遊蕩。

螯蝦還另有本領，可以鑿洞深入土中幾公尺，農藥因此對牠莫可奈何。牠還是開路先鋒，誰阻止牠的去路，就咔擦打個洞，水田田埂常因此「蝦門」大開，田中的水一夜流盡，農民為之氣結，生態學者也都「看好」牠潛力無窮的破壞力。

福壽螺後患無窮

不過螯蝦事件「方興」，對農業尚未造成大損失；福壽螺事件則至今「未艾」，農民為牠花掉的成本已頗為可觀。

在福壽螺被野放，並吃農作物上癮後，農

被引進做爲寵物的巴西龜，因爲人們「放生」，流入野外，進而威脅本土龜的生存。

Originally brought to Taiwan as pets, Brazilian Tortoises have become a threat to native tortoises after some were released by their owners.

for themselves.

"In particular, island animals, which have existed in seclusion from outsiders and are thus less aggressive, often don't know how to run away or what to do when encountering natural enemies from outside," says Yang Ping-shih, professor of plant pathology and entomology at NTU. Island ecologies are more vulnerable than continental ecologies. As a result, Taiwan gives outside species a better chance. Yang Ping-shih cites the example of the mild-mannered Taiwan tortoise, which had existed on the island for countless generations, enjoying an easy and stable existence. "How are they a match for the mean Brazilian tortoise?"

If a tough troublemaker is carelessly brought in from abroad, the native species will find it difficult to defend themselves against it. Take for example the Houttyn, which is used to clean the waste in fish tanks. It can survive even in sewers. Its skin — which is used to make files in its original home of South America — is thick and tough. In addition, its predator, the alligator, does not swim in the rivers of Taiwan. It was very expensive when first sold in Taiwan. Later, because it numbers grew and because it does not die easily, people got bored with it, and it came out of the sewers and into the rivers. "Today its traces can be seen in the Tamsui and Kaoping rivers and in Taoyuan's ponds," says Tseng Ching-hsien, a doctoral candidate in zoology at NTU.

Only One Turnip per Hole: "Island ecologies can be said to support 'only one turnip for every hole.' If there is one extra turnip, you won't be able to find a hole for it," says Yang Yi-jung, also an NTU doctoral candidate in zoology.

For a variety of reasons, outside intruders will not bring "greater variety" to the species of the island, but rather will introduce a new round of "survival of the fittest."

Those who have travelled to Sun Moon Lake will know of the *Erythroculter ilishaeformes*, which is called the president fish in honor of former President Chiang Kai-shek, who was fond of it. Because the president fish does not protect its young, the eggs it lays on water plants have become easy pickings for the Tilapia. As the numbers of president fish have declined, its price per Chinese pound (600 grams) has risen to NT$1000.

In the long run, several species of fresh water fish, such as the zacco and the smelt, can simply not compete with the Tilapia, which eats anything and protects its eggs. It is of such strength that NASA of the United States has selected it as "a fish for use in outer space."

The Coming of the Cray Fish — a Bad Situation Gets Worse: Foreign guests are also making themselves at home in the sky. A 30-year investigation on birds around the globe, carried out by the International Council for Bird Preservation, has found that "inability to compete with foreign species" is the reason accounting for 20 percent of all extinctions, a cause ranking higher than environmental pollution.

"It is not unusual for bird watchers to see mainland *hwa-mei* mating with Taiwan *hwa-mei*" says Chen Ye-wang, a senior member of the Birds Association, who is worried that Taiwan's *hwa-mei*, might soon be facing extinction.

The Phasianus colchicus formosanus, easy to raise and of opulent appearance, is also facing a crisis. Its mainland counterparts, which have been imported, are interbreeding with it in artificial or

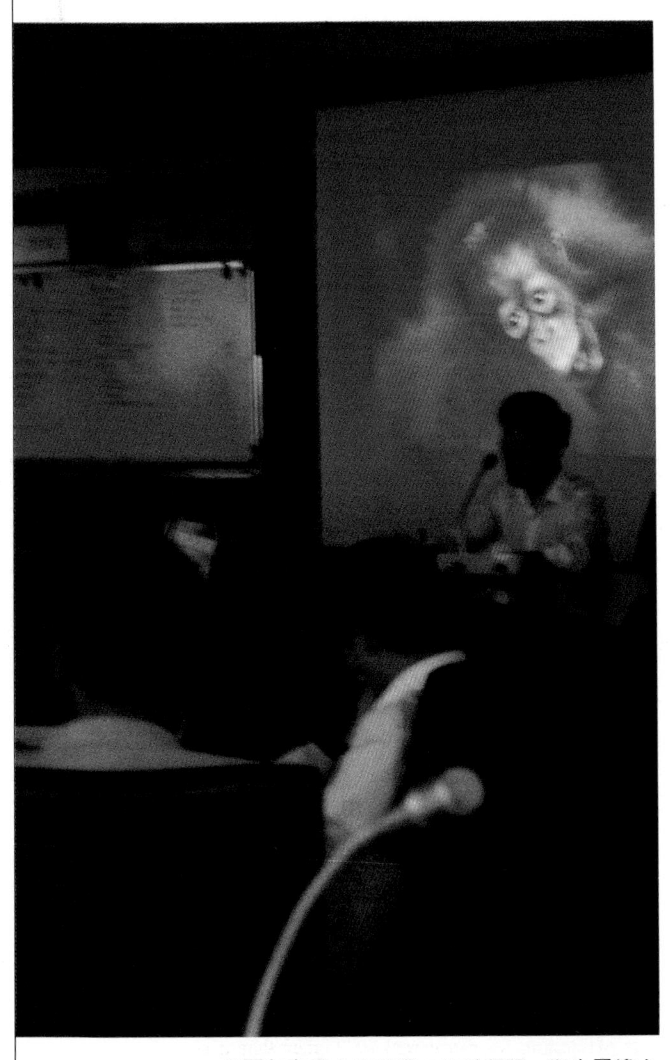

前兩年有商人走私進口紅毛猩猩，衛生署擔心牠們會傳染疾病給人們。(張良綱攝)
Two years ago businessmen started to smuggle in orangutans. The department of health is worried that they may transmit diseases to humans and has held special public announcement meetings about this problem. (photo by Vincent Chang)

業人員拚命想爲牠找天敵，沒想到除了姥姥——人們——不愛牠的味道，舅舅——鴨子、麻雀、螞蟻，各種家禽、野鳥、昆蟲等，也無一對牠青睞。

彷彿著了鐵布衫的福壽螺，直到五年前，農業商才發現有一種藥效強、價格昂貴的「三苯醋錫」對牠稍具殺傷力。一名經營農藥進口的商人就表示，原來進口五百公斤三苯醋錫要賣兩年，如今一年就有二千公斤的市場。

但殺得了福壽螺，也不免傷及無辜。已被預定徵收蓋運動公園的士林蘭雅區是大片空心菜園，過去此處的沼澤泥濘地中有許多本土魚類，文建會還一度想將之列爲「臺灣鬥魚保護區」。但在福壽螺駕臨之後，農民不得不一個禮拜祭出一次農藥，結果各水族同歸於盡，保護區就此成爲泡影，也留下土地錫污染的隱憂。

保護原生動物

不論外來種是否能廣傳後代，定居下來，卽使只是被當寵物養在深閨，也可能帶來傷害。

前年走私的藏獒傳聞帶有「狂犬病」，曾造成大衆恐慌。近來有人引進蛇類，萬一被未拔除毒牙的蛇咬傷，國內缺乏血淸，受傷者只有回天乏術。

「人們不知利害地引進某些生物造成的危害，連生態學者也幫不上忙，」臺大動物系副教授李玲玲說。

除了避免給人們自己製造麻煩，今天世界各國都希望保存自己特有的動、植物，並競相設立動物基因庫；爲避免外來種動物帶給原生種激烈的競爭，各國對野生動物進出口的防範愈來愈嚴格。

生物界也證實，當本地生物愈少時，外來生物能够入侵的機會愈大；反之，本土生物愈多，留給外來種的空間愈小。「除了非必要不引進外來種動物外；保護本土動物，更是拒絕『強龍』入主的根本之道，」師範大學生物系教授王穎說。

（原載光華八十年十二月號）

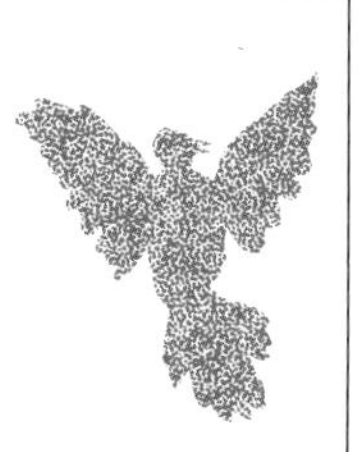

semi-natural environments, slowly eliminating the special characteristics of the Taiwanese gene pool.

Lacking in natural enemies and with startling powers of adaptability and procreation, these foreign species may also be harmful to man. An early historical example are the rats that crusaders brought back from the Middle East as spoils of victory in the fourteenth century. These rats and the fleas on them were the cause of the plague that decimated more than a fourth of Europe.

Because of a lack of rainfall, there is a shortage of water in southern Taiwan this summer. While many farmers' fields are already dry, many now face the additional menace of the American cray fish.

Introduced by pet fish stores, the cray fish is a giant among species of shrimp — growing longer than ten centimeters — and a fearsome combatant. Regardless of size, fellow occupants of a fish tank all give way to it. Even if a person gets hold of its backside, it can still lurch back and cause injury with its pincers. Following in the footsteps of the *Ampullarius canaliculatus* and the Houttyn, it has been discarded by man only to thrive in the wild.

The cray fish has another ability of note: it can dig a hole several meters deep. As a result, pesticides are of no use against it. It is also a determined builder of roads — whatever stands in its way gets a good clip from its pincers. When the ridges dividing wet fields are worked over by the cray fish, a field can go dry in one night — much to the exasperation of farmers. Ecologists stand in awe of its destructive powers.

The *Ampullarius canaliculatus* — Endless Trouble: Nevertheless, the cray fish has just started causing trouble, whereas the ever-mounting mischief of the *Ampullarius canaliculatus* has already taken a very visible toll. Since it was released and gained a liking for crops, agricultural experts have done their best to find a natural predator. They have learned that — like man — ducks, sparrows, ants, wild fowl, and insects find it entirely unappetizing.

Five years ago the pesticide industry finally discovered something that could pierce the armor-like shell of the *Ampullarius canaliculatus*: the expensive TTTA. A manager of a pesticide import company says that while it originally sold 500 kilograms of the stuff over two years, it has sold 2,000 kilograms in the past year alone. But killing this mightiest of pests also takes the lives of lots of innocent bystanders. An area of vegetable fields in the Lanya district of Shihlin that has been set aside for development as athletic park was previously a muddy swampland full of great varieties of native fish. The Council for Cultural Planning and Development had once wanted to set it aside as a "Protected Area for Taiwanese Rumble Fish." But since the arrival of the *Ampullarius canaliculatus*, farmers have had no choice but to use pesticides once a week. As a result all kinds of aquatic life have died together, and the idea of a protected area has died with them, leaving behind worries about tin pollution.

Protecting Your Own: Whether or not descendants of these foreign invaders can survive and prosper, animals just passing through as well as those coming to stay for good pose a threat to Taiwan.

The year before last, smuggled-in Tibetan Mastiffs that were rumored to have rabies caused public panic. Lately people have been introducing snakes. If an exotic snake that has not been defanged bites someone, there will be no available antidote.

"People are introducing animals unaware of the results of their actions, creating potential crises for which even ecologists will be of no help," says Li Ling-ling.

Besides preventing people from hurting themselves, countries around the world are scrambling to establish gene banks in the hope of preserving their native plants and animals. To prevent competition from foreign species, countries are imposing ever stricter controls on the import and export of wild animals.

Biologists have proven that opportunities for outside species grow more numerous as native species grow sparser. Conversely, the more abundant is native life, the smaller is the foothold available for outsiders. "Beyond limiting the introduction of foreign species to those that are really needed," says Wang Ying, professor of biology at National Taiwan Normal University, "preserving native species is an even more important step in closing the door on these unwelcome guests."

(Chang Chin-ju/photos by Pu Hua-chih/ tr. by Jonathan Barnard/ first published in December 1991)

黑白、黑白配？

——烏頭翁與白頭翁

烏頭翁是臺灣特有種，僅分佈在臺灣東南沿岸。

The *Wu-T'ou Weng* is a species unique to Taiwan's south-east coast.

Black and White Unite? —
White-Headed *Weng* and Black-Headed *Weng*

文・張靜茹　圖・郭智勇

白頭翁與烏頭翁，原本楚河漢界，一定居臺灣西部，一落脚臺灣東部；如今他們卻悄悄展開「聯姻」。人們的開發脚步，在這段「黑白配」中也扮演了要角。

「明明頭是黑色的，爲什麼稱爲白頭翁？」家住臺東的廖聖福小時候常這樣問大人。

「白頭扣仔！白頭扣仔！」看到枝頭上成羣的白頭翁，西部平原上的小孩總會興奮地大喊，但他們卻沒有機會見到廖聖福口中的「黑」頭翁。

你在東來，我在西

「黑」頭翁是臺灣特有鳥種，正式名稱爲「烏頭翁」。

根據鳥類學者調查，白頭翁是亞洲東部的特產，由中國大陸東南、海南島、琉球到中南半島均普遍可見。但由十九世紀到廿世紀中葉有關白頭翁的記錄資料則顯示，白頭翁在臺灣的勢力範圍，只及於蘭陽平原以西，向南延伸至屏東潮州、大武山一帶。

至於「山」的那一邊——花蓮、臺東，直到屏東楓港，不見白頭翁芳踪，卻是全世界僅見於臺灣東南沿岸的土產——烏頭翁的地盤。

烏頭翁與白頭翁在鳥輩中同爲燕雀目鵯科，過去一直被分爲二個不同種。白頭翁頭部白色，眼先（眼睛至上嘴間條紋面）、耳覆羽、頰爲黑色，耳羽下方有一似「胎記」的白斑；烏頭翁顧名思義，與白頭翁大唱反調，頭上羽毛以黑色出現，眼先、耳覆羽、頰則爲白色，兩腮還長著狀似八字的鬍鬚，下頷基部則有一橘黃或紅色斑點。

Originally from different places, the White-Headed* Weng *(Hartert's Chinese Bulbul) and the Black-Headed* Weng *(Styan's Bulbul) came to live in western Taiwan and eastern Taiwan respectively. Today they have begun quietly to interbreed. The steps of human progress have also played an important role in this episode of matching black with white.

It is not only the United States and South Africa that have running problems between black and white races; in Taiwan there is a color problem amongst the birds!

''It's head is obviously black, so why is it called the White-Headed *Weng*?'', was the question Liao Sheng-fu, whose home is in Taitung, would often ask of his elders when he was a child.

''Whitey! Whitey!'' yell out the excited children of the western plateau on catching sight of the gathering White-Headed *Weng*. Yet they have not had the opportunity to see the ''black''-headed *Weng* spoken of by Liao Sheng-fu.

You Come from the East — I'll Come from the West: So what is the ''black''-headed *Weng*? Its formal name is in fact the ''*Wu-T'ou Weng*,'' and it is a species unique to Taiwan.

According to investigations by ornithologists, the White-Headed *Weng* is a product of East Asia which can be seen in southeastern China, Hainan Island, Okinawa and Indochina. But evidence concerning the White-Headed *Weng* from the nineteenth and mid-twentieth centuries makes it apparent that the extent of the bird in Taiwan only ranged from the area west of the Langyang Plain south to Chaochou in Pingtung county and the Tawu range of mountains.

As for the other side of the mountains — the coast of Hualien and Taitung county to Fengkang in Pingtung — although there are no traces of the White-Headed *Weng* to be seen, Taiwan's

雖然兩者你在東來我在西，楚河漢界，嚴守分際，但除在「半身正面」照片上顯有不同外，兩者體型與橄欖褐色身體卻如出一轍，且習性、個性大同小異。也因此閩南人初到臺灣見到烏頭翁，仍以「白頭翁」稱之。

習性相同、數量衆多

由食衣住行逐項細數，除衣——長相——外，烏、白頭翁對食物的品味十分雷同，均是雜食性，昆蟲、野漿果、嫩葉、農家果園的水果概不挑剔。至於住、行，生態學者在東、西兩岸觀察到的也「同步」。兩造平時皆三五成羣，組社結黨，一起覓食，一起在疏林、空地上追逐。立春三月，繁殖季一到，則風度盡失，爲尋伴侶，雄鳥開始出現排擠、警戒、打架等行爲。待佳偶「打」成，有情鳥成眷屬，一對對各自帶開，共同建立領域及進行傳宗接代的大業。

此時在臺灣東、西部，烏、白頭翁的雄鳥有志一同，都愛獨立於領域範圍內高處樹枝上，眼觀四面的守衞；雌鳥則忙著在草地上尋找草根、榕樹氣根、相思樹葉等建材。且兩者的巢也均爲碗狀，掛在離地三公尺左右的樹枝上。

新居落成，抱孵蛋同樣需要十一、二天，蛋由二到四個不等，小鳥則九到十天卽可離巢；生殖季中皆有繁殖二窩的現象。結束天下父母心的忙碌季節，八、九月後又恢復「當我們同在一起」的團隊生活。

此外白、烏頭翁均適應力強，喜愛人工開墾過的地方，因此在各自領土上數量均頗爲可觀。在西部，白頭翁普遍分佈在一千公尺以下稀落的闊葉林和平原、農田；在其他動物都退避三尺的都市中，牠和麻雀、綠繡眼等少數鳥類則瓜分了公園、庭院、綠地。中正紀念堂榕樹下就常可見白頭翁吃果實。

常帶隊賞鳥的臺東豐里國小教師廖聖福也表示，在臺東，每天淸晨聽到的第一聲鳥叫不是麻雀，就是烏頭翁。

就連休息時頭頂羽毛會呈九十度豎立，形成鮮明的黑冠與白冠，也是烏、白頭翁共有的特徵。

白冠、黑臉是白頭翁的特徵。
White cap and black face are the marks of the White-Headed *Weng*.

兩族共和？

在臺灣一百卌多種陸棲鳥類中，表不一、裡相同，分佈又如此明顯呈地理隔離的，只有烏、白頭翁一例。而在牠們雙方有何淵源仍不爲人所知以前，白、烏頭翁互不踩地盤的佈局，已悄悄「和平演變」。

一九七〇年代，鳥類學者陸續在中橫、花蓮及楓港北部發現烏、白頭翁領域有重疊現象。八〇年代更在重疊區發現頭部羽色兼具兩者特性的「雜」頭翁，譬如白色頭冠卻有橘黃點和黑色唇鬚。

中央研究院動物所在民國七十八年針對太魯閣國家公園做調查，結果發現太魯閣峽谷內只見「中間型」，幾乎不見純種白、烏頭翁。研究人員並證實雜交後代具有繁衍能力，顯示二者在此聯姻時日已長。

southeastern coast is the only place in the world where the *Wu-T'ou Weng* can be seen.

Both the White-Headed and the *Wu-T'ou Weng* are members of the same family and they have been divided into two different species. The White-Headed *Weng* has a white head. The plumage in front of its eyes, behind its ears and on its chin is black, while there are speckles resembling a white birth mark below its ear. The *Wu-T'ou Weng*, or Black-Headed *Weng*, as it name suggests, can be distinguished from the White-Headed *Weng* by its black head plumage, while the feathers in front of its eyes, behind its ears and on its chin are white. Its two cheeks have marks that form a V-like shape, and under its chin is a speckled mark in orange-yellow or red.

Although "east is east and west is west" and the two species keep a conservative distance, apart from the obvious differences that can be seen in their "passport photographs," their physiques and olive-brown bodies could be from the same mold and their habits and personalities bear more similarity than difference. It is because of this that when the early settlers came to Taiwan from Fukien and saw the *Wu-T'ou Weng* they still called it the "White-Headed *Weng*."

Masses with the Same Habits: In their eating, clothing, habitation and movements — apart from their appearance — the *Wu-T'ou Weng* and the White-Headed *Weng* have identical tastes, being generally omnivorous and eating insects, berries and shoots, while not being fussy when it comes to choosing from the fruits of a farmer's orchard. As for habitation and movement, ecologists have observed that in both east and west the birds are in step. On both sides they normally live in groups, or organized social units, foraging together and following the crowd in sparse woods or on open ground. As soon as the nesting season arrives in March, grace is thrown to the wind and the search for a mate begins with the cocks' behavior of expelling, standing guard and fighting. When the prize is attained, the love-birds make a couple, establish their territory and put into practice the work of continuing the family line.

At this time, in east and west, both the White-Headed and *Wu-T'ou* cocks love to perch high up on a branch within their territory so as to keep a watchful eye on all sides. The hens, meanwhile, are busy searching the scrub for twigs, banyan risers and leaves that will make do for building materials. The nest will be an even bowl that hangs from a branch around three meters from the ground.

With the new residence complete, incubation of the two to four eggs will take eleven to twelve days, and the young birds will leave the nest within nine to ten, and there are generally two clutches of eggs within each nesting season. Finally, after August and September, the universal hustle and bustle of parenthood gives way to life within the larger troupe.

The adaptability of the birds and their love of built-up places means that *Weng* can be seen in large numbers in every area. In the west, the White-Headed *Weng* is spread among the sparse deciduous woods, plateaux and fields under a thousand meters in altitude. In the cities, from where other animals have retreated, the *Weng* have divided up the parks, gardens and green spaces along with the sparrows and a small number of other birds. They can even be seen eating fruits under the banyan trees at the Chiang Kai-shek Memorial in Taipei.

Liao Sheng-fu, a school teacher who often takes groups of pupils bird watching, says that every morning at dawn in Taitung, the first birds you can hear are not the sparrows but the *Wu-T'ou Weng*.

Even when resting, their head plumage stands up at a ninety-degree angle, like a brilliant white or black cap — yet another of the particular characteristics shared by both the White-Headed and *Wu-T'ou Weng*.

Two Clans Together: Among Taiwan's more than 130 species of land-bird, only the *Weng* are of different appearance while being internally the same. Whatever the reasons lying behind this, their mutual separation has already quietly been undergoing a process of "peaceful evolution."

In the 1970s, ornithologists discovered an overlap in the territories of the White-Headed and *Wu-T'ou Weng*, along the cross-island highway to Hualien and the north of Fengkang. Then, in the 1980s, two unique kinds of bird were found with plumage that was a mixture of colors, such as having a white cap but with orange-yellow spots and black whiskers.

In 1989, the zoological institute of the Academia Sinica undertook an investigation at Taroko Gorge National Park which concluded that only cross-

但調查人員觀察日久，乍看「二族共和、天下一家」的大勢，其實是由白頭翁主動攻擊，一直往烏頭翁地盤推進。如今花蓮市也到處可見黑白頭翁雜處，全臺幾乎只剩臺東維持「清一色」。鳥類學者擔心的是，如此下去，將只剩白頭翁與雜頭翁，烏頭翁可能成爲十四種臺灣特有鳥類中，目前數量最多、卻將最早絕種的臺灣鳥種。

促成白、烏頭翁相遇的主要媒介，「是人類的開發脚步，」以白、烏頭翁爲論文主題的臺大動物研究所研究生林華慶說。蘇花、北宜、北橫、中橫及南廻等公路如八爪魚般四處延伸，打通了東西部交通；加上原始密林的砍伐、崇山峻嶺的開墾，屏障盡去，白頭翁一路暢行，如今族羣不只輻射四散，分佈也已「爬高」到一千八百公尺左右。

此外，佛教信徒的放生風氣也是白頭翁的大媒人。林華慶說，兩年前臺東海山寺放生了二百多隻白頭翁，不久臺東就發現有雜頭翁出現。

白頭翁步步爲「贏」

但爲什麼白頭翁的侵略性格比烏頭翁強？這兩個不同種的鳥類爲什麼能雜交，而且還瓜藤蔓延、多子多孫？對面臨生存危機的烏頭翁，人們又是否應採取一些保育策略？爲尋求答案，目前中研院動物所、臺大動物系已合作展開研究，除習性及棲地調查外，也更上層樓——在鳥兒的聲音下工夫。

生物界已知，通常鳥唱的曲目愈多、歌聲變異度愈高，生殖成功率也愈大。「因此收集二者所有鳴音，比較不同，能增進對兩種鳥的認識，並探討鳴音在地理隔離及品種間所扮演的角色！」對烏、白頭翁曾有深入研究的張瓊文在其碩士論文中指出。按理推測，「表現強勢的白頭翁，是否牠的鳴聲更多樣，較吸引烏頭翁的雌鳥？」目前研究烏、白頭翁習性的中研院動物研究所研究助理鄭儀說。

烏頭翁，等一等

但聲音的研究卻不容易，在訓練有素、耳清目靈的人員採集之下，僅一年，在兩個地點就錄了兩種鳥的三百五十六種叫聲，又要一一比較牠們對鳴聲的偏好、頻率高低，「聽聲辨影」的工作不易在短期內見成果。另一方面，生態人員也沒有放棄傳統的生態調查，「我們希望從已知的行爲展示、求偶、驅逐狀況中更進一步找出蛛絲馬跡，了解爲何白頭翁會表現較佳？」鄭儀說。研究人員已發現：白頭翁平均體型比烏頭翁大，烏頭翁繁殖季節比白頭翁晚半個月。但有何意義，仍不敢遽下定論。

問題是：研究需要時間，而在人們爲主導的這個動物演化過程中，白頭翁步步爲「贏」，烏頭翁還能撐持多久？够不够讓人類解開這兩種動物間的謎團？ S

（原載光華八十年十二月號）

西部平原長大的人大多只知白頭翁，不知有烏頭翁。(陳永福攝)

Most people who grow up on the western plain do not know about the *Wu-T'ou Weng*. (photo by Chen Yung-fu)

breeds could be found there, with almost no pure White-Headed or *Wu-T'ou Weng* in sight. The researchers also proved that the offspring of cross-breeding had the ability to proliferate, so it was apparent that the marriage between the two was already long-standing.

But after further observation by the researchers, what had originally been seen as the great situation of two families living together as one, was in fact discovered to have arisen from incursions made by the White-Headed *Weng* into the territory of the *Wu-T'ou*. Thus, today, White-Headed *Weng* can be seen all over Hualien, while Taitung is almost the only place left in which a "pure color" can be seen. What troubles ornithologists about this is that, if it continues, then there will only be White-Headed and Mixed-Headed *Weng* left. *Wu-T'ou Weng* could become the most numerous of Taiwan's unique species at present — but the earliest to become extinct.

The most important matchmaker in facilitating the marriage between the two types of *Weng* has in fact been "the march of human progress," says Lin Hua-ching, a graduate student in the department of zoology at Taiwan University who wrote a thesis on the *Weng*. The numerous cross-island highways have reached all areas in connecting up the east and west while the felling of the original dense forests and development of the high mountains has meant that the White-Headed *Weng* can move so easily that it is not only spreading in all four directions but is also climbing up to heights of around 1800 meters. The practice of Buddhist releasing animals has also acted as a matchmaker for the White-Headed *Weng*. Lin Hua-ching says that, two years ago, the Haishan Temple in Taitung released a flock of more than two-hundred White-Headed *Weng* and it was not long after that the cross-bred type of *Weng* was discovered.

Victory by Stealth for the White-Headed *Weng*: Yet why is the White-Headed *Weng's* expansionary character stronger than that of the *Wu-T'ou*? Why can these two different species of birds cross-breed and continue to do so for generations? And should we adopt a policy to preserve the *Wu-T'ou Weng*, which is facing this crisis of survival? To find the answer the zoological institute of the Academia Sinica and the zoological faculty of Taiwan University have already begun joint research. Apart from investigating habits and demography, they are even working hard on the voices of the birds.

Ecologists already know that, the more melodies there are in the songs of a bird and the higher the fluctuations in pitch, the more likely it is to have a high rate of successful breeding. Through collecting the songs of the two species of bird and comparing their differences, we can deepen our knowledge about them and further explore the importance of song to geographical distribution and separation and the entire role this plays between the two species!" points out Chang Chiung-wen in his master's dissertation, after carrying out in-depth research into the *Weng*. "Is it that the more varied songs the White-Headed *Weng* have, the more attractive they are to the *Wu-T'ou* hens?" asks Cheng I, research assistant at the zoological institute of the Academic Sinica.

Hold On, *Wu-T'ou Weng*: But audio research is not easy. Well-trained personnel with sensitive ears and eyes have recorded 356 different types of song from the two birds within barely a year. Moreover, within this short period, such work as comparing the songs one-by-one to find out which is preferred and at what frequency they occur does not easily yield readily tangible results. Another aspect is that ecologists have not yet given up their traditional ecological investigations: "We hope to find clues in what we already know about displays and courtship that will tell us why the display of the White-Headed *Weng* is more attractive," says Cheng I. The researchers have already discovered that the White-Headed *Weng* is on average larger than the *Wu-T'ou* and that the latter's nesting season comes about two weeks later than the White-Headed *Weng*. Yet just exactly what the significance of this might be, no one has yet dared to pin down.

The problem is: research necessitates time, and in the human-led evolutionary process of animals, the White-Headed *Weng* has been step-by-step victorious. For how long can the *Wu-T'ou Weng* hold on? Will it be long enough for mankind to solve the riddle of these two species? S

(Chang Chin-ju/
photos by Kuo Chih-yung/
tr. by Christopher Hughes/
first published in December 1991)

（李淑玲繪圖）(drawing by Lee Su-ling)

趕緊來，火金姑

文・張靜茹　圖・張良綱

Fireflies: Brilliant-Bummed Ladies of the Night

When the video* Tomb of the Firefly *was selling fast and furious and the lyrics to the popular "Firefly Song" were on the lips of children everywhere, it occurred to people that they hadn't seen a real firefly for ages.

"The shower hasn't let up. The gold carp is taking a wife, his brother the *gong* fish is beating the gongs and drums, and the *tousa* fish is the match-

當日本卡通「螢火蟲之墓」的錄影帶充斥市面，小朋友均能琅琅上口近來流行的「螢火蟲之歌」；人們卻發現，好久沒看到眞正的螢火蟲了。

「西北雨，直直落，鯽仔魚，要娶某，
鮕呆兄，打鑼鼓，媒人婆，土虱嫂，
日頭暗，找無路，趕緊來，火金姑，
做好心，來照路，西北雨，直直落……」

當西北雨再度直直落下，森林溪畔重新上演了一場鯽仔魚娶親。

鮕呆兄仍然打頭陣，敲鑼打鼓，架勢十足；媒人婆土虱嫂更早早粧點好，扭腰擺臀而來，大夥促擁著志得意滿的新郎倌上路。迎過新娘，又一路鑼鼓歡暢回程，結果吵嚷的隊伍誤了時辰，天色已暗，又迷失了路。鯽仔魚急著完成終身大事，其他人累得想回家歇息，個個急得跳脚。

於是，有人在夜裏聽到遙遠的森林裏，傳來微弱的呼叫：火金姑……火金姑……。

只要你的屁股亮晶晶？

螢火蟲，閩南人稱爲火金姑；古人也稱爲耀夜、熠燿。在電燈尙不普及的農業社會，螢火蟲卻「屁股會點燈」，成爲許多人嚮往、利用、吟誦的對象。浙江民間就有這樣的歌謠「螢火蟲兒，飛落來，給你三個銅錢買草鞋，不要你的金，不要你的銀，只要你的屁股亮晶晶。」

亮晶晶的屁股，除了可以爲鯽魚娶親「好心照路」，讓家貧好學、卻買不起燈油的車胤「囊螢夜讀」，還可以「照爹爹，犂大坵；照哥哥，到杭州」；「照媽媽，好紡紗；照姊姊，去織麻」。當然，在生物學家看來，光憑幾隻螢火蟲的光源，苦讀未成，恐怕早已老眼昏花，就更別提犂地紡紗、織麻照路了。

古人也留下不少藉螢遣懷的詩詞歌賦。

「巫山秋夜螢火飛，簾疏巧入坐人衣，
忽驚屋裏琴書冷，復亂簷前星宿飛；
卻繞井欄添箇箇，偶經花蕊弄輝輝，
滄江白髮愁看汝，來歲如今歸未歸。」

流螢不愼誤闖位在林木密佈的草堂，除了陪公子讀書，還惹得懷才不遇的文人滿腹愁悵。尤其午寐忽醒，黃昏已盡，文人百無聊賴，忽見停於帳上的流螢，更近感自身、遠懷家國，魏晉時期傅咸的「螢火賦」就如此感懷：「不以姿質之鄙薄兮欲增輝，乎太清雖無補於日月兮，期自照於陋形；不競於天光兮，退在晦而能明，有似於賢臣兮於疏外而盡誠……」

提著燈籠找老伴

說來只怕要大掃古人的雅興，螢火蟲「提著燈籠」四處飛行，其實是在對異性「打電報」，互通款曲。至於牠們「退在晦而能明」的賢臣美德，是由於白天日光對螢光干擾太大，因此牠們和蝙蝠、貓頭鷹一樣，屬於「夜行性動物」，在黑夜活動，白天停棲在樹上。

不同種類的螢火蟲所發出的光，有不同的頻率、波長。就像每種鳥類靠特有的鳴叫聲吸引異性，流螢不同的發光訊號，可以防止牠們「搞錯對象」。牠們的螢光以綠色最多，但因種類而異，也有紅、橙、黃、藍等色澤。

東南亞與大陸東南部的濶葉林中，有一種「齊爍螢」，會上萬隻在同一株樹上同步明滅，步調絲毫不差，像個發報臺。如此可避免落單時，微弱的光源被密林遮蔽，「發報器」失靈，求偶不成。

螢光不燙人

螢火蟲體內含有「螢光酶」酵素與發光質，二者經連串生化反應，產生氧化作用（燃燒作用）而產生光，再由腹部末端的發光器發射而出。至於發光過程中百分之九十的能量是耗在光能上，只有極少的部分轉爲熱能，因此螢火蟲發出的是不灼人的「冷光」，可以說是一種「低溫氧化」，不似一般氧化作用會產生高熱。

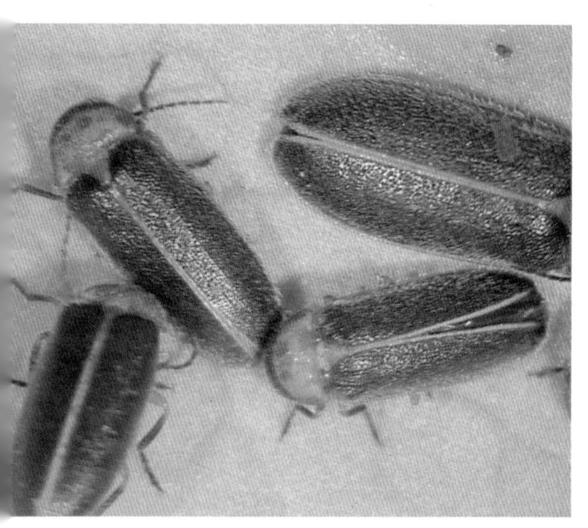

你看過眞正的螢火蟲嗎？圖爲水生螢火蟲黃緣螢。(陳仁昭提供)

Have you ever seen a real firefly? The photo is of a water-bred firefly. (photo courtesy of Chen Jen-chao)

maker. It's dark already and we can't find our way. Come quickly firefly, be nice and light the road. The shower hasn't let up.''

As the rain falls on the banks of a forest creek, a gold carp is once again taking a bride.

In full regalia the groom's brother is beating on the drums and the gongs. The matchmaker, all dolled up, is swaying her hips as she walks, and the self-satisfied groom, surrounded by pressing throngs, is walking proudly along the road. After fetching the bride, there is another joyous march with drums and gongs returning to the groom's home. But they are behind schedule. It turns dark and they lose their way. The gold carp is anxious to perform the ceremony, and the others are tired and want to go home to rest.

Thenceforth, at night people would hear soft calls in the distant woods: ''firefly. . . firefly. . . .''

If only your bum could light up: The Taiwanese call fireflies ''golden fire ladies,'' and in ancient times they were called ''night brighteners'' or ''dazzlers.'' In pre-industrial times, before electric lights were common, fireflies ''with bums that light up'' were widely used, praised and longed for. As the Chekiangese sing, ''Firefly fly down. I'll give you three pennies to buy some straw shoes. I need neither gold, nor silver, just your sparkling tail.''

According to a folk song, besides lighting up the road for the gold carp's wedding procession and providing light for the studious but poor Che Yin, these brilliant bottoms also ''give father light to plough the fields and brother light to go to Hangchow . . . give mother light to spin and sister light to weave.'' Of course, a biologist can tell you that even when straining you can't read by the light of a few fireflies — let alone plough, spin, weave or take a journey — and you might hurt your eyes trying. The ancients also left behind many lines of verse in which fireflies are used to dispel one's sorrow:

On a fall evening the fireflies take flight on Wu Mountain.
Flying through the thin curtains they land on my clothes,
Startling the chill among my books and instruments.
Before the eaves, they shoot like stars.
A growing horde, they circle the well and the railing,
Brightening the occasional flower.
White haired, I look on with longing,
Still not home after so many years.

A passing firefly mistakenly enters a humble cottage in the dense forest. Besides accompanying the literary scholar as he reads, it also fills the scholar with melancholy. In particular, when he wakes from his afternoon nap to find that the sun has already set, the exiled scholar feels as if life no longer has meaning. Suddenly, he sees the firefly resting on the mosquito net and he is further gripped with longing for home. ''Though your light is thin, adding nothing to the sun and moon, you still illume your simple shape. Lost in the bright of day, you are clear to see in a darkened place — like a loyal official in the hinterland.''

A lantern for your spouse: The ancients might have found what follows disappointingly prosaic. Fireflies ''holding up their lanterns'' are actually ''strutting their stuff'' to members of the opposite sex. Because daylight greatly interferes

臺灣目前已被發現的螢火蟲有三十多種，其中不乏體型碩大者。圖下兩隻「臺灣筒螢」幼蟲，體長超過五公分，屬稀有螢火蟲。

More than 30 species of fireflies have been discovered on Taiwan, and among them are larger sized species. On the right are two young fireflies of a rare Taiwanese species that grows longer than five centimeters.

若由人類發電照明的方式來看，螢光是一種極有效率的發光技巧。因爲人類發電過程，大部分的能量都轉爲熱能消耗掉，卽使是時下流行的冷光燈，也只是借助玻璃纖維降低燈泡溫度，發電過程耗掉的資源一樣多。根據估計，如果以螢光方式發電、照明，可以減少十倍電力。

至今學界仍然不斷研究螢火蟲的發光機制，國際上還有所謂「國際生物發光會議」不定期展開，爲尋找螢火蟲發光的「更深一層意義」而皓首窮經，論辯不斷。

尤其在會發光的動物中，烏賊、水母、水螅與許多發光菌，大部分都集中在海裏；陸地與淡水中與螢火蟲具有相同特質的，只有少數眞菌和蠕蟲。螢火蟲卻廣泛分布在熱帶與溫帶，目前全世界已知就有兩千多種，臺灣出現的紀錄有卅幾種。

寵物昆蟲

由於數量衆多，螢光不燙手，成蟲又不咬人、不螫人，幾乎沒有攻擊性，遂成了人們最愛的「寵物昆蟲」。

今天的中生代，談起兒時捉螢經驗，人人手舞足蹈；但古人其實「玩」得更凶。西方婦女曾流行將活生生的螢火蟲，挾在衣襟上當裝飾品。

隋朝奢靡、放縱而喪國的隋煬帝，曾由民間蒐羅了數斛螢火蟲，由洛陽景華宮一起放走，螢火輝映，蓋過皎月。根據清朝顧鐵卿所著的「淸嘉錄」記載，蘇州人常挖空鴨卵，圖上色彩，再鑿孔放入流螢，成爲「螢火蟲燈」，供小兒玩耍。

可惜的是，今天的孩子們雖然仍在小學課本裡讀到「囊螢夜讀」的故事；從百科全書、動物全集裡看到火金姑精美的照片、解剖圖，却已少有追撲流螢的親身體驗。

民國六十年次，出生於臺北的曾珍琦，在進屏東技術學院植物保護系念書之後，生平第一次見到火金姑，「比圖片上畫的小多了！」她的答案可以總結今天部分同齡以下小朋友對螢火蟲的經驗。

螢火蟲的繁殖力強，不易因爲人們好玩捕

with the light fireflies produce, they are nocturnal animals like owls and bats, active at night and resting during the day.

Different kinds of fireflies produce lights of different frequency and wave length. Just as species of birds rely on their unique calls, fireflies use different wavelengths of light to ensure that they don't find "an unsuitable partner." For the most part they give off green light, but there are also species of firefly that emit red, orange, yellow and blue light.

In the broad-leaved forests of Southeast Asia and mainland China, fireflies of the species *Pteroptyx malaccae* can gather in the tens of thousand on one tree, flashing nearly simultaneously, like a giant neon sign. In this way, they avoid having their lonely solitary lights obscured in the dense forest and losing out in their search for a spouse.

Luciferase enzymes in fireflies go through a series of bio-chemical reactions in which they are oxidized for use as fuel to create light which is sent out from the end of their abdomen. Because 90 percent of the energy consumed becomes light and very little of it is turned into heat, fireflies produce a "cold light" that doesn't burn. It could be called "low-heat oxidation," differing greatly from most forms of oxidation which create great amounts of heat.

A light that won't burn: Considering the human method of using electricity to generate light, in which most of the energy is consumed in the generating of heat, the method the fireflies use is remarkably efficient. Even in the now-popular "cool lights," the energy consumed is just as high. Glass fibers are used only to reduce the heat of the light bulb and not to increase its efficiency. According to estimates, if the firefly method was used to generate electricity, the energy needed to create light could be reduced to a tenth of what it is used now.

With scholars constantly researching the light-producing technique of fireflies, "International Biological Lighting Conferences" are held every once in a while for debate and discussion of the matter.

The vast majority of animals that emit light — such as cuttlefish, jellyfish and hydra — are sea creatures. Among land and fresh water animals, the firefly's ability is extremely rare, shared only by a small number of fungi and helminth. Spread over the tropics and temperate zones, there are 2,000 varieties of fireflies, 30 of which can be found on

恒春農場主人張國興希望能在農場這一角小小河域重圓螢夢。
Hengchun farmer Chang Kuo-hsing hopes that the dream of fireflies can be realized once more in a small area by this stream.

捉，而消「光」匿跡。那麼，今天螢火蟲盛況難再的原因是什麼？

林塘、清風，碧梧、綠苔

「碧梧含風夏夜清，林塘五月初飛螢，」古詩裡，已明示了欲見點點螢火飛光，需要的是什麼樣的環境。

現代人經營都市生活時，既未有心保留一些林塘、碧梧，又缺乏清風，火金姑逃之夭夭自是必然。問題是，即使野外如今也已難見「微雨過時松路黑，野螢飛出照綠苔」的景致。

鄉下長大的兒童都知道，過去稻田中最容易捕捉到螢火蟲。當農藥使用量愈來愈大，有些甚且是針對蝸牛而噴灑，造成了幼螢食物來源匱乏。楊平世就舉臺大舟山路上的實驗農田爲例，不過幾年前，一個晚上還可抓上四、五十隻螢火蟲，後來學校在稻田裏做農藥實驗，「今天將水稻田翻遍，也找不到半隻螢火蟲，」他說。

五彩霓虹裏難覓芳踪

水生幼蟲利用土堤化蛹，「新式」的溝渠、水圳，全部以水泥襯底，蟲兒再厲害，也無法在水泥地上打洞。「以前由烏來瀑布沿著北勢溪，進入信賢村，一路上，每晚可見上百隻臺灣窗螢，這幾年路拓寬後就不見了，」從高中就跟著陳仁昭做昆蟲觀察的海洋大學養殖系學生張譽馨解釋說，「光害，也是螢火蟲消失的重要原因」。

利用光源配對的成蟲，在路燈四處林立，五彩霓虹大放光芒的今天，恐怕也眼花撩亂，衆裏尋「牠」千百度，却沒法尋到對象，也只有在「無處覓芳踪」的感嘆聲裡，打光棍終老一生了。

廿年前，日本已開始針對日漸稀少的螢火蟲進行人工繁殖與復育，尤其原本數量繁多的平家螢、源氏螢，更被特意保護在嚴格控制水質的幾個河段裡。

兩年前，對兒時撲螢經驗「懷念特別多」的恆春農場主人張國興，找到陳仁昭教授，希望在農場中找一塊乾淨的水域，復育螢火

Taiwan.

Pet Insect: Though lacking economic value like butterflies and bees (which are collected as specimens or used to make honey), because the fly is numerous and harmless — it doesn't burn, bite, prick or sting (see "The Life of a Firefly") — it has become a beloved "pet insect."

When those who are now in middle age speak of their experiences collecting, they become very expressive, but in ancient times people were even crazier for the bug. Among Western women, it was once fashionable to wear a live firefly on the lapel as a kind of broach. Emperor Sui Yang, who led a lavish life of debauchery and ended up losing his crown, once had enormous numbers of fireflies collected and let go in the Chinghua Palace of Loyang. Their light was brighter than the moon. As recorded by Ku Tieh-ching, people living in Soochow would often empty duck eggs, paint them, and make a hole to put in fireflies, creating "a firefly lamp" for children to play with.

It's unfortunate that although the children of today read the story "Reading by the Light of Fireflies" in their elementary school textbooks and see pictures and anatomical drawings of fireflies in encyclopedias and books about animals, few have the chance to experience what it is to chase after them.

Tseng Chen-chi, a 21-year-old native of Taipei, didn't actually see a firefly until she enrolled in the Plant Protection Department of National Pingtung Polytechnic Institute. "They're much smaller than in the pictures." Her reaction can speak for the experience of most of those her age and younger.

Fireflies multiply rapidly, and they're not easily affected by people catching them. Why then are their numbers declining?

The wind in the woods: "The breeze rustles the leaves of the green firmiana on a cool night in early May. Fireflies start to fly around the woodland pond." From these lines of ancient verse we can see what kind of environment fireflies require. Since there is no will to preserve wood ponds and green firmiana and since there is a lack of fresh air in modern cities, the fireflies have had no choice but to flee. The trouble is that even in the country, it is difficult to find scenery where "after the light rain, by the road through the dark pines, fireflies fly out and illuminate the green moss."

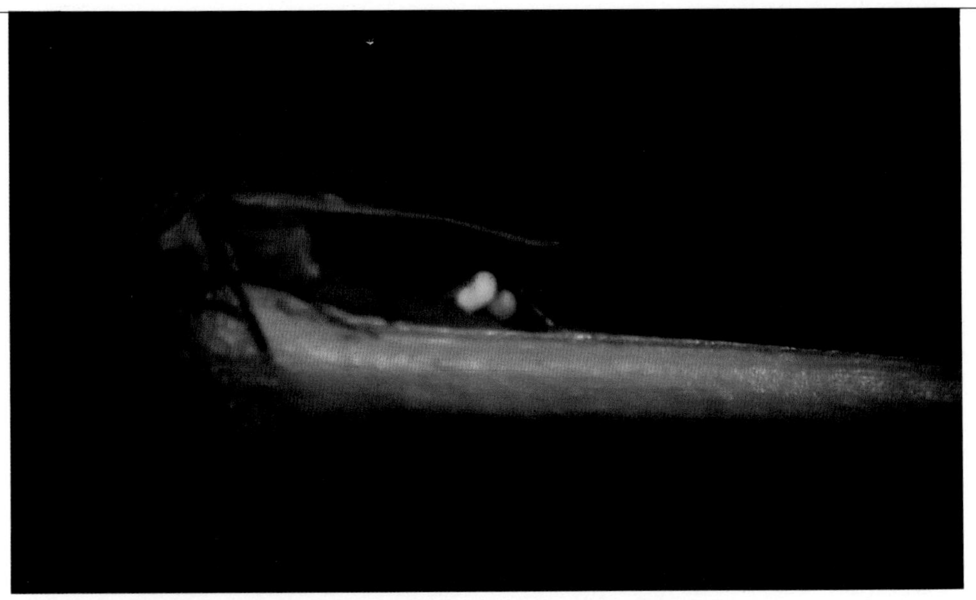

螢火蟲的屁股亮晶晶，是因爲牠的發光器位於腹部末端。(陳仁昭提供)
Fireflies have "brilliant bums" because their light comes from the end of their abdomens. (photo courtesy of Chen Jen-chao)

蟲。

雖然主要的工作仍在實驗室裡進行人工繁殖，陳仁昭已憂心忡忡。因爲螢火蟲的食物——蝸牛，已不容易找；如果大環境繼續惡化，螢光再現的情景，也只能局限在一個小小角落裡。

何時重圓流螢夢！

從事螢火蟲全臺調查的楊平世表示，雖然目前仍不易提出螢火蟲對水溫與溪水溶氧量要求的數據是多少。但由國外情形和全臺調查可以確定，螢火蟲可說是溪流中，對新環境適應很慢的昆蟲。

許多溪流仍可見蜻蜓點水，竹節蟲漂浮，却獨獨缺少火金姑光臨。「環境稍有劣化，螢火蟲是最早遭殃的，」楊平世說，日本爲了繁殖螢火蟲，不惜巨資，一隻成本多達一萬多臺幣，而接下來環境的維護與管理更難，因此日本也只集中火力在短短的幾個河段裡執行。

開發自然，輕而易學；恢復舊觀，重圓螢夢，卻困難重重——這是學者的心聲。「如今想要看流螢，恐怕得遠溯溪流源頭，探訪人跡難至之處，」楊平世說。

致力推動螢火蟲復育的農委會技士陳超仁則以爲，復育螢火蟲，更大的意義，是提醒我們恢復舊觀，學習尋回一個良好的生活環境。

螢火蟲，螢火蟲，慢慢飛

「螢火蟲！螢火蟲！
你有一個小小的手電筒，
你在草地上慢慢的飛，
我們笑著在後頭追，
追啊！追啊！追到了河邊，
看你閃啊！閃啊！飛到了對岸……」

詩人的生活經驗能否傳遞下去？夜裡，聽不清遙遠的森林裡是不是有：火金姑……火金姑……的微弱呼叫；鯽仔魚娶親的隊伍，到底會不會平安到家呢？ S

(原載光華八十一年十二月號)

Those who spent their childhoods in the country know that rice fields used to be full of fireflies. But as the use of pesticides has grown — to the point where even snails are being targeted — there has been a resulting lack of food for young fireflies. Yang Ping-shih cites the example of National Taiwan University's experimental fields. Not long ago, you could still catch 40 or 50 fireflies in one night. Later the school made experiments with pesticides there. "Now, even if you turn the field upside down," he says, "you won't find half a firefly."

Lost amid the neon: Water bugs make use of earth dikes for pupation, but newfangled gutters and drainage ditches are lined with cement, and no matter how tough these bugs are, there's no way they can burrow into cement. "Previously, as soon as you got on the road from the Wulai Waterfall along Peishih Creek to Hsinhsien Village, you could see hundreds of Taiwan window fireflies," says Chang Yu-hsing, a student at National Taiwan Ocean University who has observed insects with Chen Jen-chao since high school. "In the past couple of years, since the road has been widened, there are none to be seen. Damage from light is also a major reason they have disappeared."

Perhaps adult fireflies, which use their light to match up with their spouse, have grown dazzled and dazed by the streets lined with lights. With potential mates lost in the glare, they live celibate lives.

Twenty years ago, Japan had already started proceeding with artificial propagation and repopulation of gradually disappearing fireflies. In particular, the originally numerous *Luciola cruciata* and *L. Latoralis* have been specially protected in a few stretches of river with strict controls on water quality.

Two years ago, Chang Kuo-hsing, a Hengchun farmer who has fond memories of chasing fireflies as a child, contacted Chen Jen-chao, saying that he hoped he could find a clean water area on his farm to raise fireflies.

Although his main work is still in the labs carrying out experiments on artificial propagation, Chen Jen-chao is very worried about the scarcity of snails, on which fireflies feed. If the environment continues to deteriorate, the light of fireflies will be confined to only a few small areas.

The dream of the fireflies re-realized: Yang Ping-shih, who has made an island-wide investigation of the firefly, says no one knows what water temperature fireflies require and how much oxygen they need dissolved in the water. But from foreign test results and the island-wide investigation, one can say with certainty that the firefly is an insect living by running water that adapts very slowly to environmental change.

While you can still see dragonflies skipping along the water and stick insects floating on the surface of many streams and rivers, the light of the fireflies is missing. "When the environment deteriorates a little," Yang Ping-chih says, "the fireflies are the first to meet misfortune." Japan has spent no small sum trying to breed fireflies. But the cost to produce a single bug has exceeded NT$10,000, and the next hurdle of maintaining and controlling the environment is even more difficult. As a result, the Japanese are concentrating their efforts along the stretches of a few rivers.

It is easy to open up the wilderness, but it is extremely difficult to bring back old scenes and re-realize the dream of the firefly: This is what scientists know in their hearts. "If you want to see fireflies," says Yang Ping-chih, "I am afraid you've got to head to the source of streams, places where people go with difficulty."

Restoring firefly populations to remind people of how things used to be and to teach people how to bring back a pristine environment, says Chen Chao-jen, a technician at the Council of Agriculture who is working to restore populations of fireflies, is more meaningful than just the return of fireflies.

Firefly, Firefly, Fly Easy

Firefly! Firefly!
With your small flashlight,
You fly slowly in the grass.
Laughing, we chase you from behind.
Chase! Chase! Chase to the river's edge.
We watch you flash! Flash! Flying to the other side. . . .

Can we pass along the life experience of this poet? Can you hear the muffled call in the distant woods: "Firefly. . . firefly. . . ." Will the wedding procession of the gold carp make it peacefully home?

(Chang Chin-ju/photos by Vincent Chang/ tr. by Jonathan Barnard/ first published in December 1992)

最後的巡禮

The Last Pilgrimage

文／圖・徐仁修

此刻我在推土機的嚨嚨聲中，對這美麗的野池做最後的巡禮。何時，人類才會珍惜自己原本擁有的？……

在龍潭靠山的小丘間，有一方隱藏在防風竹林裡的小池塘，四周青草葳蕤，岸樹垂掩；池水清幽，波平如鏡，難得一陣穿林而來的涼風，微微弄縐這反映著北臺灣夏日的藍天白雲。幾波漣漪，池面立刻恢復映照著清明的福爾摩沙天空。

一方小池，生機無限

池裡緊靠著岸草的水中，長著一簇簇有如睡蓮的浮葉植物——臺灣萍蓬草，一朵朵如

爲了灌溉，客家先民在桃園臺地開鑿了無數的池塘，它不只滋潤著農田，形成美麗清幽的景觀，也是許許多多水生動植物的天堂。
For irrigation, the Hakka pioneers in the Taoyuan Tableland dug numerous ponds, which have not only provided water for the fields and beautiful scenery but have also served as islands of paradise for numerous species of water plants.

Amid the sound of bulldozers, I have made my last pilgrimage to this beautiful wild pond. Why is it that we wait until times like these to treasure what we should have treasured all along?

In the hills of Longtan, there is a small pond hidden among the bamboo wind breaks in lush surroundings, where the trees on the banks stretch their green tentacles out above the water. The secluded pond is as flat as a mirror, and only rarely does a gust of cool wind passing through the woods wrinkle the white clouds and blue sky of the northern Taiwan summer day reflected on its surface. Soon the ripples subside, and it once again reflects the clear Formosan sky.

In a single pool, the life is boundless: At the edge of the pond, near the grass on the banks, grows a cluster of Taiwan yellow water lilies. Bursting forth from the green leaves, the beautiful flowers make the pool seem as if it is the abode of some nymph, far away from the world of men.

In this clear water, paradise fish surface from among the reeds, making bubble nests. The resplendent color of their scales and beautiful long fins can

黃金打造的金花，自那田田的葉間穿水而出，不只使這池子美麗脫俗，也使人相信它是沼澤仙子幽居的家園。

在這清淨的水中，蓋斑鬥魚時時從水草間浮起，吐著小泡沫營造牠的愛巢。那斑爛的體色以及飄飄多姿的長鰭，能使最沒有美感的人也著迷。

幾隻黃黑相間的斑蜓，在青葱的岸草間追逐嬉戲，細小的豆娘在臺灣萍蓬草的黃花間飛飛停停，偶爾一隻鮮麗的擬蛺蝶，低低掠過水面，飛越到對岸。這些小小的生物，使這一方小池子變得生動活潑起來。

我坐在岸樹的涼蔭下，陶醉於這美麗的風景裡，也爲這些臺灣特有種生物的存在而感動。在這小小的池子，我窺見了福爾摩沙原貌的一斑。

可是，這小池子在一位年僅十九歲、中興大學昆蟲系二年級的顏聖紘眼裡，它有更深層的意義。

他幾乎走遍了臺灣

八年前，他不過是一個十一歲的小朋友，一次野外的旅行，他爲水生植物的美妙迷住了，就此開始了研究臺灣的水生植物。

年復一年，他利用課餘，在臺灣的濕地沼塘搜尋、觀察，幾乎走遍了臺灣。他跟隨水生植物專家楊遠波博士跑了不少野外，而雙親也爲他從國外訂了大量的生物雜誌與參考書，現在，他幾乎是少數對臺灣濕地沼塘最了解的人。

在這嬌小、美麗的野池邊，他爲我介紹了許多臺灣特有種生物，以及名字充滿異國風味，甚至邊荒風情的植物，例如日本簣藻、印度水豬母乳、烏蘇里聚藻……。

這些植物都別具意義，它們大多是由候鳥從外國攜帶過來的，正如水韭、東亞黑三稜一樣，這些屬於溫熱帶的植物，竟然能在亞熱帶的臺灣落戶，讓國際的植物學家驚奇不已，其中不乏是植物地理學上的南限植物。

臺灣萍蓬草更別具深意，因爲萍蓬草屬於溫帶性水生植物，主要分布在北美洲及歐亞大陸北部，而臺灣萍蓬草可能是世界萍蓬草

全世界只有一屬一種的田葱。
The species Phylidrum lanuginosum is the only member of its class.

這是日本學者發現的植物，但顏聖紘發現它被誤認爲其他種植物，有必要重新命名。它的葉片有紫蘇的芳香。
This is a plant discovered by Japanese scholars, but Yen Sheng-hong realized that it was misidentified as another plant. It needs a new name.

entrance even the least aesthetically minded.

Some yellow and black spotted dragonflies chase each other along the green banks. Several tiny demoiselles rest and alight among the yellow flowers of Taiwan yellow water lilies. Occasionally, a beautiful and brightly colored Precis lightly skims over the water's surface across to the opposite bank. These little creatures make this small pond a site of great life and activity.

In the cool shade of the bank, I sit intoxicated with the beauty of the scenery and am greatly moved by these species unique to Taiwan. In this small pool, I get a glimpse of the original face of Formosa — "the beautiful isle."

But Yen Sheng-hong, a 19-year-old sophomore studying entomology at Chunghsin University, can see deeper levels of significance.

Combing the island: Out on a hike eight years ago as a child of eleven, he became entranced with the wonder of water plants and began researching the varieties on Taiwan.

Year after year, when not in school, he went all over the island to find marshes and swamps where he could collect and observe. He went on numerous treks in the wild with Dr. Yang Yuan-po, a water plant specialist, and his parents bought him reference books and gave him subscriptions to various foreign plant magazines. Now, he is one of only a few who understand much about Taiwan's marshes and swamps.

By the side of this small and beautiful pool, he introduced me to many plants unique to Taiwan as well as some exotically named species from abroad — even plants from remote wild regions, such as the *Blyxa japonica*, *Rotala indica*, *Myriophyllum ussuriense*. . . .

These plants are all interesting in their own ways. Usually found in the temperate zone, most of them were brought from abroad by migratory birds like the quillwort and the *Sparganium fallax*. Their taking root in the subtropical environment of Taiwan has greatly startled international botanists. Among them are plants found at their southern most geographical limit.

The Taiwan yellow water lily is especially noteworthy because it belongs to temperate zone family of water plants found mostly in North America and the northern parts of the Eurasian land mass. Taiwan's variety represents the southern limit of the family's range, a keepsake the island took from the glacial epoch. What makes it remarkable is that very few areas of the world have unique species of water plants. As a species unique to the island, the Taiwan yellow water lily has attracted the attention of international scholars.

Where are all the adults? Just as I was becoming intoxicated with the pride I felt for the glory of Formosa, loud noises jarred me from my state of enrapture. The noises came from behind the wind breaks on the opposite bank. Yen told me that the bulldozers were hurrying to finish work on a major construction project.

With helpless resignation, he said, "This pond is near its end. In a few days the bulldozers will fill in the pond, and the beauty of all of these plants will be no more."

The pond's tragic fate touched me deeply, bringing tears to my eyes. There are so many beautiful spots on this island that we destroy before we have a chance to discover them. For even so large a construction project, no investigation was made to determine how much special scenery would be destroyed or how many unique plant species would be uprooted.

I remember several years ago, when France was building a famous high-speed railway line, biologists discovered that one stretch of the line would obstruct a unique species of frog from returning to swamps where it mated and propagated. Eventually, for the sake of this group of frogs, the track was elevated at an expense of tens of millions of francs.

Such a concern for nature truly displays a nation's culture and civilization.

At the edge of the construction, Yen was constantly rescuing precious species of water plants about to be buried. As I saw him grabbing, pulling and taking photographs in the already half-buried wetlands, I couldn't help but be moved. Why was it that a kid like this was willing to fight a struggle he knew he would lose? Where were his elders?

Beautiful ponds — bulldozed and awaiting sale: Anticipated prosperity has wreaked even more destruction than actual construction projects. Looking forward to sales, citizens of the villages and townships along the edge of this construction project have drained or filled wetlands and ponds that had existed there for centuries. Even beautiful ponds like the one I have described cannot get listed

分布的南限和冰河期遺留在臺灣的孑遺植物。而更難得的是，在世界各地區裏很少有特有種水生植物的情形下，獨獨臺灣萍蓬草是臺灣特有種，而引起國際學者的重視。

大人哪裏去了？

正當我沉醉在福爾摩沙光榮的驕傲裏，一連串的巨響驚醒我的沉緬，那聲響來自對岸防風林後面，顏聖紘告訴我這是新建工程的推土機正在趕工。

他無奈地說：「這池子就要跟我們永別了，過幾天，推土機就會填平這個池子，所有的動植物與它的美麗都將就此消失……。」

一股悲傷自我心底湧起，直冲我的眼睛。這個島上有多少美麗的角落，在我們還沒有發現它就被我們摧毀了。這麼大的工程，竟然沒有調查過它將破壞多少有特色的風景地形，毀滅多少特有種或珍稀生物。

記得好幾年前，法國在建造一條著名的高速鐵路時，生物學家發現，有一段高速鐵路會妨礙一羣特有種青蛙回到沼澤去交尾繁殖。最後鐵路爲了這一羣小青蛙而改爲高架，工程費也多了好幾千萬法郎。

這種尊重自然的精神，眞正表現出了一個國家的文化、文明。

美麗沼塘，塡平待沽

顏聖紘在工程沿線上不斷地搶救一些即將被掩埋的珍稀水生植物。我看他在一些已經被半埋浸的濕地裏，撈、拔、拍照，使我感動不已。爲甚麼是一個這樣的孩子知其不可爲而爲，大人哪裏去了呢？

除了工程所破壞的濕地外，因它所帶動繁榮地方的期待，造成的影響更大。沿線的鄉鎮居民，紛紛把已有上百年歷史的濕地、池塘抽乾或塡平待售。這麼豐富美麗的沼塘，竟然沒有人願意將它列爲保護區。

在楊梅一處行將乾枯的沼塘裏，我拍到了有名的田葱、水杉菜、石龍尾、微果草，以及尙待命名、我戲稱它爲「顏氏紫蘇菜」的水生植物。它的豐富可見一斑。

看著日漸乾竭的池水，以及逼近的工廠，

小雨蛙的鳴叫，使濕地熱鬧非凡。

The call of the little Hyla arborea rings through the wetlands.

我知道它已來日無多，就正如它方圓一公里以內已經被塡平的十五個沼塘一樣。

臺灣雖然是蕞爾小島，但水生植物卻多達五十餘科二百多種，目前有幾十多種面臨絕滅的危機。以臺灣原產的五種睡蓮科植物爲例，其中芡（Euryale ferox）、 午子蓮（Nymphaea tetragona）、 紅花睡蓮(Nymphaea lotus var. pubescens)早已絕種，而蓴菜(Brasenia schreberi）亦已瀕臨滅種，只有臺灣萍蓬草（Nuphar shimadai）在桃園的一些老池塘裏一息尙存。

至於難得一見的長葉毛菁菜，踏破鐵鞋的顏聖紘只在工程塡土邊緣的幾寸濕地上，找到一株已枯乾的「屍體」。

水盡淚乾

濕地沼塘的生態遠比森林脆弱，但也比森林容易保護；只是它常被忽略，甚至輕視。

墾丁國家公園的南仁山沼澤濕地，原本是一大片各種水草叢生的美麗濕地，過去，顏聖紘幾乎每年都到這裏探望它們。

但自從那戶滯居沼地旁邊的人家建堤攔水，使湖水變深，以便行使他的「遊筏」之後，濕地消失了，無數的水生植物與候鳥也逐年減少了。而這卻是唯一劃入國家公園的濕地。

此刻我在推土機的嚨嚨聲中，對這美麗的野池做最後的巡禮。

望著這些即將被毀滅的池塘與形形色色的各種生物，我不禁要問：「何時我們才會珍惜自己擁有的！何時人類才會學到尊重大自然、尊重生命？如果我們生活的環境變得醜陋不堪，我們的水不再澄淸、天空不再湛藍；空氣充滿惡臭、耳中盡是噪音，我不知道文明的定義究竟爲何？……」

（原載光華八十一年十月號）

as protected areas.

In a pond which is almost dry in the vicinity of Yangmei, I photographed the well-known *Philydrum lanuginosum*, *Rotala hippuris*, *Limnophila trichophylla kamarou*, *Microcaipaea minima*, and a plant which I playfully called ''Yen water basil.'' One can see how abundant the water plants in this pond are.

Looking day by day at the ever-dryer pond and the ever-closer factories, I knew that the pond's days were numbered — just like the 15 ponds within a square kilometer which had already been filled in.

While Taiwan is a small island, there are some 50 families and over 200 species of water plants. Currently, more than 80 species are endangered. To take the five native species of water lily as an example, *Euryale ferox*, *Nymphaea tetragona*, and *Nymphaea lotus var phbescens* have long been extinct, the *Brasenia schreberi* is also endangered. Only the Taiwan yellow water lily is still hanging on in a few old ponds in Taoyuan.

As for the hard-to-find drosera indica, in the few inches of wetlands still remaining at the edge of the construction, Yen found only a few dry ''corpses.''

Bone dry: Although the ecology of wetlands and pond is more fragile than the forests ecology to begin with, it is also easier to protect. It's just that it is usually overlooked or even regarded as unimportant.

The Nanjen Mountain swamps and wetlands of Kenting National Park were originally a large expanse of beautiful wetlands where various water plants grew in abundance. In the past, Yen could find some new species brought to Taiwan by Anatidae ducks virtually every year. But ever since work began on a nearby major construction project, wetlands have been lost as countless water plants have disappeared into the mouths of the grass carp and the number of migratory birds have gradually declined as well. And this tale of demise took place in the only wetlands found in a protected area.

At this moment, amid the sound of bulldozers, I am making my last pilgrimage to this pond in the wild. Looking at the soon-to-be-destroyed pool and the colorful abundance of life in and around it, I can't help but ask, ''When will we treasure what we have? When will humanity learn to respect nature and life? When our environment becomes an ugly abomination, when our water is clear no longer, our sky never blue again, when our air is permeated with smog and our ears are ringing from the ever-present clamor, how will we go about defining civilization?''

(photos & text by Hsu Jen-hsiu/ tr. by Jonathan Barnard/ first published in October 1992)

土動物、新生機

New Life for Native Farm Animals

文•蔡文婷　圖•張良綱

藍瑞斯豬、努比亞羊、白羅曼鵝、愛拔益加鷄、聖達牛……，如果我們進一步追問平日所吃的家畜禽，會發覺全是些洋名洋姓、混血改良的「西式大餐」。

曾幾何時，土裡土氣的黃牛、黑豬、土山羊漸漸被淘汰或雜交，現在的小孩可以說是「沒吃過土豬肉，也沒見過土豬走路。」

為了搶救這些和我們同在一塊土地生活的土畜禽，也留下牠們身上可貴的「土本領」基因，行政院農業委員會在民國七十六年展開農林漁牧保種計畫。經過三年的努力，總算繁殖了相當的數量，保住這些土畜禽的一縷香煙。

Duroc pigs, Nubian goats, White Roman ducks, Santa Gertrudis cattle . . . anyone who tries to find out what kind of domesticated animals we eat every day will discover they're almost all improved, crossbred varieties for "Western cuisine." For some time now native Chinese cattle, black pigs and Taiwan goats have steadily been displaced from the marketplace or crossbred with other varieties. Children today may well have "never eaten native pork or seen a native pig walk down the street."

To save these native animals, which have lived on the same piece of earth with us for so many years, and to preserve their genes with all their native characteristics, the Council of Agriculture (COA) in 1987 initiated an agriculture, forestry, fishing, and livestock breed preservation program. After three years of effort, a considerable quantity of native species have been preserved and propagated, thereby maintaining the continuity of their genetic lines.

種畜繁殖場內，一群土山羊從牧場要回家，圖右的母羊正懷了一對雙胞胎。

A herd of native Taiwan goats at the propagation station are about to return from the pasture. The female at right is carrying twins.

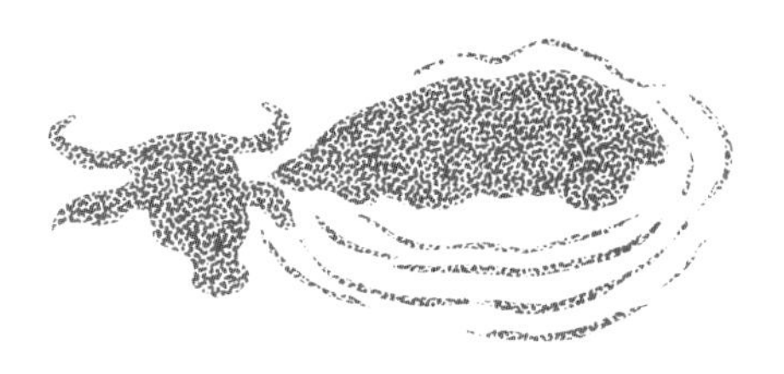

「我很醜，可是我很會生」，這是桃園豬最拿手的招牌曲。

"I may not look like a movie star but I've got what it takes (to have a lot of babies)" is the signature song of the Taoyuan pig.

說起現在流行的畜禽，眞是非常「優秀」，像是——

現居世界肉鷄銷售量排行第一的白肉鷄，每吃兩公斤飼料就能長一公斤肉；成長速率也快，平均五十天就可以上市，不像土鷄需要一百天才能上市。

專門生蛋的蛋鷄年產二百八十個蛋，比土鷄每年多出一百八十個以上，每個蛋重量還多上十公克左右。

餐桌上的常客——雜交洋種豬，成長速度不但是土種桃園豬的一倍，而且皮薄骨頭小，噸位相同的，身上精肉也多出百分之十。

奶牛就更明顯了，水牛、黃牛的奶量不過剛好餵飽自己的一個寶寶，哪還有多餘牛奶供應人類。

家畜禽一直是人類動物蛋白的主要來源，人類飼養畜禽的唯一目的是「裹腹」，因此在追求商業效益原則下，經雜交改良的新品種，成爲生長效率較高、繁殖能力較強的「優等生」，日漸得寵，廣受歡迎。相形之下，未經雜交的純種動物，也就是一般所稱的土種，則日漸散失，瀕臨絕滅。

土動物「鹹魚翻身」

從野生動物變成家畜禽那天開始，人類就不斷以「好還要更好」來改良牠們。十八世紀中葉產業革命後，畜禽商品化，專業養殖開始取代農家飼養，各種檢定制度提供育種學家豐富資料，長得快、吃得少、生得多的各式合成品種不斷上市。

到了廿世紀，所有畜禽的經濟效益幾乎都提高了一倍。改良的脚步卻無法再邁進了，西元一九四〇年加拿大萊康貝豬改良後，豬的生產就再也沒有任何突破，跟著鷄也走到同樣的地步。

一九六〇年，當世界育種專家正苦於無法再提升豬的繁殖力時，一羣法國學者卻在土種的中國梅山豬身上，找到一胎十八至廿頭的高產子率。土種於是鹹魚翻身，引起世界矚目，也掀起了各國對一些固有品種的研究高潮。

不同品系的同種動物交配，原也是自然演化的方式之一，但在人類介入以前，一個品種能不能留存下來，取決於牠適應環境的能力。也就是能不能通過大自然「物競天擇，適者生存」的考驗；晚近的人工配種則將遊戲規則改變，單憑經濟價值高低；大自然的考驗，人類已替牠們排除。

天生我材必有用

俗話說得好，「天生我材必有用」，土種動物的整體表現——經濟價值——雖不及改

Some of the most popular breeds of livestock and poultry these days are quite something. For instance:

White broilers, the best-selling type of chicken, grow one kilo in weight for each two kilos of feed consumed. At this rate of growth, they are ready for the market in just 50 days, unlike local varieties which require 100 days before they're prepped.

White leghorn pullets, which are bred specially to lay eggs, can produce 280 a year. This is at least 180 more than native chickens, and the eggs are each about 10 grams heavier to boot.

Duroc pigs, a common guest at the dinner table, not only grow twice as fast as the native Taoyuan pig, they have thinner skin and smaller bones, yet a comparable body size, so that they have ten percent more prime meat.

The discrepancy in cattle is even clearer. Water buffaloes and Chinese cattle produce only enough milk just to feed one calf, so don't even think about having extra for people.

Domesticated livestock and poultry have always been a major source of animal protein, with the sole point of raising them being to ''fill our bellies.'' Given the demand for ever-greater commercial efficiency, new and improved varieties with rapid growth, greater fecundity, or better taste are constantly being raised and are very popular. In contrast, pure strains that haven't been interbred — what are called ''native'' types — have steadily declined in popularity or even face extinction.

Native breeds find their niche: Ever since the day wild animals were first domesticated, people have constantly tried to ''take what's good and make it better'' by improving them. After breeding and agriculture was commercialized during the industrial revolution in the mid-18th century, specialized breeding began to replace family farm animal husbandry. With all manner of measurement systems providing specialists with comprehensive data, many new varieties were developed that grew faster, ate less, and produced more offspring then their forebears.

By the 20th century, the economic efficiency of domestic fowl and livestock was nearly doubled. But then the pace of improvements was stalled. There have been no further breakthroughs in swine production since 1940, following the chicken along a similar dead-end path.

In 1960, as animal husbandry specialists labored futilely in their inability to raise hog reproductive ability, a team of French scholars discovered in the Chinese Meishan pig a rate of reproduction of 18-20 head. Native breeds thus turned their fate around, drawing the attention of the world, and generating a wave of studies of some special varieties from various countries.

Interbreeding has always been one of the methods of natural evolution. But before man's intercession, whether or not a strain could survive depended upon its ability to adapt to its environment. This is to say, it depended on whether or not it was able to pass the rigors of ''natural selection and survival of the fittest.'' With artificial breeding, the rules of the game have been changed, with the sole criterion for survival now economic value. Man has already substituted for nature's tests.

Heaven must have put me here for something: The saying goes, ''Heaven must have put me here for some purpose.'' While native breeds may not be able to compare with improved strains in terms of economic performance, they still possess one or two special characteristics, such as a high tolerance for sun and heat, powerful resistance to disease, and the ability to eat coarse fodder, that have enabled them to pass nature's tests and survive.

''These innate endowments were given to them by the Creator and are gifts to mankind as well. If we don't preserve them today, they could disappear forever. If by chance the improved strains should run up against an especially virulent illness, or die off because of a sudden change in the environment, people would only be able to regret not having kept around the tougher native breeds,'' says Dr. Tai Chien, director general of the Taiwan Livestock Research Institute (TLRI).

Improved breeds are like delicate hothouse flowers, protected under a myriad of safeguards, that produce beautiful blossoms. Native breeds are like wildflowers in the field. They may not be much to look at, but they can survive the frost and the wind.

良種完美，但生存在天地間，身上總有一、二個特點，像是不怕曬、抗病力強、耐粗食等，所以能通過大自然的千錘百鍊。

「這些土本領是造物主贈予牠們，也贈予人類的禮物。今天不保存，便可能永遠失去。萬一有一天，改良種遭惡疾纏身、或因生存環境驟然改變而消失，又沒有生命力頑強的土畜禽爲後盾，人類只怕要後悔莫及了，」臺灣省畜產試驗所所長戴謙表示。

改良種就像溫室裡的花朵，在萬般保護下，開著最美麗的花朵，土種動物則像野地裡的小野花，雖不起眼，卻能在寒風霜雪中，存活下來。

像現在的改良種豬，吃慣玉米、大豆等合

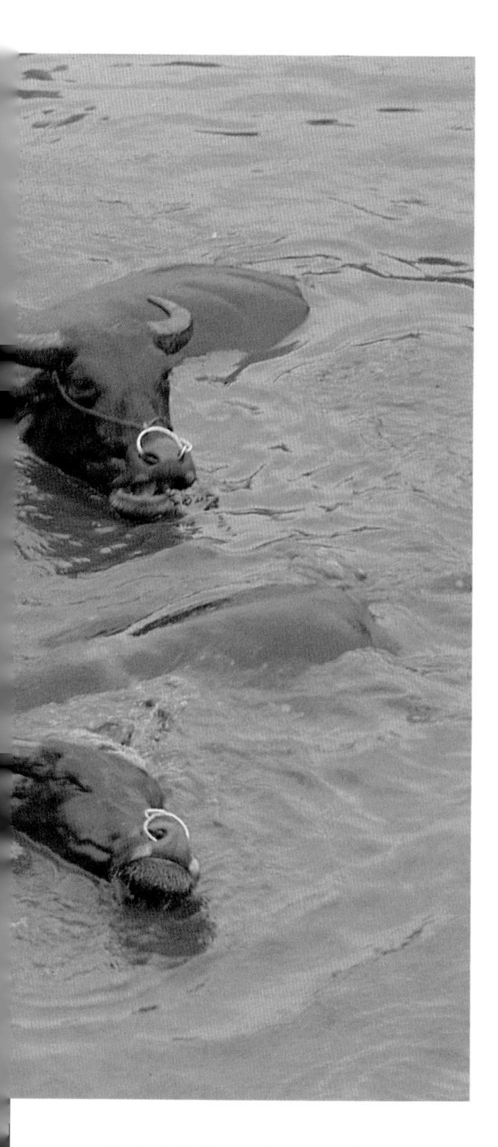

洗個泥水澡，退暑氣又驅蚊蟲，眞是不亦快哉！

A mud bath beats the heat and drives away insects. What a feeling!

Today's improved pig, for example, is accustomed to eating a mixture of Western feeds such as maize and beans. If an emergency like a war or an oil crisis ever broke out, and feed factories had to shut down, improved breeds would find it hard to survive. The Taoyuan pig, on the other hand, with just waste water, sweet potato leaves, and grass roots, is as happy as, well, a pig in slop. And the Lanyu small-eared pig can even live off human excrement, according to Dr. Sung Yung-yi, a professor of animal husbandry at National Taiwan University who has done quite a lot of research on pigs.

Improved breeds are not only coddled and spoiled, they've even lost some of their innate abilities under mankind's remolding. For example, in order to prevent Leghorn hens — specially bred for egg production — from stopping to incubate previous eggs, man has chosen to select females who seemed disinclined to incubate as the breeders. Today, these Leghorns only know how to lay eggs, but the "natural instinct" of brooding has been taken over by incubators. That means that if the electricity ever went out, the chicks would probably die in their shells.

The great preservation plan: In 1987, in view of the importance of preserving genetic resources, the Council of Agriculture (COA) initiated a multi-faceted agriculture, forestry, fishery, and livestock breed-preservation program under the direction of the TLRI, located in Hsinhua City in Tainan County. Under this flag, five breeding and propagation stations were set up in Ilan, Hualien, Taitung, and Changhua counties and in Hengchun Township in Pingtung County, run in conjunction with the Department of Animal Husbandry at National Taiwan University.

Just which breeds are being preserved?

From the standpoint of utility, the more strains preserved the better, because the greater the variety of types, the easier it is to cull the outstanding characteristics of each and cultivate a new breed in corporating the best features of all. However, each country in the world still tends to concentrate on

成的「西餐」，一但發生緊急狀況，例如戰爭、石油危機，飼料工廠停擺，改良種就難以活命。而土種桃園豬，只要餿水、地瓜藤、草根照樣甘之如飴。蘭嶼系的小耳迷你豬甚至只要吃人的糞便即能生存——對豬頗有研究的臺大畜牧系教授宋永義舉例說明。

改良種不僅嬌生慣養，甚至在人類的「改造」下，連本能也消失。例如產蛋用的來亨鷄，爲了不要牠因孵蛋而停止產蛋，人們便留下不愛孵蛋的母鷄爲種鷄，時至今日，來亨鷄只知下蛋，孵蛋的「天職」則交由「孵蛋器」包辦。要是遇上電力斷絕，只怕小鷄全要悶死蛋中。

保種大計

民國七十六年，行政院農業委員會有鑒於「種原」的重要，於是展開農林漁牧多項「保種計畫」，畜禽保種則由位於臺南縣新化市的臺灣省畜產試驗所爲總指揮，旗下有宜蘭、花蓮、臺東、恒春、彰化五個分所或繁殖場，及臺大畜牧系共襄盛擧。

保「種」，到底保的是哪些種？

站在利用觀點上，保種原是保越多的種越好，庫存品種越豐，自不同種源取得優良特性就越便利，得以培育「集百家精華」於一身的新品種。不過世界各國仍偏向保護自己固有的土種畜禽。一來是財力有限，二來是原有地較適合品種保存。因此現階段農委會以「本地固有種」、「臺灣改良特有種」爲保留對象。

像是本地水牛、山羊、土鷄、桃園豬、中國鵝、褐色菜鴨這些畜禽，由先民拓荒時自大陸帶來，雖非本地原有，卻能入境隨俗、落地生根者，稱爲本地固有種或土種，以相對於外來種及改良種。

此外，一些經我們自己以本地種再選拔出的新品種，如由褐色菜鴨選拔出的白色菜鴨；或是由土種小耳豬和外來藍瑞斯種交配的李宋豬。這兩種品種在選育出來以後，以多代繁衍觀察，令特徵、性狀穩定，經國際認可「定名」後，才取得文憑，成爲被珍藏的一員。

比較特殊的是黑色番鴨。「番鴨」顧名思義來自番地，是十六世紀荷蘭人帶進來的，本是黑白皆有，然而在西方人嗜白，中國人認爲黑色動物比較「補」的文化差異下，形成了歐洲只有白番鴨，而臺灣保有黑番鴨。牠的存在可是深具「中國特色」。

土法相種

要保種，先得要知道種長得什麼模樣，方能引種。

農村社會，對畜禽沒有建立太多資料，更不用說是遺傳基因、生理血型檢定；土種的模樣也只能從老一輩農民的形容中，勾勒外形、特徵，或待各地農會四處搜集來時，請老農夫親去「相相看」。

「相親」難易程度不一。其中水牛因染色體對數與其它牛種不同，沒有雜交，不需辨識。本地山羊則對伴侶「來者不拒」，各種「混血兒」也特別多，因此有一點小白毛的、頸下有肉垂的、大耳朶的都可能是別「羊」的孩子。

「有時候，遠遠看見一大羣山羊好不高興，仔細一看卻沒一隻合格，」花蓮種畜繁殖場場長施義章無奈地表示，引種工作是「寧可殺錯一百，也不能錯判一個」。

踏遍千山萬水

爲了尋找土種畜禽，各繁殖所眞是煞費苦心，跋山涉水往較封閉的地方去。因爲越是封閉的地方，越可能存有土種畜禽的行跡。

負責土鷄保種的新化總所助理鍾秀枝就跑遍了臺中、嘉義、臺南、臺東各縣鄉村，還深入秀姑巒溪上游的山區。

恆春分所的本地山羊則是渡海到金門找著的。因爲金門是離島、又是戰地，先天便是個閉絕的環境。加上爲了戰地安全，金門早期推行植樹運動，爲了不讓山羊吃掉嫩芽，一度禁止養羊，在居民只能偷偷飼養本地山羊的情況下，杜絕了和外來羊的雜交機會，留下這一批土山羊。

「不過，因農家賣羊都是挑大的賣，剩下的小羊反成爲種羊，加上近親交配，使得繁

conserving its own native breeds. For one thing, finances are limited, and for another the original area is best suited to preserving types native to it. As a result, the COA at this stage has designated "endemic breeds" and "Taiwan improved breeds" as objects for preservation during the current stage of the program.

Take for example many types, such as water buffaloes, goats, chickens, Taoyuan pigs, Chinese geese, and brown tsaiya ducks, which were brought to Taiwan by early pioneers from the mainland. Although they are not actually native to Taiwan, they have been able to adapt to the surroundings and set down roots here. Thus they are considered native breeds, in contrast to imported or improved strains.

In addition, there are new strains that have been developed from native breeds, such as white tsaiya ducks bred from the brown variety, or Lee-Sung pigs which are a cross-breed of Lanyu small-ear pigs and a foreign variety. Only after been selected out, after generations of breeding and observation allowed their special features to become stable elements, and after international recognition, did these two types "get their certificates" and become treasured members of the program.

The black "barbarian duck" is a very special case. The name "barbarian duck," as the breed is known in Chinese, seems to have a certain alien ring to it. Both white and black types were brought to Taiwan by the Dutch in the 16th century. Given the cultural difference of the Westerners preferring white while the Chinese found black to be more "tonic," it evolved that Europe was left only with the white version while the black was preserved in Taiwan.

Faces in the crowd: If you want to preserve a breed, you have to know what the real thing looks like before you can start collecting them.

Farmers in traditional society didn't systematically keep a whole lot of information on domesticated creatures, much less their genes or biological blood types. The distinguishing characteristics of native breeds can only be roughly sketched out from the descriptions of old-timer farmers; otherwise the local farmers' associations have to round up as many types as they can and then ask the older farmers to "pick them out of a lineup."

The degree of difficulty in identification varies. There is no need for on-site recognition for water buffaloes, for example, because they have a different number of chromosome pairs than other cattle so they don't cross-breed. Native Taiwan goats, on the other hand, are less discriminating in their tastes, and there is an especially large number of mixed varieties. Thus some small lamb with a little white on it, or with a little extra meat at the neck, or with big ears, could very well be the child of some "casual outside relationship."

"Sometimes you're really excited when you see a flock of goats from a distance, but when you get close you find that not a single one is up to standard," says Shih I-chang, head of the Hualien Livestock Breeding Station, with a helpless expression. In gathering samples, he says that "it's better to reject 100 by mistake than to misjudge one."

Hitting the dusty trail: In order to find the native varieties, the workers at each breeding station really rack their brains and bodies, trekking over hill and dale to even the most remote locations. This is because the more isolated a location is, the easier it is to find pure traces of native breeds.

Chung Hsiu-chih, an assistant at the Hsinhua station in charge of the preservation of native chickens, has been through every rural township and village in Taichung, Chiayi, Tainan, and Taitung counties, and even into the mountainous areas of the upstream portion of the Hsiukuluan River.

The local mountain goat at the Hengchun Station has actually come "across the sea" from Kinmen Island. Because Kinmen is a lonely island and is also a restricted war zone, it is a naturally isolated environment. Add to this that Kinmen initiated a tree-planting movement in the early days for security reasons and, in order to prevent goats from eating the buds, banned domesticated billies. With the local people only able to covertly nurture the local mountain goats, with all chances for interbreeding with outsider sheep preempted, a group of pure native mountain goats have been left behind.

改良鴨在人為刻意改變下，早失去孵蛋本能，番鴨就因「會孵蛋」而被列入保種群英譜。

Improved duck strains have been deliberately deprived of their ability to brood. Native ducks have made the preservation honor roll because they know how.

殖力漸減，在如此反淘汰的飼養方式下，這些土山羊體形已較一般羊兒小，」找到這批山羊的恆春分所副研究員黃政齊有點惋惜地說。

種源搜尋不易，有的甚至歷經三年，連「類似」的都找不到，如美濃豬、頂雙溪豬。也有的看來很像，但血統「純度」可疑，這時就須以親兄妹交配、淘汰模樣差太多的下一代方式，進行「純化」。

分寸難捏拿

保種，首要目的是保存此種動物的特性，所以，尋到種之後另一個重點工作是讓牠們依照天性，自然生長。最理想的飼養方式為維持大量族羣，給予自然環境，不刻意選拔，讓牠們自由交配。

但實際情況往往不那麼理想，像鷄族羣多、辨認不易，保種前必須先進行「純化」。如此不可避免地要以人工過濾；而為了建立保種族羣的「祖譜」（系譜），確知每個蛋的親生父母，只得以樊籠將對對鷄羣隔開。

「如果不先確定牠們是不是土、純，只是保了一窩身分可疑的鷄，人類還是無法了解牠，就更不可能去運用牠身上的遺傳基因了。」負責編輯保種手册的畜試所研究員吳明哲指出。

"However, because farm families always choose the biggest rams when they sell, the remaining small rams have become the breeders. On top of this add incestuous breeding and the result is a decline in fecundity. Given these circumstances, these native mountain goats have become smaller than ordinary ones," says Huang Cheng-chi, the associate researcher at the Hengchun Station who discovered this flock, with a tinge of regret in his voice.

It's not easy to find "genetic originals." Sometimes staff can't even find something approximate after three years of searching, as with the Mei-nung pig or the Ting-shuang River pig. Sometimes the animals look all right, but the purity of the blood line is questionable. In this case they can only use the technique of interbreeding siblings and weeding out the offspring that are too far off the mark to undertake "purification."

Fine-line judgements: The first priority of breed-preservation is to save the special characteristics of the given breed. Therefore, once a breed in obtained, another important task is to let them grow and reproduce naturally according to their innate natures. The ideal method is to maintain a large flock or herd in natural surroundings, and let them interbreed without deliberately arranging things for them.

But things aren't always so ideal in practice. For example, there are many varieties of chickens, and differentiation is difficult. It is necessary to do "purification" before preservation. For this human screening is unavoidable. And in order to maintain the "pedigree" of a species, and to know the parents of each egg, the only way to be sure is to separate each pair off in cages.

"If you don't first certify that they are indeed native breeds and thoroughbred, then you're just collecting a henhouse full of chickens of doubtful lineage. People won't be any closer to understanding them, much less putting to use their unique genes," says Dr. Wu Ming-che, a research physiologist and geneticist at TLRI in charge of putting together the breed preservation handbook.

This is the difference between preserving domesticated and wild strains. The main point of preserving domesticated creatures is still "resource utility," so not only do you want to preserve the animals, it is even more important to utilize the special genetic characteristics they have, and to

stabilize them. There is no other way but to undertake "artificial" controls.

Moreover, a natural environment by no means need be a primitive environment. We call them "domesticated" precisely because they have lived for so long together with humans. "Simulating the surroundings in which they were once raised is not very possible, practically speaking," says Dr. Wu. "In fact, sometimes those surroundings can actually blur their special characteristics and make them less like the way they should be."

The Taoyuan pig has just this problem. Like the Chinese Mei-shan pig, the Taoyuan pig has the special characteristic of multiple breeding. According to some old-timers, a Taoyuan pig can give birth to 15 or 16 head at a time. But because most farmers lack sufficient fodder, or the fodder is inadequately fibrous, hog litters average only 11-12 head in practice. Faced with this problem, those trying to preserve the breed features have to raise the amount of fiber in the feed, and propagate from the relatively more fertile females. They can thus restore their original capacity because of man's intercession.

爲了建立系譜、純化血統，工作人員只得以籠飼及人工授精，來確定誰是小鷄的親生父母。

To establish correct genealogies and purify bloodlines, researchers have to keep hens in separate cages and artificially inseminate them so they know who the real parents of the chicks are.

這也是畜禽保種和野生動物保育不一樣的地方，畜禽保種主要目的仍在「資源利用」，因此不只要保存畜禽，更要找出遺傳特性的基因，加以固定，那就不得不有「人爲」的操縱了。

而自然的環境，並非原始環境，我們說「家」畜「家」禽，就因爲牠們一直和人們生活在一起，「而模擬早期的飼養環境，不僅實際上不太可能，有時這些環境反而使牠的特性逐漸模糊，越來越不像牠們自己，」吳明哲說。

桃園豬就有這樣的問題。桃園豬和中國梅山豬一樣有多產特性，據老一輩說，一胎總可生個十五、六頭，然而一般民間飼養，由於種豬有限、或食物纖維素不足等因素，每胎平均只有十一、二頭。針對這個問題，保種人員便需在食物中加重纖維素、挑選其中較具「興家旺子」潛力的母豬來繁殖。透過人爲方式，使牠們恢復原有的特性。

但人爲的力量可以介入多深呢？如何確定那只是協助牠們恢復本性，而非另一次改造？這其間火候的控制得非常嚴謹，所以現場人員必須定時觀察牠們、視實況而修正脚步，每種畜禽的方式又不盡相同，眞可說是「知易行難」了。

鷄飛狗跳、人仰馬翻

平時照顧這些「活標本」的工作人員，有

But just how far should humans interfere? How can we be sure that we're simply helping them restore innate traits, and not remolding them? This requires meticulous control of the process, so field staff must make regular observations and adapt the process according to the current situation. The methods for each breed are also not completely identical, so you could say its a case of "easy to say, hard to do."

All things lazy and dangerous: Those who are charged with the continuous care of these "living samples" have a lot of basic homework to keep up with. For example, they have to weigh the animals regularly, take their measurements, and figure out their rate of growth. When recorded behavior begins, maturity and fecundity are determined by the number of head they (re)produce. Ultrasound testing is also done to see whether their skins are thick or thin; there are also blood tests and stool samples. These tasks might seem easy at first glance, but things can get pretty ugly when you have to perform them on beasts who have no regard for others.

Just to get the weight of a Taoyuan pig takes three people -- one to grab him, one to handle the scale, and one to record the data. "When you run up against their basic laziness, no amount of threatening or inducements can move them, and then you have to get eight people to carry them to get the show on the road," says Cheng Yu-hsin, director of the TLRI farm.

Things have gotten even hairier at the Hualien propagation station. Once when station chief Shih Yi-chang was doing a prenatal checkup on a cow, he put a shackle on the back legs of the animal. Who knows how, but that cow got out of the shackles and gave a "mule kick." Lucky for Shih, he had jumped back in surprise, and all he was left with was a souvenir hoofprint on his stomach. Thinking back on that incident, Shih says straight out, "that was really scary." If he really had taken the full force of the kick, he wouldn't be around to talk about it.

Breeds of preservation: When you try to look after unruly, jumpy creatures, lots of nasty things can happen. But animals can only produce one generation at a time, so not only is taking care of them not easy, management costs are high. Moreover they only have a fixed life span, and are not like easy-to-care-for plants where you just need to freeze one seed and you can keep it for a decade or so. If "seeds" could be similarly preserved for animals, that would save a great deal of trouble.

"Right now up-to-date methods for freezing sperm or embroyos are the cheapest and easiest ways of preserving genetic resources," points out Liu Jui-chen, an associate researcher at TLRI.

For example, in just one cc of chicken sperm, there are five billion individual sperm. This sperm can be preserved for ten or twelve years in liquid nitrogen at 196 degrees below zero.

But retaining frozen sperm can only be considered preserving half a breed. If in the future there are no pure females to be inseminated, then everything will have been for nought. Thus only by preserving an intact, already fertilized egg can one say one has preserved an actual life. When needed, this can be implanted in a female to produce an animal genetically identical to the originals. If by chance there were some terrible disaster so that a strain was totally eradicated, a breed could be extended through "surrogate motherhood."

Local specialists are operating at the same level of technology in this area as the rest of the world. Right now there is basically no problem with sperm preservation, and embryo preservation has been achieved with the Chinese cattle, mountain goat, and pig, giving the maintenance of native livestock yet another form of guarantee.

I'm not very pretty, but I may come in handy some day: Times change, and so do the needs of a society. Genetic resources today have even more diversified roles to play. With progress in genetic engineering, gene applications have become even broader. Some things which in the past were seen as dead ends have become attractions today.

For example, the Taoyuan pig has creases like a Chinese Shar-pei on its body. These make the skin thicker, meaning less meat inside; they are also tougher to slaughter. Thus their market value is

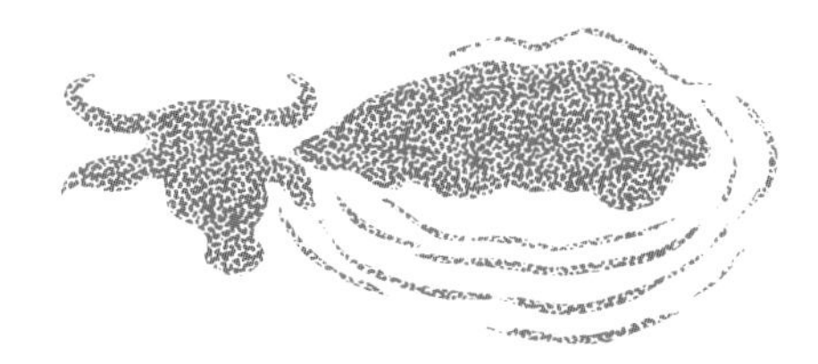

不少的基本功課要做，譬如定時爲牠們測量體重、量三圍、了解牠的生長速度；記錄性行爲開始時間、生產頭數以推定其成熟年齡、繁殖能力，及以超音波測量他們皮厚不厚（背脂厚度）、抽血、檢查糞便等，這些工作看似輕鬆，但是放在「六親不認」的畜牲身上，就有一堆糗事了。

單是爲桃園豬量體重，一次就得三員人手伺候——一個抓、一個秤、一個記錄。「碰上豬的懶性一發，怎麼威脅利誘都賴著不走，也只有請來八人大轎（拖車），擡牠上路了，」畜試所畜牧場主任鄭裕信表示。

而花蓮種畜繁殖場發生的事就更驚險了，場長施義章有一回在爲母牛產檢時，明明以鐵鍊鎖住了母牛後腿，不知怎麼的，那母牛硬是掙脫了鐵鍊，還來一記「牛後踢」，所幸施義章機警地往後跳，只在肚子上留下一個大牛蹄印，回想起那一幕，施義章直說「好險！」若眞紮實受牠一腿，只怕要一命歸陰了。

保種之種

活蹦亂跳的動物照顧起來，總是有著許多驚險鏡頭。而且動物只能一代傳一代，不但照顧不易、管理費用高，又有一定壽命，不若植物保護方便，只要冷凍一顆種子，便可保存十年左右不變。要是動物能像植物那樣保留「種子」，就可省去不少麻煩事了。

「現在最新的畜禽精子、胚胎冷凍保存法，便是既經濟又實惠的保存種源方式。」畜試所副研究員劉瑞珍指出。

像是一西西的鷄精液中，便有五十億隻精蟲，將精液放在攝氏負一百九十六度的液態氮之中，就可保存十一、二年。

保存冷凍精液只算保了一半的種，萬一以後沒有純種的母動物可以交配，就功虧一簣了。所以要保留已受孕的胚胎，才算是保存了一個完全的生命，需要時取出移入母體，即可生出和以前相同基因的動物來。萬一遭不可抗拒的災變，該品種完全消失，也還能「借腹生子」，延續種源。

國內專家在這方面的技術與世界同步，目前精子保存大多沒問題，而黃牛、山羊、豬則已達胚胎保存的境界，使得畜禽延育有更深一層的保障。

我很醜，可是我有新用

時代不同、社會需求不同，種源在今天也扮演更多元化的角色。隨著遺傳工程的精進，基因的用途更廣，有些以往被視爲致命點的，現在往往反成爲吸引力。

例如，桃園豬身上那一身像沙皮狗的皺摺，使得牠皮厚肉少、屠宰不易，售價只有市面上商用肉豬的六分之一。而腦筋轉個彎，想想，桃園豬小時候也是細皮嫩肉的，長大後才出現一身皺紋。「如果我們找出桃園豬身上引起皺皮的基因加以控制，那桃園豬就不皺皮了，更重要的是，應用到人身上，太太、小姐也可永保青春，不需整容拉皮了，」宋永義打趣地說，那將是愛美女性的一大福音。

而黑番鴨不愛戲水，屬於半陸性水鴨，如果能使番鴨只在陸上時間才排洩；那便可減少養鴨事業對河川的污染。

「保種原本只是保留收藏動物特性，以備不時之需的目的，在基因工程的營造下，保種在今日也有了更積極的作用，」所長戴謙表示。

世世代代傳香火

說起保種，保的是動物的種，爲得卻是人類的享用，似乎都盡是些「利用」的目的。

其實保種一個最大的意義是「承先啓後」，「將祖先交給我們的資源傳給下一代，可以說是項『歷史使命』，」農委會畜牧處處長池雙慶語重心長地說。

「如果你看過桃園豬，你就會知道西遊記裡的豬八戒是怎麼來的，」戴謙表示。那大耳、垂腹、王字臉、大鼻孔全是中國豬的特點。如果未來的孩子沒有機會一睹桃園豬的眞面目，只以小耳蜷尾的洋種豬爲豬，大概要以爲古人的想像力眞是天馬行空，創造出豬八戒這麼一號怪物來了。 S

（原載光華七十九年十一月號）

only about one-sixth that of the commercial pig. But when you think of it from another angle, the Taoyuan pig is also thin-skinned and delicate when young, and only shows a body of wrinkles after growing to adulthood. ''If we can isolate and control the gene in the pig which determines the folds, then the Taoyuan pig will no longer have creased skin. Even more importantly, if this can be adapted for human application, people might be able to maintain their youthful skin without resorting to plastic surgery,'' muses Sung Yung-yi. That's good news for beauty-worshipping women.

Or take the black ''barbarian'' duck, which doesn't like the water and is ''semi-aquatic.'' If it can be made that the ducks will only defecate when on land, this can greatly reduce the water pollution from the duck-raising industry.

''Breed-preservation originally was just storing up special characteristics in order to achieve occasional objectives. Given developments in genetic engineering, breed preservation has an even more positive function today,'' says institute head Tai Chien.

Passing the torch from generation to generation: When you talk about breed preservation, what's being preserved are the animal's characteristics, but these are for the use of man. It seems just a bit ''utilitarian.''

In fact, one of the most significant things about breed preservation is ''insuring the continuity of the family line.'' ''Passing along the characteristics our ancestors gave us to the next generation is something of an 'historical duty,''' says Chyr Shuang-ching, director of the Animal Industry Department of the COA, with utmost seriousness.

''If you've seen a Taoyuan pig, then you know where the character Pigsy in *Journey to the West* came from,'' states Tai Chien. The big ears, protruding gut, facial shape, and large nostrils are all features unique to Chinese swine. If in the future children have no chance to see a real Taoyuan pig first hand, and just take the small-eared foreign varieties to be the image of a pig, then they'd probably assume that the ancients had rather extravagant imaginations to make Pigsy such a bizarre creature.

(Ventine Tsai/photos by Vincent Chang/ tr. by Peter Eberly & Phil Newell/ first published in November 1990)

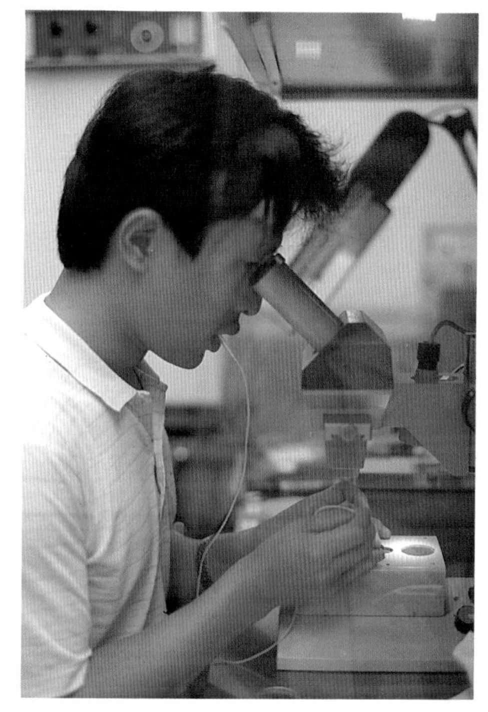

高科技化的精子、胚胎冷凍保存法，是最經濟、安全的種源保存法。

High-technology freezing of sperm and embryos is the safest and most economical way of preserving genetic resources.

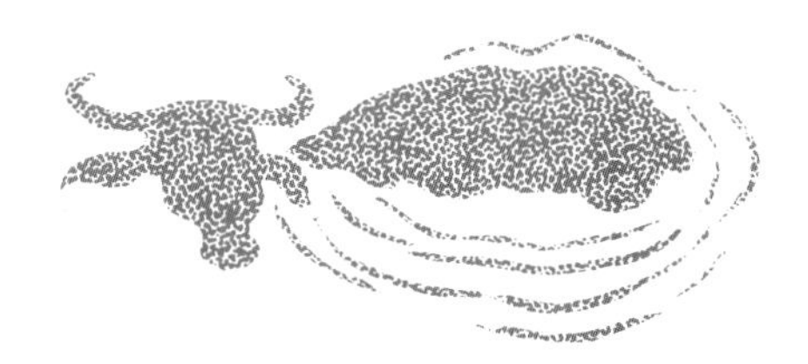

臺灣水牛悲喜調

Soliloquy of a Water Buffalo

文・蔡文婷　圖・張良綱

「老牛、水田、白鷺飛」，這樣的景致，記憶中並不陌生。

在這次農委會列名保種的畜禽裏，大多屬於「只要牠長大，我便宰來吃」的食用動物；唯有黃牛、水牛爲役用，往往和農民朝夕相處十年以上。在以稻米爲主食的長江以南地區，水牛更成了農家最忠實的一份子。

我們常說「誰知盤中飧，粒粒皆辛苦」，感謝的是農夫；對爲我們耕田服役千年的水牛，又了解多少呢？

牛群吃草，白鷺齊飛。這等悠閒的農村景致，在「鐵牛」耕田的今日已不多見。

A herd of oxen graze as egrets take wing. Tranquil bucolic scenes like this are rarely seen now that paddies are cultivated by mechanical tillers, or "iron oxen."

水牛爲人耕田幾千年，人們對牠又了解多少呢?

Water buffalos have tilled our paddies for millennia, yet how much do we really know about them?

前一陣子報上說，一羣臺北市的小朋友到市立動物園上課，經過黃牛家時，有的小朋友竟然指著牠們叫「水牛」。唉！枉費我和中國人同甘共苦四千年，如今卻落得「兒童相見不相識」，眞叫「牛」不得不唱起思想起了。

我的祖先早在西元前二千年，就從印度到了中土。自此我們世世代代在這兒土生土長，成爲人類忠實的僕役，和中國人一起踏遍爛泥水田，共同期待金黃稻穗。

隨著人們渡海墾荒，我們也從廣東、福建一帶移民到臺灣。明崇禎年間，閩南起大旱，地方父母官便頒布「三人一牛」政策，每人給銀三兩，三人再給牛一頭，大夥一起到臺灣開創新天地。這可不是我水牛胡謅，「臺灣通史」寫得明明白白的。

異族不通婚

在目前動物界開放的婚姻觀下，我們是禁止異族通婚的少數民族。即使黃牛、洋乳牛、肉牛等牛友高唱世界一家，生了一堆混血兒，我們也只和同族婚配。

太保守了？那可不，我們是「非不爲也，不能也」。

癥結在於「染色體」。一般牛種的染色體全是卅對，水牛卻天賦異秉，有廿四與廿五對兩種。

世界上水牛族分爲沼澤型和河川型兩支。沼澤型是苦力派的役牛，有廿四對染色體，我們臺灣水牛就屬於這一類。河川型則是又能使力、又供牛奶的乳役兼用牛，有廿五對染色體。這兩種水牛，雖然染色體對數不同，內涵卻一樣，仍可結婚生小寶寶。

除了生理上的特異，我們沒被亂點鴛鴦譜也是基於人類和我們的情感。臺灣大學畜牧

"A** **n old water buffalo, a rice paddy, and an egret on the wing" is a scene deeply embedded in the Chinese consciousness.

Most of the farm animals listed by the Council of Agriculture for inclusion in its breed preservation program are animals which fall under the old Chinese adage, "if it's reached adulthood, then I'll eat it" — creatures raised for food. Only the water buffalo and the Taiwan zebu (Chinese or yellow ox) are animals which are kept primarily for labor, working with farmers day in and day out, often for ten years or more. In the region of China south of the Yangtze, where rice is the main staple of the diet, the water buffalo has long been a faithful member of farming families.

We often cite the old proverb, "Who knows how

much hard work has gone into each and every grain of rice" as a way of expressing gratitude to farmers. But how much do we know about the water buffalo, which has toiled away in our rice paddies for thousands of years?

It said in the papers a while back that a group of Taipei City schoolchildren went on a field trip to the municipal zoo. When they passed by the Chinese oxen, wouldn't you know that several of them would point and say, "Look, water buffaloes!" Has it come to that, then? We've shared life's ups and downs with the Chinese people for 4,000 years, and now it's at the point where "kids don't even know us when we're staring them right in the face." It really makes you wonder.

My ancestors came to China from India as far back as 2000 B.C. We've lived here generation after generation ever since, becoming mankind's faithful servants, plodding shoulder to shoulder with Chinese through muddy rice paddies, waiting together for the golden tassels to ripen in the sun.

When people from Kwangtung and Fukien crossed the sea to open up the wilderness in Taiwan, we came right along with them. There was a terrible drought in southern Fukien during the Chung-chen reign (that's 1628-1644 by your calenders) of the Ming dynasty. Local officials then announced a "three person, one water buffalo" policy. Anyone who would emigrate to Taiwan to open up new land was given three taels of silver, and every three were given one water buffalo. And that's no bull, either: it's written in black and white in *A Comprehensive History of Taiwan*.

We frown on intermarriage: With mixed marriages pretty common in the animal world these days, we are one of the few groups that forbid getting involved with outsiders. Yellow oxen, dairy cows, and beef cattle can all moo till they're blue about how the world is one big happy family and have as many half-breed offspring as they like, but we only link up with our own kind. Too conservative? It's not what you think. It's not that we wouldn't want to, it's just that we can't.

The crux lies in our chromosomes. Most bovine species have 30 pairs of chromosomes, but water buffaloes have 24 or 25.

Water buffaloes the world over come in two types: swamp and river varieties. The swamp variety, with 24 pairs of chromosomes, is the "coolie" among draft animals. That's the type we Taiwan water buffaloes belong to. The river variety, with 25 pairs of chromosomes, is an animal that can serve as both labor and in the dairy service as a supplier of milk. Even though the two varieties have different numbers of chromosomes, we're really the same inside, and can marry and have families if we want.

Besides our biological characteristics, another reason we are set apart from other animals is the sentiment that people seem to hold for us. Dr. Sung Yung-yi, a professor of animal husbandry at National Taiwan University, explains that "water buffaloes were brought over from the mainland by our forebears and have toiled away in the fields with farmers for years and years, so the feeling is that they're almost part of the family." As for other livestock, they are seen just as dumb beasts, good for nothing but filling a belly, so people have continually been crossing them with "imported" breeds.

Actually, way back in 1959 mechanical tillers were already pretty commonplace. Seeing that we weren't much needed anymore, the government arranged a marriage for us with some Southeast Asian murrah buffaloes, which yield a lot of milk, hoping to give us a new "career opportunity." But it turned out that farmers didn't care much for the twisty-horned little calves that resulted, and the murrah wives, rejected by their in-laws, quietly packed up and went home. This was the only time any of us Taiwan water buffalo ever had "married outside the clan."

Can't tell a water buffalo when you see one: I said before that now children can't even recognize us, but actually it's very easy to tell us apart from the rest of the herd. Shih I-chang, an old friend of ours from way back, and head of the Hualien Animal Breeding and Propagation Station, describes us this way: "Water buffaloes are black or dark brown in color, with a crescent of white under the throat and neck, a little like that of the

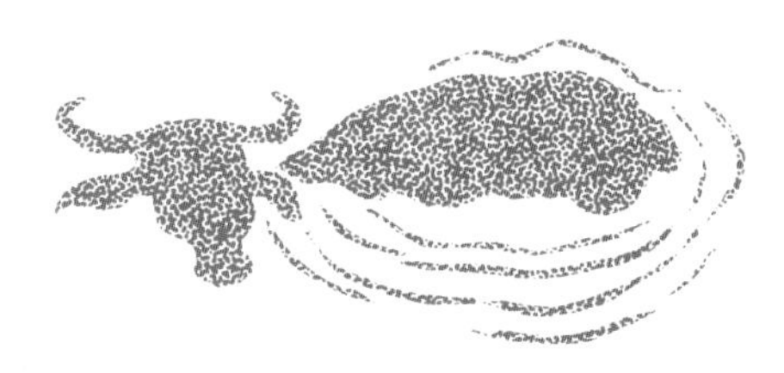

系教授宋永義便表示，「水牛是先民自大陸帶過來的，加上長年和農民一起耕耘，培養出類似一家人的感情。」至於其他家畜禽，則被視爲畜牲一類，飼養只爲下肚，人們也不斷拿牠們和外來種交配。

但是，民國四十八年時，耕耘機已經普及，政府鑑於我們的役用功能漸減，便引進來自東南亞、產乳量大的摩拉水牛和我們完婚，希望爲我們增加另一種「謀生能力」。結果農民對那蜷角的混血小牛大大排斥，摩拉姑娘在不得大家長的歡心下黯然離去，這也是臺灣水牛唯一的一次異族婚姻。

不識水牛眞面目

前面說過現在小朋友不認得我們了，其實我們的外型很容易辨認。和我們很接近的花蓮種畜繁殖場場長施義章這麼形容：「水牛的毛色爲黑或黑褐，最特別的是咽喉下及頸下各有一道月牙型白線，和臺灣黑熊有點相似，配上那一雙大而平行後彎的角，非常優雅。」

除了外型，我們的膝蓋、脚踝（繫節）柔軟，脚丫子（蹄）的面積又大，在泥水田行走，不但不會深陷其中、不可自拔，也不易受傷。

在吃食上，我們從不忌口，不論是纖維素較高或品質較差的芻料，都能消化，這得歸功於我那副「鐵胃」中的微生物。我們還有潛入水中吃水藻的絕技。

泥巴澡最健康

我們的毛髮稀疏，毛孔只有黃牛的六分之一，加上一身黑褐皮膚，吸熱快卻又散熱慢，照理說並不適合熱帶氣候。不過，路是「牛」走出來的，爲了調節體溫，我們非常喜愛泡水。施義章場長最愛看我們沒入泥水中，搖耳擺尾的快樂模樣。

洗泥巴澡除了涼快，還有驅除體外寄生蟲、防止蚊叮的功能，就和人們用「沙威隆」的意思一樣。當我們泥浴完畢，身上自然形成一件泥衣，待陽光將泥水曬乾，身上的牛蝨也就和乾硬的泥巴一塊脫落。別以爲我們很髒，這一件就地取材的自然泥衣可是妙用無窮呢！

人們常會罵我們「大笨牛」，其實我們哪兒笨，我們和馬一樣會「識途」，在農村長大的小孩應該都看過，我們將農作物和累得睡著的主人一起自田中拖回家。

走上餐桌

我們來到臺灣也有三百多年了，一部水牛史也可說是一部農業發展史。曾經我們也有數量多到卅三萬口牛的聲勢，被人們悉心照顧，不少老一輩的農民不吃牛肉，爲表示謝意和尊敬，甚至還爲死後的水牛埋葬立碑。

然而在民國五十年後，耕耘機幾乎完全取代水牛，佔有水田天下。沒被金髮碧眼的洋牛取代，卻敗在銅筋鐵骨的「鐵牛」手中，也無話可說了。目前我的族牛只有二萬多頭了。

失去了耕作的職位，我們的用途也被改爲肉用，比氣力我們沒被挑剔過，比好吃卻常被冠上乙等。雖然被宰來吃不是我的本願，但爲了名譽，我可是要大聲疾呼：我的味道絕不比進口牛肉差。

施義章曾表示，「過去爲了保持耕作力，防止屠宰商濫殺水牛，不但每頭牛有張貼照片、註有出生年月日的身分證（牛籍），年齡未滿十三歲的還不准屠殺。」

你想想嘛！二歲的牛已經性成熟，那十三歲的少說也有人類的六、七十歲了。這樣老態龍鍾的老牛肉當然不好吃。現在沒有這些規定，我們也可展現壯年期的一身精肉，供人類品嚐了。

花蓮種畜繁殖場以前還舉行過水牛品嚐會，在不知牛肉種類的情況下入口，一般人根本分辨不出誰是誰。

據臺灣省農林廳的統計，臺灣每年自國外進口六萬七千多公噸的洋牛肉，約合四十八萬頭全牛的總合，足足多出我們總牛數的二十幾倍。唉！爲了延續水牛在臺灣的光榮歷史，我水牛只好自賣自誇，希望人類在青睞進口牛肉之餘，也別忘了再愛我一次。

（原載光華七十九年十一月號）

Taiwan black bear, and have a pair of large, parallel, sweptback horns which are quite elegant.''

Besides the outside appearance, our kneecaps and heels are soft, and our cloven hooves are large. This means we can walk through the mud, and not only won't we get stuck in anything we can't get ourselves out of, we also don't get hurt easily.

In terms of food, we aren't picky eaters. We can digest any kind of fodder no matter how rough the fiber or how poor the quality. The credit here goes to the microorganisms in our ''iron stomachs.'' We also have a knack for sticking our heads under the water to snatch up a mouthful of water plants.

Mud baths are still the healthiest: Our hair is sparse and thin, with only one sixth the number of pores in our skin as the Chinese ox. Add to this a layer of black-brown skin, and we absorb heat quickly but dissipate it slowly. It seems to reason that we shouldn't be suitable for a tropical climate, but a buffalo has got to go where it is told. To adjust our body temperature, we just love to soak in the water. Shih I-chang really enjoys watching the way we slip into the water, shaking our ears and wagging our tales with delight.

Besides cooling us off, mud baths also get rid of parasites and prevent mosquito bites, just like spraying on that insect repellent you people use. After we've finished bathing, a mud coat naturally forms on our bodies that dries in the sun and drops off piece by piece, taking the lice and vermin with it.

People sometimes curse us as ''stupid oxen,'' but we're not so slow. We know our way around as well as horses. Anyone who grew up in the country must have seen us hauling home crops and farmers so tuckered out that they've fallen asleep on top, completely on our own.

Headed for the dinner table: We've been on Taiwan for more than 300 years, and a history of water buffaloes could practically serve as a history of agricultural development. At one time our numbers reached 330,000, carefully looked after by the people on the island. Many older farmers wouldn't touch beef, as a way to express gratitude and respect, and some even buried their buffaloes in proper graves with headstones.

Ever since the 1960s, though, we've been almost completely replaced by mechanical tillers. They literally have won the field. We weren't defeated head to head by foreign cattle, but suffered defeat at the hands of clanging ''iron bulls,'' and now there's nothing left to be said. Today there are only about 20,000 of us left.

Now that we've lost our jobs in the field, our usefulness comes in our meat. Yet, while no one ever complained about us before, when our strength was all that mattered, now that flavor is all that matters, we're ''Grade B.'' Although being slaughtered for the dinner table isn't my number one career choice, I still demand respect, and will tell anyone who will listen that we're not one whit inferior to those fancy foreign imports.

Shih I-chang once said, ''in the past, in order to maintain an adequate supply of water buffaloes for tilling, butchers were restricted in slaughtering them. Each water buffalo had its own ID with photo and date of birth. It wasn't permitted to slaughter any that hadn't reached the age of 13.''

Just think of that! A two-year-old buffalo is sexually mature, so a 13-year-old would be the equivalent of a person in their sixties or seventies, at least. It's no wonder meat that old tasted lousy. Now that there are no such regulations, we can give people a taste of our meat in its prime.

At the Hualien breeding station a while back, they held a meat-tasting event where people were given different kinds of beef without being told which was which. Most people couldn't tell them apart.

According to statistics of the Taiwan Province Department of Agriculture and Forestry, Taiwan imports more than 67,000 metric tons of beef each year — that's 480,000 whole head of cattle, or more than 20 times our entire population. Alas! If we want to extend the glorious history of the water buffalo in Taiwan, we'll have to start tooting our own horns and hope that, despite their infatuation with imported beef, people won't forget to love us once more.

(Ventine Tsai/photos by Vincent Chang/ tr. by Peter Eberly & Phil Newell/ first published in November 1990)

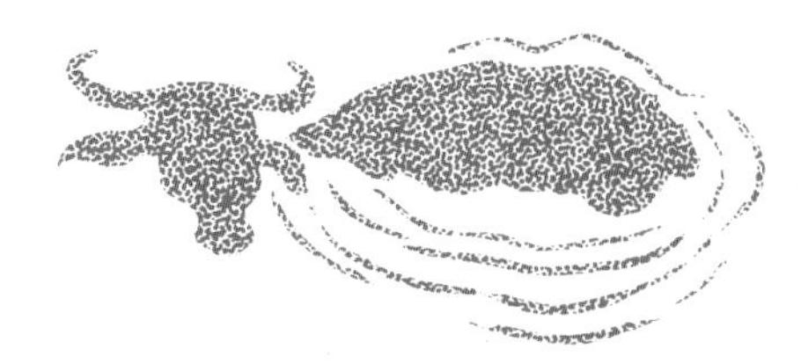

復育大自然的「少數民族」

Restoring Nature's Dwindling "Minorities"

文・張靜茹　圖・張良綱

「特稀有動物」近來成了代表身價非凡、人人耳熟能詳的名詞。

各國政府急著要保護他們特有及稀有的動物；學者專家忙著爲牠們做生態調查；在中華民國臺灣地區，近來民間更爭相想爲牠們做人工繁殖。

爲什麼要保護特有及稀有的動物？是不是只因爲物以稀爲貴？怎樣才是對待牠們最好的方法？是不是幫助牠們交配，繁殖出很多小特稀有動物？

Rare and unique species have become a hot topic of conversation just about everywhere.

The governments of countries around the world have been taking urgent steps to protect these animals. Scholars and specialists are busy at work doing ecological research. On Taiwan people are striving to develop means of artificial reproduction.

Why should we protect these unique and rare animals? Is it because what is rare is good? How do we best treat them? Should we help them mate and multiply?

五彩繽紛的蝴蝶被大量用來做標本——資源爲我所用；但濫用却使人類可掌握的資源愈來愈少。

Colorful butterflies have been captured in great quantity for use as specimens — an example of mankind's indiscriminate exploitation of precious animal resources.

許多美國科學家和鳥類學家花了八年時間，才使一種名叫「寶芬」的海鳥回到北美緬因州居住。

寶芬鳥原來居住在大西洋岸的加拿大、美國緬因州。牠身材像隻小企鵝，嘴喙五顏六色，喜歡挖土洞為巢撫育幼鳥。本來緬因州寶芬鳥數量非常多，但因獵人無限制的獵殺和取走鳥蛋，終於完全絕跡。

寶芬鳥回家

絕跡八十年後，新一代的美國年輕人才發現自己國土原來也有寶芬鳥。他們於是展開了「寶芬鳥回家」的復育計劃。

首先他們向加拿大求救，由加國每年提供他們一百隻小寶芬鳥，帶回緬因州撫育，但辛辛苦苦養大的鳥，一能出海覓食後，却毫無留戀地一去不回。這樣的情形一直持續到八年後，一位漁夫在海上捕魚時才發現第一隻覓食後轉頭回來的寶芬鳥。

鳥雖然回來了，牠們却仍不願留下來求偶、築巢、繁殖，原因是寶芬鳥要達到某一數量聚集在一起後，才會發生求偶行為。於是科學家又忙碌起來，後來人們製造了許多唯妙唯肖的假寶芬鳥，希望引誘牠們留下來交配、繁殖……。

「人類要破壞大自然的生態往往不費吹灰之力；但想還原，付出千百倍代價還不一定能成功，」一位從事保育工作的人士說。這也幾乎是所有自然保育學者的共同看法。

那人類為何還要花力氣去還原它呢？

失去「婚姻自主權」

人類利用野生動物的歷史雖無法確實考證，但根據英國牛津大學動物學博士莫理斯在「裸猿」一書中的說法，人類在一萬年前已懂得馴養動物，山羊、馴鹿在那時已被有組織的飼養與選種改良。

四千年前，鳥類中的雞、鵝、鴨與魚類中的金魚，也加入馴養的名單。農業社會以後，更多的動物如牛、豬、狗、鴿子也開始成了家禽、家畜。

除了做為人類主要的食物，牠們的功能更擴及到觀賞、科學實驗，甚至加以訓練，以補人力之不足，如信鴿、警犬等。

十八世紀工業革命後，人類育種技術大大提升，被人視為有經濟價值的動物更失去了「婚姻自主權」，依人類需要被育種改良。透過人工繁殖、基因選汰等技術，被留下來或改良出來的肉牛，是長得快又多肉的品種，乳牛則留下產奶多的，而其它不符合人類需要的品種日漸消失。

動不動就「全軍覆沒」

「依物競天擇的原理，人類既進化為現階段的萬物之首，利用周遭的東西來改進人的生活品質並不為過，」臺灣大學畜牧研究所教授宋永義接着話鋒一轉，指出問題：「但生物的品種、遺傳基因趨於單一的後遺症是，生物一經病菌傳染、出現新的天敵或天候等環境改變，這類生物就『全軍覆沒』。」

過去發生在植物上的許多例子，更令畜產、生物學者覺得不得不小心提防。十九世紀中葉，曾有北美葡萄蟲傳到歐洲，使全歐葡萄園盡毀；後來發現美洲原產葡萄樹種可以抵抗葡萄蟲，將歐洲葡萄株嫁接在美洲葡萄砧木，才挽救了歐洲葡萄的生產。

「對於動物，我們所知道、所能掌握的比植物更少，」宋永義說：「人類想依自己意願，隨時育出動物新品種以應付突來的變故，還早的很。」

大自然是最好的種源庫

「人類已經意識到不能只保存目前需要的動物種類，」臺灣大學動物系副教授李玲玲指出，保存動物種類多樣化成為各國的重要工作。

畜產動物中漸被淘汰的品種也紛紛有人搜集。例如美、法兩國向中國大陸要中國原種豬飼養，因為牠雖長得慢，不合現代快速生產的需要，但牠一胎可生個十六、七隻小豬，誰也不知這項特點未來是否會變得很重要；在中華民國臺灣，政府也急着要保護漸被耕耘機取代的水牛，因為牠的勞動力、耐力非其它牛種可比。

U.S. scientists and ornithologists spent eight years striving to repopulate Maine with the puffin.

Originally native to Maine and the Atlantic provinces of Canada, the puffin resembles a small penguin with a multicolored beak. It digs a hole as a nest to rear its young. There used to be a thriving population of puffins in Maine, but because of unrestricted hunting and snatching of the birds' eggs, the puffins there were completely wiped out.

The puffin comes home: Upon learning eighty years later that puffins had once lived on their nation's soil, a new generation of Americans unveiled the "Puffin Come Home" restoration plan.

First they asked Canada to give them 100 puffin chicks a year. But after painstaking steps were taken to raise the birds in Maine, it was discovered that the adult birds would fly away after they went out to sea to seek food. It went on like this for eight years until a fisherman saw the first puffin return.

Yet though the birds had come back, they weren't willing to wait around to find a mate, build a nest and propagate. The reason was that puffins will only seek mates if they are gathered together in sufficient numbers. And so the scientists got busy again, making a flock of decoys in the hope that the real birds would be attracted to stay around, mate and produce offspring. . . .

"Without working too hard at it, people are destroying natural ecologies everywhere," says someone working in environmental protection. "We can spend a thousand times as much energy to bring them back, but it won't necessarily work." This is the view shared by just about everyone in the environmental protection field.

Then why are we making efforts to bring them back?

Losing the right to pick one's spouse: Man has a long history of using wild animals. As early as 10,000 years ago, goats and reindeer were already being raised and bred in an organized fashion.

Four thousand years ago birds like the chicken, goose and duck and fish like the goldfish joined those being raised by man. After the agricultural revolution, even more animals — cows, pigs, dogs and pigeons among them — were domesticated.

Besides being an important source of food, they also provided enjoyment, were used for scientific experiments and even trained to help out where man himself was insufficient, as in the case of carrier pigeons or watch dogs.

After the industrial revolution of the 18th century, great advancements were made in breeding. Animals of economic value lost the right to pick their own spouses as species were bred to meet man's needs. With breeding reaching the point where even particular genes could be selected or rejected, cattle grew quickly and yielded more beef and cows produced more milk. Species that did not meet man's requirements gradually disappeared.

Easily extinct: "According to the law of natural selection, with man having progressed to become lord of the beasts, it isn't outrageous that he would make use of the things around him to improve his quality of life," explains Sung Yung-yi, professor of animal husbandry at National Taiwan University, before turning to the problem: The result of the narrowing of the gene pool is that it leaves species much more vulnerable to extinction in the face of disease, new natural enemies or changes in climate.

With many such past examples among plants, animal husbandmen and biologists feel that they must be remain extremely vigilant. In the middle of the 19th century, for example, North American grape vine pests that were brought to Europe destroyed all of Europe's vineyards. When it was discovered that American vines were resistant to these pests, European vineyards imported stock from America to save Europe's grape production.

"What we know and can control about animals is much less than for plants," Sung says, "Mankind is far from being able to breed animals at will to cope with sudden and unexpected problems."

Nature — the best storehouse of species: "Mankind has already become aware that it cannot preserve only the species it now needs," says Li Ling-ling, an associate professor of zoology at National Taiwan University. Preserving the diversity of animal species has become important work everywhere.

People are also beginning to collect breeds of domesticated animals that have been found wanting. The United States and France, for example, have asked mainland China for a breed of pigs native there. Although slow growers — modernity demands that porkers grow portly promptly — a sow of this breed can give birth to 16 or 17 piglets in

因貿易利益而過度濫捕，臺灣已難見野生梅花鹿蹤跡。

Because of overexploitation, Formosan sika have practically disappeared in the wild.

比起經人類飼養的畜產動物，克服大自然各種考驗、自行生存下來的野生動物，種類更是豐富、多樣。而且人類對其未知的事也更多，許多野生動物的用途還不斷在有意、無意中被發現。

例如原產在南美洲的犰狳，是目前已知除了人類以外唯一會罹患痲瘋病的動物，對治療痲瘋病的研究，提供了許多幫助。

近年來科學家也發現北極熊的毛是極佳的吸熱材料，除用來製禦寒衣物外，可能還可做爲吸收、利用太陽能的材料。

有的野生動物更是人類尋找自己根源的重要證據。例如冰河時期留在臺灣的臺灣鱒，就被視爲大陸和臺灣曾經相連的證據；少了牠，地質史也少了一項佐證。

斧斤以時入山林

「野生動物這麼『好用』，因此人們始終沒有停止利用牠們。」師範大學一位生物研究所的學生說：「但不合理的捕獵，加上人類的開發活動，破壞了牠們棲息的環境，使野生動物一直在消失。」

據野生動物專家統計，目前全世界百分之四十的脊椎動物有滅種危機；若加上昆蟲，評估結果是，到本世紀末將有五十到一百萬種生物會遭到滅絕的命運。

野生種不斷消失的危機，使得保育觀念興起。例如歐美許多國家爲免竭澤而漁，規定捕魚網目不可小到「大小通吃」、產卵季限制釣魚，及釣到稚魚必須放走等，而且捕獵某些動物要處以罰款或徒刑；目前被各國視爲美味、大量撈捕的南極蝦，也是五種稀有鯨魚的食物，爲避免與鯨魚爭食，國際上已訂立公約限量捕抓；我國的野生動物保育法也已草擬完成，正在立法院審議中。

特稀有種列爲優先

除以法令保護，各國野生動物專家也積極以人工繁殖、人工孵化，或將年幼動物帶回實驗室養育，待其成長再野放等人工復育的方法，來達成保存野生種的目的。而因特殊生存本能或對環境的要求，成爲某一地區的

a litter. Who knows, some day this proficiency might come in handy. In Taiwan, the government is eagerly protecting the water buffalo, which has gradually been replaced by power tillers, because it can outmuscle all its bovine brothers.

In comparison with animals bred by man, wild animals, which have overcome nature's tests to survive on their own, are even more numerous in diversity, and man knows even less about them. Whether by accident or design, new uses for them are constantly being found.

For example, the South American giant armadillo is the only known animal besides man that can suffer from leprosy, and so it is of great help in researching the disease.

In recent years, scientists have also discovered that the fur of polar bears is excellent for absorbing heat. Besides being used as thermal clothing, it could be employed to absorb and make use of the energy of the sun.

Some wild animals serve even more as important evidence in man's search for his own origins. The Taiwan Trout, which has been in Taiwan since the ice age, is taken as evidence that Taiwan was once connected with Mainland China. Without it, the history of geology would be missing a piece of its puzzle.

The time is ripe: ''Because animals are so useful, people haven't stopped making use of them,'' says a graduate student in zoology at National Taiwan Normal University. ''But overhunting coupled with encroaching human development is destroying habitat and causing wild animals to go extinct one after another.''

According to wildlife experts, 40 percent of all vertebrates are on or close to inclusion on the endangered species list. If invertebrates are added, it is estimated that by the end of the century 500,000 to one million animals will have become extinct.

This crisis of increasing extinctions has given rise to the concept of conservation. In order to prevent overfishing, many European and North American countries, for instance, have stipulated that fishing nets cannot be woven so tight that all fish are caught, and during the egg laying season hook fishing is prohibited. If a young fish is caught it must be released, and the hunting of prohibited species is punishable by fine or imprisonment. The Antarctic shrimp, whose delicious flavor has

特有動物，一旦在當地亡種，就等於是在地球上消失，因此數量稀少的特有種更常是各國優先選擇的復育對象。

國際保育聯盟也將稀少的特有動物，依瀕臨絕種程度分爲四類，列在「紅皮書」上，例如我國的帝雉、藍腹鷴都在其中。除名列書上的項目，各國仍視自己的情況，選擇其它動物復育。

先保護棲息環境

我國在民國七十四年行政院核定的「臺灣地區自然生態保育方案」中，指明將選擇合適地點對帝雉、藍腹鷴做復育，還選出現已無野生種的梅花鹿，和冰河時期留下來的臺灣鱒、山椒魚做人工復育。

如何進行人工復育呢？

臺灣大學動物系教授林曜松認爲，人工復育的目的是要保存野生動物原有的特性，維護原有基因，最好不要有人爲的力量直接介入幫助復育。

對於國內民間一直有人從事帝雉人工繁殖，並表示想放回野外，帝雉專家謝孝同博士也一直持保留態度。因爲人工繁殖常有近親交配情形，他擔心這樣產生的新一代會隱藏不良基因，如覓食能力差、行動緩慢等，野放後牠們若與野生種交配，下一代遺傳到不良基因，最後可能連野外種一起滅亡。

成立大武山自然保護區

據國際自然保育聯盟研究調查指出，自然環境的破壞是導致野生動物滅亡的主因。「棲息環境若不加以保護，用人工復育再多的野生動物，放出去後可能仍是走向滅亡一途，」美國蒙他拿州大學野生動物博士 Chris Servheen 最近來臺接受訪問時表示。

美國在前年曾發行一套郵票，呼籲大家搶救各種瀕臨危亡的野生動物和牠們棲息的森林、沼澤、高山、草原……，因爲自然環境破壞了，生活其間的動物也就消失了。

而動物一旦消失，並非人類想爲牠們復育就能做到。例如住在地洞，以白蟻爲食的穿山甲，除非人類可以不斷供應牠白蟻，否則不可能將牠帶到實驗室復育。再以臺灣久未出現的雲豹來說，動物學家花了九牛二虎之力走遍臺灣山區，找都找不到，如何復育起？但學者調查發現，大武山區可能還有雲豹，行政院農業委員會在今年初將大武山區列爲自然保護區的做法，被認爲是一種最佳的選擇。

「如果一種動物的數量還不至於在短期內滅絕，只要不再有人爲破壞，就可以自然復育，這時保育工作應以禁止捕捉、保護其棲息環境爲主，」師範大學生物系副教授王穎說：「這種讓野生動物『奪回失地』的保育方式，是我們最樂意見到的。」

如果原始野性被「稀釋」……

沒有自然復育條件的野生動物，只好多管齊下，除保護棲地，還得借助人力。像美國寶芬鳥的復育，是由加拿大引種回美國，將鳥養大後再野放；阿拉斯加的麋鹿，棲息地在冬季常缺乏食物，得用大批人力每年冬天將牠們趕往較溫暖、食物充裕的南方；有些名列「紅皮書」上的動物，更是由專家在實驗室裏育種。

但無論是人工繁殖、人工育種、人工飼養、人工……，最後都希望把費盡心機照顧的動物放回自然環境，因爲唯有這樣才能維持野生種的天性和特質，也才眞正達到復育的目的。因此，在復育過程中也就有更多需要注意的地方。

例如做鳥類的復育，爲提高小鳥孵化的成功率，實驗人員可能需要把鳥蛋從鳥巢拿到沒有天敵的實驗室孵化，然後把小鳥放回巢中；但取卵時若嚇跑母鳥，那就算小鳥孵化成功，放回原有巢中，沒有母鳥引領，恐怕也無法完全回歸自然了。

「人工復育是以『人擇』要來做『天擇』，要達到百分之百何其容易，」宋永義說。

前幾個月，美國加州成功地以人工孵育了一隻小禿鷹，研究人員不和牠接觸，而以人造的鷹嘴機器餵牠。因爲若不如此，禿鷹可能會與人發生情感，原始野性被「稀釋」，以後放回山林可能見了獵人也不怕，無異羊

caused it to be fished on a large scale, is also food for five rare species of whale. In order to prevent competing with the whales for food, the international community has agreed to public limits on fishing of this prawn. And the R.O.C. Wildlife Conservation Law was passed several years ago.

Rare and unique species take priority: Besides providing legal protection, the wildlife experts of various countries are also actively moving ahead with artificial propagation and incubation and with raising young animals in laboratories with the hope that they can be freed in the wild as adults. All these methods have been used to attain the goal of wildlife preservation. Once an animal that is unique to one place because of its special survival skills or its need for a certain environment goes extinct in that place, it is extinct everywhere on earth. Hence, these unique and rare species have become the priority of conservation efforts. The International Wildlife Federation has put these rare and unique animals into four categories in its "Red Book," according to level of endangerment. The mikado pheasant and Swinhoe's blue pheasant are among the species listed from Taiwan. Individual countries are also looking at their own situations and selecting other animals for restoration beyond the ones listed in this book.

The "Taiwan Area Natural Ecology Preservation Directive," made by the Executive Yuan in 1985, ordered the reintroduction of the mikado pheasant and Swinhoe's blue pheasant in appropriate areas. For restoration, it also selected sika deer, which no longer exist in the wild, as well as the Taiwan Trout and the Formosan salamandar, leftovers from the ice age.

How does one proceed with restoring a species? **First preserve their environment:** Lin Yaosong, a professor of zoology at National Taiwan University holds that since artificial propagation is aimed at preserving the special characteristics and gene pool of the species, it is best not to interfere directly in helping species repopulate.

Dr. Hsieh Hsiao-tung, an expert on Taiwan's mikado pheasant, has always had reservations about plans to release into the wild animals that have been artificially bred and raised under carefully controlled conditions. Because human-directed breeding often means that animals mate with close relatives, he worries that this will result in a new generation of inferior genes, leaving animals slower afoot or less adept at seeking food. If artificially bred mikado pheasants mated with wild ones after being released, future generations would inherit their inferior genes and the end result might be that the bird would die out altogether.

Establishing the Mt. Tawu Nature Preserve: According to statistics of the International Wildlife Federation, destruction of the natural environment is the reason behind the extinction of wild animals. "If greater protection is not given to the living environment of animals, the number of wild animals will continue to diminish despite man's efforts at artificial propagation and release," said Chris Servheen, grizzly bear recovery coordinator at the University of Montana, when he visited Taiwan.

Several years ago, the United States published a set of stamps calling for everyone to save endangered species and the environments (the forests, wetlands, high mountains, grasslands, etc.) in which they live. If natural habitats are damaged, after all, the animals that live in them will likewise disappear.

And once animals have gone, it's not just a matter of people wanting to restore their populations. Take, for instance, the pangolin, which lives in holes in the ground and eats termites. Without a steady supply of termites, you can't go about restoring their population in laboratories. As for the clouded leopard, which hasn't been spotted for a long time, zoologists spent great efforts searching through the mountainous areas of Taiwan but came up with nothing. If you can't find it, how can you revive it? When a study determined that the Mt. Tawu area might have some of the clouded leopards left, the Council of Agriculture of the Executive Yuan designated Mt. Tawu as a nature preserve — a decision hailed as the ideal choice.

"If a species isn't facing extinction in the near future, you can naturally revive it by simply stopping destruction of its habitat. At such times conservation work ought to aim at stopping poaching and protecting habitat," says Wang Ying, an associate professor of biology at National Taiwan Normal University. "This method of conservation — letting a species 'recapture lost ground' — is what we are happiest to see."

If the "wildness" is diluted: Wild animals

臺灣鱒雖然人工繁殖成功，但還有很多復育工作要進行。(蘇焉攝)

Although some success has been achieved on Taiwan in reviving endangered species through artificial propagation, much work remains to be done. (photo by Su Yen)

入虎口，徒讓研究人員白辛苦一場。

要爲野生動物配對也須避免亂牽紅線——把脾氣好，容易抓的都抓來配對，使脾氣較凶的失去配對機會；或加入經濟觀點，選體型大的配對……。

做紅娘，大不易

「人工繁殖出來的東西，外表可能很像回事，但內在——基因却可能隱藏了問題，」臺灣鱒復育計劃主持人、也是中山大學海洋科學院院長張崑雄說。因此在前年臺灣鱒繁殖成功後，他表示，這並不代表已經復育成功，還有待放流後追踪調查、觀察其生長情況與環境變化等。

「人工繁殖是幫助野生動物復育的一種過程、一種手段，是不得已的做法。未來放流後的臺灣鱒如果能自然繁殖，就不要再使用人工繁殖，」張崑雄強調。

由於人類對大自然的了解仍太有限，大自然生態互動的關係又太複雜、精微，把一個動物趨趕或野放到新環境，牠能不能適應、對當地生態體系會不會造成傷害……，雖然可以借重事先的研究評估，但「失控」的狀況仍隨時可能發生。美國野生動物管理法中有一條針對「人工繁殖」的進行、野放訂定，就是怕沒有規範，會擾亂野生種。

一加一不等於二

人工養殖的動物往往覓食、逃避敵人、適應天候轉變的能力較差；可能野放出去就死亡；也有可能出去後，當地植物一下子被這一「新來的」動物吃掉幾百種，造成「環境浩刼」，也影響了其它野生動物的覓食。

「野放不是原來有一百隻，再放進去一百隻就等於有二百隻這種單純的邏輯，」農委會保育科陳超仁說。

因此要進行野放前，必須先劃出一塊最像牠自然棲息環境的地方，看牠在裡面的生長情形。觀察一段時間，確定沒問題再放到更大的模擬環境，直到最後放到野外。

「要等牠在模擬的環境中適應、適應成功、野放、找到配偶、願意交配，再評估牠對環境的影響、要不要繼續做人工復育……，可能是一段無限期，」從事梅花鹿復育計劃的王穎表示。

集各方人才於一堂

在這段「無盡期」中，必須集合植物、土壤、氣候，甚至其它種動物的專家一起進行，如果要爲被復育的動物養些食物野放，如老鼠，還得老鼠專家研究如此會對當地造成什麼影響，該不該做。

目前在墾丁國家公園社頂進行的梅花鹿復育，三年前開始進行時早就沒有野生種了。因爲梅花鹿鹿角有利用價值，原來有人養殖，復育人員就由其中找出較具原始天性的鹿來復育。

但人工養殖的鹿可能早被鹿農進行過「異國婚姻」——將之與日本、韓國種梅花鹿，或和不同品系的水鹿、紅鹿交配過，因此梅花鹿復育計劃中還找了動物品系鑑定專家宋永義鑑定，以防復育的其實是「混血兒」，却以爲是自己的原生種。

that lack the conditions to revive themselves need extra support — not just preservation of habitat but also more direct assistance from people. Take puffins, for instance. People brought them from Canada to America, raised the chicks and released the grown birds. Alaskan moose often lack food in the winter, and it takes massive human effort to encourage them to move to warmer climes to the south where food is more plentiful. A few species listed in the Red Book are even being artificially bred in the laboratory.

But whether this human assistance comes in the form of breeding, feeding or whatever, the hope is that these well cared for animals will eventually be released into the wild. Only in this way will a wild animal's nature and special qualities be preserved, and only by so doing will the species really be restored. In the process of restoring a species there are numerous considerations.

For example, in order to increase the rate of eggs being hatched when restoring bird species, researchers may need to move the eggs from the nest, where natural enemies lurk nearby, to the laboratory. After the chicks hatch, they can then be returned. But if the mother is frightened away when the eggs are taken, then even if the eggs are successfully hatched and the chicks put back in the nest, without a mother's protection they won't stand a chance in the wild.

''In artificial procreation, 'human choice' is substituted for 'nature's choice' to bring an easy success rate of 100 percent,'' says Song Yung-yi. A few years ago, researchers in California succeeded in hatching a young condor. To ensure that it could survive in the wild, it was fed by a mechanical condor so as to avoid human contact. Otherwise, it might not be wary of human hunters when released in the wild, and all of the researchers' efforts would be for naught.

In order to get wild animals to find mates, one should perhaps avoid drawing arbitrary red lines. If one mates the relatively docile, easy-to-catch members of a species, this means that the ill-tempered have lost the chance to reproduce. Or for economic reasons one might be inclined to select those with larger bodies.

It's not easy playing matchmaker: ''Artificially bred animals may look perfectly normal on the outside, but there could be trouble with the genes on the inside,'' says Chang Kun-hsiung, dean of the College of Sea-Ocean Science at National Sun Yat-sen University and the coordinator of the plan to revive the Taiwan trout. And so, several years ago, when attempts to propagate the fish artificially were successful, he made clear that it was too early to celebrate. Further study would have to be done, he said, to observe their growth and environmental changes.

''Artificial propagation is a method of last resort,'' Chang stresses. ''In the future, if Taiwan trout can do the job themselves, then artificial propagation will no longer be needed.''

People still have limited understanding of nature, of its complicated and profound interconnected web of relationships. If you release or drive an animal into a new environment, you may be able to make scholarly estimates about whether it will adapt to or damage the ecological structure, but at any time the situation may get out of control. The U.S. wildlife management law has an article that sets rules about ''artificial propagation'' and introducing animals to the wild. It was feared that a lack of guidelines might hurt wild species.

One plus one doesn't equal two: Artificially bred animals might be ill-equipped to seek food, evade predators and adapt to changes in climate, dying out only shortly after being released. Or they might cause an environmental disaster by devouring hundreds of species of plants and affecting what other wild animals eat.

''For releases in the wild, you can't use the simple logic of 'there were originally 100; by releasing another 100, you get 200,''' says Chen Chao-jen of the conservation office of the Council of Agriculture.

Before a release into the wild, you've got to first find a site that closely resembles the animal's original habitat and observe its prospects there. After a period of observation and making sure that there are no serious problems, a release can be made in a larger simulated environment. Only then can a release be made in the wild.

''You've got to wait for them to get used to the simulated environment, successfully adapt, be released in the wild, find a spouse and procreate willingly,'' says Wang Ying of the sika deer re-introduction plan. ''Then you can reassess their effects on the environment and consider whether

野生動物的生態調查是人工復育的第一步。圖爲臺大動物系在陽明山做臺北樹蛙的調查。

Surveying the habits of animals in the wild is the first step in artificial propagation. Here a team from the zoology department of National Taiwan University studies Taipei tree frogs on Yangming Mountain.

即使是原生種，民間飼養時也往往因開始時只有幾頭，繁衍下去多是近親交配，而影響了遺傳基因。

此外，未受過專業訓練的人，可能選擇鹿角大的做繁殖，但對叢林型的鹿而言，鹿角大容易被林木絆住，因此角愈大反而愈吃虧，若有人將這樣繁殖出來的鹿野放，天敵來了，逃都逃不掉。

勞命傷財也得做

去年日本曾送我們一批過去由臺灣「引渡」到日本，經過人工繁殖成功的帝雉，我們很樂意地接受了，但却安置在木柵動物園。因爲這些帝雉體型比野外種小，可能是近親交配，爲防止牠們影響野生種的血統，一直沒有野放。

「人力、物力、時間，誰說人工復育不是『勞命傷財』？可是讓我們的特有種紛紛消失行嗎？說得過去嗎？」林試所生物系一位副研究員嘆口氣，重複許多學者說過的話：「破壞容易，復原太難」。

「人工復育不是養動物、讓牠們交配繁殖就可以，」李玲玲說，現在國內許多民間人士有心想做人工繁殖，技術水準也不錯，但沒有學術與各方面的支持，又通常具有想展示、買賣等商業企圖，徒令人擔心。

陳超仁也對目前國內想做人工復育者大多興趣遠高於專業能力的情形，引以爲憂。他假設：如果有人善意地引進泰國雲豹繁殖野放，萬一該種雲豹的捕獵能力比我們的雲豹強，那我們其它野生動物面對這個從天而降的天敵，只有遭殃的份。

「爲防範這類問題發生，我們必須着手制訂一套人工復育的法令，」李玲玲呼籲。

也有人想將特稀有動物大量繁殖，豢養供做食用。多數保育學者認爲這亦無妨，但他們擔心的是，大量馴養的動物流出去會破壞野生種；或爲了做繁殖，到野外去搜捕，加速該種野生動物的絕跡。

留得青山在，不怕沒柴燒

「大自然種源庫裏的東西不是不能用，只是有一些快被用光了，有些我們還不知道怎樣才是最好的用法，」李玲玲說：「我只是希望大家有耐心，利用野生動物，不要趕盡殺絕；人工復育不要急功近利地湊熱鬧，讓這些資源能在最合理的使用方式下，永久被利用。」

如果能拋棄私利，道理其實很簡單，不過是一句國人都耳熟能詳的俗諺——留得青山在，不怕沒柴燒。

（原載光華七十七年六月號）

or not you want to continue artificial propagation . . . It may go on forever.''

A team of experts: During this unending process, one has to work with experts of plants, soil, climate and even of other animals. If one wants to release prey for the animal being revived to feed on — like mice — then you've got to consult with experts on that animal to determine what effects such a release will have and whether or not it should be done.

Currently the Sika Deer Repopulation Plan is being undertaken in Kenting National Park. Three years ago, when the plan began, there were already no deer in the wild. But because the antlers of sika deer have economic value, there were people who raised them. The members of the repopulation team picked deer that seemed most like their originally wild ancestors.

But the deer farmers may have already cross bred the deer with Japanese and Korean varieties of the sika deer or with such varieties as the Formosan sambar or red deer. And so the repopulation team also brought in Song Yung-yi, an expert in determining animal lineages, to distinguish the real Taiwanese sika deer from the mixed breeds.

But even if the domesticated animals are pure bloods, because human stocks are usually descended from a small group of ancestors, this also has an effect on the genes.

In addition, people who have not received professional training will perhaps select larger antlered animals for breeding. Yet in the forest, deer with larger antlers easily get entangled in the trees. And so the larger the antlers, the more disadvantaged the animals. Deer so bred will not be able to run away from their predators in the wild.

It's worth the trouble: Last year Japan gave Taiwan a group of mikado pheasants descended from stock that had originally been introduced from Taiwan. Though happily accepting the birds, we put them directly in the Mucha Zoo. Perhaps overbred, the birds were smaller than the mikado pheasants in the wild. To prevent them from affecting the wild population, they have not been released.

''Human effort, money, time. . . who says restoring animal populations isn't costly? But can we just let our unique species go extinct one by one?'' says an assistant researcher at the Forest Research Institute, who then echoes what many scholars say: ''It's easy to destroy, very hard to restore.''

''In restoring animal populations, it's not enough just to raise animals and let them find mates and breed,'' says Li Ling-ling. Currently there are many people in Taiwan interested in artificial propagation whose techniques are up to par, but they lack the needed wide-ranging support from scholars and others. Their aims are often commercial — breeding the animals for display or sale — and this too makes people nervous.

Chen Chao-jen worries that many people's interest in artificial breeding far exceeds their abilities. Suppose, he says, ''people with good intentions introduce Thai clouded leopards for breeding and release.'' If they are much better than the native leopards at catching prey, then pity the poor animals that become their dinners.

''In order to prevent this kind of problem from happening,'' says Li, ''we must make up a set of regulations governing species restoration.''

There are also others who want to raise unique rare animals for food. Most conservationists don't find this too objectionable. Yet they worry that if large numbers of these man-bred animals find their way back into the wild, they may taint the gene pool of the species. And they worry that these breeders might accelerate the extinction of the wild animals by going to the wild to catch them.

Keeping the forests for the wood: ''It's not that the things in the great storehouse of Mother Nature can't be used, it's just that some of them are almost used up and others we still don't know how best to make use of,'' says Li. ''I just hope that everyone will have patience, and that in using wild animals we won't use them to extinction. People who want a quick buck shouldn't look for it in artificial breeding. Managed reasonably, these resources should be able to be used forever.''

If we can cast aside private gain, the logic becomes very simple. It's clearly stated in this frequently used Chinese idiom: By keeping the mountain forests, you'll never lack for firewood.

(Chang Chin-ju/photos by Vincent Chang/ tr. by Jonathan Barnard & Kirby Chien/ first published in June 1988)

「我的爸爸是獅子，媽媽是老虎，我却是彪，傷腦筋！」
My father's a lion, my mother's a tiger, and I'm a ''liger.'' You figure it out.

喬太守亂點鴛鴦譜

Playing with the Animal Kingdom

文・張靜茹　圖・張良綱

十幾年前，臺灣省農林廳曾向農民推廣將進口乳牛與水牛交配，使產下的牛不僅可耕作，又可產牛奶。但在農民心中，水牛是同甘共苦的伙伴，豈可與門不當戶不對的「番牛」亂配。更令農民無法接受的是，如此產下的小牛簡直「四不像」。這項推廣工作終於宣告失敗。

風水輪流轉，近來「四不像」却大受歡迎。

A dozen or so years ago the Taiwan Department of Agriculture and Forestry urged farmers to import a cross between a water buffalo and a dairy cow that was said to produce milk as well as to work in the fields. Now to farmers on Taiwan the water buffalo is more than a beast of labor — it's a faithful companion and friend — and they refused to accept the idea of its being crossed with a ''foreign cow.'' The result, they felt, would be a ''monstrosity that was neither fish nor fowl,'' and the promotional campaign finally had to be called off.

But times change, and ''monstrosities'' have recently become popular attractions.

不久前，民營的臺南動物園展示了一種罕見的動物「彪」。牠是將二隻不同種的動物——獅、虎交配繁殖而成的，目前有五隻。這幾隻世上稀有的動物頗受大衆傳播媒體青睞，還上過電視亮相，臺南動物園也因此吸引了不少遊客。

多年前，國外也曾有人嘗試將這二種不同的動物交配，公獅配母虎產下的子代稱爲「獅虎」，公虎配母獅產下的稱爲「虎獅」。

畜產與野生有別

臺灣大學動物系副教授李玲玲表示，由遺傳學上來看，生物長久以來自然演化結果，各種動物有其特有的習性、構造……，不同種動物不太可能自然交配、繁殖。

即使加入人爲因素，將牠們關在一起，彼此也不見得會看上對方，就算被湊和成功，但血統相差太遠，如獅、虎交配，結果通常有幾個可能：根本不會生育、生下不良的下一代、或是第二代不孕。因此，這種人類「代替」上帝創造出來的動物，通常只是曇花一現，很快又絕跡。

但有人以爲，老祖先也曾將馬、驢交配，產下的騾子兼具二者優點，更有耐力和負重力，成爲很好的運輸工具，也沒看到什麼不良影響；獅、虎的配對又何嘗不能試試看？

事實上，爲了人類利益，畜產科學家常會把不同的動物配對，如水牛配乳牛、山羊配綿羊，「但那限制在同一種動物下的不同品系，而且是在人類已飼養、觀察及研究很長一段時間的畜產動物，不是用在人類還不够了解的野生動物上；」臺灣大學畜牧研究所教授宋永義強調：「這樣做的目的是爲大多數人的利益，不是爲滿足少數人的好奇。」

牛頭對馬嘴，對出什麼？

「獅子、老虎是野生動物，把牠們關在籠子中已不够人道，還要強迫他們不自然配對，實在說不過去，」農業委員會保育科陳超仁的看法傾向於人對動物應講人道。

臺灣大學動物系教授林曜松則以爲，即使不以道德論此事，這樣做仍令人擔心。

馬配驢雖是早有的事，而且現在因爲人們已有更便捷的交通工具，已很少「製造」騾子，也未發現騾子曾對環境產生什麼不良影響，但這並不表示不同種動物交配就都不會有不良結果，很可能只是因爲人類對大自然的了解仍太有限，還不知道而已。尤其這類情形多了，後果就更難預料。「因此最好不要輕易去攪亂自然狀態，」最近應農委會之邀來臺演講的美國動物學家Bart Ogara 博士也說。

李玲玲則表示，若是爲了學術上的驗證，想知道不同種動物配出來是怎麼回事，或是爲了動物科學上的研究，這樣的行爲才算有意義。

龍配龍，鳳配鳳

現代人在生活富裕之後，喜歡與衆不同的東西，而且有餘力去追求。因此有動物園以擁有稀有的動物招徠遊客，例如把雄山豬和母老虎關在一起，希望牠們「完成好事」；去年底還有動物園把黑熊染色，說是「黑白熊」；許多人養寵物，也以擁有別人沒有的動物爲傲。

「如此也造成許多世界瀕臨絕種動物非法進口，」農業委員會主管動植物進出口的保育科技士吳冠聰說，而且因爲有人肯出高價，就有人到野外去抓，使野生動物數量不斷減少，再加上亂交配野生動物，使保存各種動物品種的希望也愈渺茫。

「動物世界可做的事還很多，」一位生態學者舉例說，像不久前成立的貓狗收容中心就很好；動物園則可把動物在自然環境中的生活情形、動物與人的關係介紹給大衆，不只是將牠們關在水泥地、鐵欄杆中，或去製造「四不像」。「不要做本末倒置的事，」師範大學生物系副教授王穎也強調。

中國人傳統有「龍配龍、鳳配鳳，老鼠生的兒子會打洞」的說法。大自然有其自我演化、汰選的能力，人類如果扮演喬太守，在自然界亂點鴛鴦譜，會爲自己帶來什麼樣的後果呢？ S

(原載光華七十七年六月號)

Not long ago the privately owned Tainan Zoo displayed a rare animal that is called a *piao*, or a cross between a lion and a tiger. These unusual creatures, of which the zoo now has five, were the darlings of the local media, and the attention they received attracted a great many visitors to the zoo.

In fact, the two species have been crossbred for many years overseas, where the offspring of a lion and a tigress is called ''liger'' and that of a tiger and a lioness a ''tiglon.''

Wild animals are different: According to Li Ling-ling, an associate professor of zoology at National Taiwan University, the different physical structures and habits that animals of different species develop as a result of evolution make it difficult for them to mate together and reproduce.

And even if they are artificially mated, the female may not give birth, or the resulting offspring may be infertile. The creatures that result when man tries to play God are often sorry pieces of work indeed and short lived.

Yet some people point out that mankind long ago crossed the horse and the ass to produce the mule, which combines useful qualities of both. So what's wrong with crossing a lion and a tiger?

In fact, livestock breeders have often mated different kinds of animals, such as a water buffalo and a dairy cow, or a sheep and a goat, to produce improved breeds. ''But crossbreeding has been limited to different breeds of the same species and to animals that have been domesticated and studied for a long time. It hasn't been practiced on wild animals, which we know little about,'' says Sung Yung-i, a professor of animal husbandry at National Taiwan University. And the purpose has been in the greater interest of mankind and not merely to satisfy the curiosity of a few.

Jumbling genes: ''Lions and tigers are wild animals. Cooping them up in cages is inhumane enough, without forcing them to couple with each other unnaturally. That's really going too far,'' says Chen Chao-jen, a member of the Wildlife Conservation Division in the Council of Agriculture.

Lin Yao-sung, a zoology professor at National Taiwan University, believes that, ethics aside, the practice is still troubling.

Even though asses and horses have been crossbred for thousands of years and the mules created have produced no deleterious effect on the environment and have been replaced by more advanced means of transportation, that doesn't mean that crossbreeding different species won't entail negative results. It may be that our understanding of nature is still too limited. Especially if there are more cases, the consequences are hard to predict. ''It's best not to toy with nature,'' says Dr. Bart Ogara an American zoologist who was recently invited to lecture in Taiwan by the Council of Agriculture.

Li Ling-ling says that the only situation in which crossbreeding is justified is scientific verification and research.

Dragons and phoenixes: People today have the time and money to crave things out of the ordinary. Zoos attract visitors with rare species and hybrids, by trying to cross a boar and a tiger, for instance. Late last year one even dyed a black bear to produce what it called it a ''black-and-white bear.'' Many people pride themselves on keeping an odd or unusual pet.

''That's what leads to illegal trafficking in endangered species,'' says Wu Kuan-tsung, a colleague of Chen's in the Wildlife Conservation Division. Because people are willing to pay a high price for rare animals, others go out in the wild to catch them, constantly reducing their numbers. If some people are thoughtlessly crossbreeding wild animals at the same time, that further reduces the hope of preserving them from extinction.

''There are many positive things that can be done in the animal world,'' an ecologist says, citing the Cat and Dog Shelter that was set up recently and suggesting that zoos introduce the public to animals in their natural state instead of locking them up in iron cages with concrete floors and trying to produce ''monstrosities.'' ''Don't play topsy-turvy with the animal world,'' emphasizes Wang Ying, an associate professor of biology at National Taiwan Normal University.

''Dragons mate with dragons and phoenixes mate with phoenixes, while the offspring of a rat will always dig a hole,'' a traditional Chinese saying goes. Nature has its own methods of propagation, selection and evolution. If man tries to play God, what will be the consequences?

(Chang Chin-ju/tr. by Peter Eberly/ first published in June 1988)

我的長相就是這樣，不難看吧！陣陣漣漪是吐氣時造成的。 Here's what I look like — not bad, huh?

海龜回家

A Sea Turtle Goes Home

文／圖・鄭元慶

爬過沙灘，方一觸到清晨凉冽的海水，半個多月來的懷疑和疲憊，驟地消失。然後我開始認眞思考所謂「成見」的問題。

或許眞的和年齡、閱歷有關；有了這兩次經驗，我對人類的看法的確有了改變。

As soon as I crawled over the beach and touched the cold morning seawater, the doubts and fatigue I had felt over the past few weeks suddenly vanished. Then I began to think about what I had been through.

Maybe it has something to do with age and experience, but with these two experiences behind me, my idea of humans has really changed.

我是隻「海龜」，作學問的稱作「綠蠵龜」，身長九十公分，寬六十公分，體重目前總在一百廿公斤上下。

海中歲月長。閒著也是閒著，我常常一邊游、一邊想些事情。據說，在二億五千萬年前的龜類已在地球上這麼游著了。除了海龜，還有陸龜、淡水龜，細分種類有二百五十種之多。最小的只七公分長，最大的有六百八十公斤……。不知道牠們是不是也和我一樣，聽、嗅、視覺靈敏，能辨別顏色却看不遠？並且愛思考？

我好比那淺水龍困沙洲

今年四月十八日，天氣有些悶熱，我游著、想著，忍不住打了個盹兒，不想醒來却跟一羣傻魚兒困在張大網裏。挨到日落，漁民收網時，看到我縮頭縮尾（怕嘛！）地躺在網裡，就有人眉開眼笑地嚷著可論斤出售，還有個年紀大些的叫著說：「海龜肉比牛肉好吃哪！」

曾經聽同伴們說過，日本南方小笠原羣島的居民，有吃海龜肉的習慣，難道這裡的人也一樣？我對人類感到失望。

我仍然把腦袋和四肢都緊緊縮在殼裡。海龜類壽可百歲，除了天賦，就靠這身雖薄却硬的甲冑了。相信嗎？它貌不驚人，却經得起鱷魚大咬一口呢。

還是得想個辦法，慌亂之中倒也心生一計；當下就盡量伸展前脚，「秀」出腿上的不銹鋼「護身符」。一位眼尖的漁民看到了問：「這是什麼？」一陣議論，他們終於承認，我是隻不平凡的海龜，也就不敢隨便「處理」。

I'm a sea turtle, *Chelonia mydas japonica* to be exact, 90cm long, 60cm wide and about 120kg in weight.

We've got plenty of spare time in the sea, and I do a lot of thinking while I'm drifting about. They say turtles have been around for some 250 million years and come in over 250 different species, including land and freshwater varieties, ranging in size from 7cm long to 680kg in weight. . . . I wonder if they can all see, hear and smell as well as I can? If they can see colors but are nearsighted like me? And if they like to think?

Netting a whopper: April 18th was warm and sunny. I was drifting around, thinking like I do, when I wound up dozing off. When I woke up, I found myself caught in a net with a bunch of stupid fish. The fishermen hauled us in at sunset. They let out a whoop when they found me, curled up in my shell (I was scared), and started jabbering about how much they could fetch for me. Why, one old geezer even cried out, "Turtle meat's tastier than steak!"

I had heard a buddy of mine once say they eat sea turtle on the Ogasawara Islands in southern Japan. Don't tell me people here do that, too, I thought. I was disappointed in the whole human race.

I stayed huddled up in my shell. We seas turtles can live up to a hundred, you know, thanks largely to our hard, thin armor. You don't believe me? Our shells may not look like much but they'll withstand a big chomp from a crocodile.

I had to think of something fast. Then it came to me. I stretched out a foreleg and flashed one of my stainless steel "lucky charms." A keen-eyed fisherman spotted it and said, "What's that?" After some discussion, they finally realized that I was no common garden-variety sea turtle and shouldn't be "disposed of" quite so summarily.

A blessing in disguise: These "lucky

大難不死、必有「後福」

說起這道「護身符」，那也是不知多少個日出日落、潮來汐往之前的事了。那時我還是隻幼龜，在美國夏威夷羣島附近給人打撈了起來。一陣虛驚，這些夏威夷大學的海洋生物研究人員並沒有太折騰我，只觀察一番，在我兩隻前腿各套了一片號碼牌後竟把我放了。

感謝上帝，我邊游邊想，原以爲此命休矣，沒料到能夠大難不死，至於前腿上的牌子也不算太礙事。更沒想到的是這回，它又成了我的「護身符」。

但他們終究沒有輕易放掉這筆發小財的機會，爲了想探聽探聽「行情」，他們找來一個叫陳恆裕的中年人。

聽他們說起來，這陳先生是位貝類專家，當年龍宮貝殼和白蝸牛就是他發現的，他也是省立博物館的榮譽標本製作專家。自從老伴中風後，就在此地（臺灣南端的恒春）買塊地，做起進口貝殼的生意。

陳恆裕果然是個內行人。看了名牌後面的英文字，他立刻知道我是夏威夷大學用來作研究的，我有責任在身，他主張把我放掉。

(上)美國夏威夷大學研究人員替我掛的名牌，成了我的護身符。

(above) The tag that the researchers put on me is my lucky charm.

(右)龜甲很硬但不重，它護衛了我一生，我喜歡這個甜蜜的負擔。

(right) My shell's protected me all my life; it's a burden I don't mind bearing.

海龍王的恩賜？

漁民們却認爲我是海龍王的恩賜，至少要「意思」一下才能放我走，承蒙抬舉，我的身價還眞不低，陳恆裕搖搖頭，走了。

我暫住在其中一位漁民家裏。

過了半個多月，漁民們終於明白耗下去也不是辦法，決定削價賣給陳恆裕。五月六日。我永遠也不會忘記的日子，這位陳先生帶我回到他家，歇在有打氣幫浦、住來還挺舒服的池裏，接著他打電話給墾丁國家公園管理處，請他們協助放生，並寫信給夏威夷大學告知此事。

當天就有報社的駐地方記者聞風而至。第二天，我的新聞和照片果然見了報，小出一陣鋒頭。那天傍晚，還有一家雜誌社的編輯遠從臺北趕來（眞快！四百多公里的路，我得爬多久！），慷慨的陳恆裕殷勤接待，又爲他解釋我的來歷。他說：

「海龜的成長速率很慢，夏威夷的海洋生物專家，曾經捉了六百隻年幼的海龜，標示後放生。過段時間捉回七十隻，但多數沒有明顯變化。他們推測要一、二十年，才達到成熟階段。跟大多數動物比較起來，海龜的成長眞是够慢的了。」

到底是專家，由於上回「大難不死」的經歷，知道得比我還淸楚，原來，我還對研究萬物生靈的學術界有點貢獻呢！

老祖宗占卜用

他又說：「海龜具有廻游特性，多數都固定在一個地方覓食，然後到另一處產卵。有一種海龜會集體從巴西海岸，游過一千六百公里，每年二月左右到達非洲和南美間的亞

charms'' of mine go back a long, long way. When I was a youngster, some marine biologists from the University of Hawaii plucked me out of the sea near Hawaii. It was scary, but they didn't mistreat me. They just looked me over, attached a numbered tag to each of my front legs and then let me go.

Thank God, I thought as I swam away. I'd figured I was a goner for sure. As for the tags, they didn't bother me very much — and they even turned out to be my ''lucky charms.''

The fishermen didn't give up their get-rich-quick scheme so easily, though. To find out what my ''going rate'' was, they called in a man named Chen Heng-yu.

From what I heard them say, Mr. Chen is a crustacean expert who discovered two new species of crustaceans and a taxidermist emeritus for the Taiwan Provincial Museum. After his wife suffered a stroke, he bought some land in the area (at Hengchun, on the southern tip of Taiwan) and started up a shell-importing business.

Mr. Chen really is an expert. He took one look at the English on the back of my tags and found out that the University of Hawaii was using me for research. I was on a mission, he said, and urged them to let me go.

Special gift?: But the fishermen seemed to think I was a special gift from the Dragon King of the Ocean. And they deserved a little ''consideration'' for letting me go. Mr. Chen just shook his head and left.

I stayed in one of their houses for a while. After a couple of weeks, they realized they were getting nowhere and cut a deal with Mr. Chen.

May 6th. That's a day I'll never forget, the day Mr. Chen took me to his house and put me in a comfortable, aerated pool. He called Kenting National Park Headquarters and asked them to help let me go and write to the University of Hawaii to notify them about what had happened.

The same day some local reporters turned up, and the next day my story and pictures were all over the papers. That evening a writer showed up from a magazine in Taipei. (He was really quick! I hate to think how long it would have taken me to crawl all that way!) Mr. Chen hosted him warmly and explained my background this way:

''Sea turtles have a very slow growth rate. Marine biologists at the University of Hawaii marked 600 young sea turtles and let them go. After a while, they caught 70 of them again. Most had hardly changed. They suspect it takes 10 or 20 years for sea turtles to mature, a very slow growth rate compared with that of most other animals.''

It turned out he knew more about it than I did. It seemed I was making some small contribution to scientific research, at any rate!

Ancestral divination: ''Sea turtles migrate long distances, from fixed feeding areas to fixed spawning grounds,'' he continued. ''One kind

松森羣島的沙灘上產卵。它每隔十二天產下幾十個蛋，六月之後又集體游回巴西海岸覓食海草。這種海龜每隔二到三年就廻游一次。」

這我就知道得比他更清楚了（家務事嘛！），但他提到另一件事却令我驚訝。他說，在古代中國的商朝，人們若有疑問不能解決，就用龜甲占卜算卦，以定凶吉，並將結果刻在龜甲上。這還發展成中國最古老的文字「甲骨文」。我縮了縮頭，想不到咱先祖還有這段歷史。

第三天上午，我又認識了一個來客。這人叫鄒燦陽，畢業於臺灣大學海洋研究所，是墾丁國家公園管理處保育研究課的工作人員，他開了一部車，帶了幾個幫手——一看就知道是來送我回「家」的！

重返大海

這鄒燦陽，帶著一身南臺灣燦亮的陽光，笑咪咪地指著我說，海龜是國際自然保育聯盟公布禁止捕殺及販賣的動物。墾丁國家公園管理處成立後，已明令漁民不得捕捉，被捉的多半是像我一樣，自己卡在定置網中的。除了少數知識不足，又貪點小便宜的漁民，大多數網到海龜的漁民，多會通知管理處處理，「海龜放生，我們最拿手了」，他說：「已經放過好幾批啦！」

說著他們把我搬上車，行前在我身上寫了幾個字，就像前腿的「護身符」一樣，這也證明了我不平凡的際遇。往海邊的旅途雖然顛簸，但一想到卽將重返大海，也就忍了下來。

行程中，我又開始思考：海中多年，也慣見弱肉強食的生存本質，但畢竟大自然裏，除了優勝劣敗的生存競爭，還有些別的。

車子停在一個叫「風吹沙」的地方，大夥兒合力把我放在沙灘上。

嗅著海風的味道，看著水面的反光，我急切地向前爬去。一步步的脚印留在身後，算是我曾「到此一遊」的見證。

然後，我重返大海。

（原載光華七十六年六月號）

要爬回海裡，還有一小段路。慢慢爬吧！
Still a ways to go yet; just keep up the pace.

哇！還剩幾步路就可重返大海了。
Wow! Just a few more steps.

travels 1,600km from the Brazilian coast to the beaches of Asunción Island, where they arrive in February. The females lay dozens of eggs every 12 days. After June, they return en masse to Brazil. Each turtle migrates once every two or three years.''

Now *that* I happen to know more about than he does (family business, you know!). But he mentioned something else that surprised me. He said turtle shells were used by the Chinese of the Shang Dynasty (16th to 1th centuries B.C.) to record the answers to questions about divination. These ''shell-and-bone inscriptions'' are the oldest form of writing preserved in China. I never imagined my ancestors had a history like that.

On the morning of the day after that, I met a new visitor. His name was Tsou Tsan-yang. He works at Kenting National Park and graduated from the Institute of Marine Science at National Taiwan University. He brought along several assistants — as soon as I saw them, I knew they had come to take me home!

Return to the sea: Smiling, Mr. Tsou pointed to me and said that the capture and sale of sea turtles is prohibited by many wildlife protection organizations as well as by the park. Most of those that are caught have usually gotten themselves stuck in nets like I did. Most fishermen, except for a few greedy ones who don't know any better, notify the park if they net one. ''We've let a bunch of them go,'' he said. ''It's our specialty.''

They carried me to the car and wrote a few words on my back, like the ones on my ''lucky charms,'' to protect me. The road to the seashore was bumpy, but I put up with it when I thought of where I was going.

Along the way I pondered: In the ocean it's strictly a dog-eat-dog world, but there seems to be something more in life than just the survival of the fittest.

The car stopped at a place called Fengchuisha, and everybody pitched in and carried me onto the beach.

Sniffing the salt sea air and glimpsing the sparkling waves, I crawled as fast as I could. The footprints I left behind me were my ''Kilroy was here.''

And then, I was back in the deep. S

(text & photos by Cheng Yuan-ching/ tr. by Peter Eberly/ first published in June 1987)

擬虎鯨浮沈記

One Whale's Story

文／圖・鄭元慶

午後開始漲潮，大部分的旁觀者都退到海堤上，只剩幾個人無奈地守在它身邊。海水漲到他們腰際的時候，又退了下去。

海潮一寸一寸地退，也慢慢帶走擬虎鯨的生命。

During the afternoon, as the tide came in, most of the onlookers retreated to the sea wall, leaving only a few behind. When the water was up to their waists, they too withdrew.

The tide ebbed inch by inch, slowly stealing away the life of the false killer whale.

碼頭邊人影晃動，目光集中在擬虎鯨身上。　All eyes are concentrated on the sight below.

「這是海豚嘛！」「不，報上說是鯨！」
「我看它有五公尺長！」「不只，有八公尺！」
「別說了，要噴水了！快拍！」
岸邊，慕「鯨」而來的人羣比劃著說，目光全集中在水裡的擬虎鯨上。
軀體龐大的擬虎鯨，無畏於「人言人語」，悠然地在水中游著，不時還浮上水面，噴出晶瑩四射的水柱。

大牌訪客驚動臺中港

「我最早看到這條鯨，」臺中港公共關係課的林憲宗回憶說，元月十一日下午，他帶著宏周機械公司的員工到港區參觀，在十二號碼頭附近的航道中，看見一條「大」魚，但是由於距離遠，弄不清楚究竟是什麼魚。
兩天以後，這條大魚終於決心要露面；它一直游近碼頭邊，港務組的工作人員這才從它龐大的腰身，在水中有規律的潛潛升升，和不時噴出的水花，證實是條鯨。
消息傳來，工作人員發現茲事體大。臺中港開港九年，進進出出的大小船隻不下八千艘，但是鯨類擅自闖關，却還是頭一回。
以往，臺灣附近海域的鯨類被人碰上，大多未能得享天年。例如去年五月，一條七公尺半的藍鯨，在桃園縣大園鄉竹圍漁港外海，被漁民捕殺。
因此站在保育的立場，臺中港務局局長沈霖一聽有「稀客」來訪，不敢怠慢，立刻指示港警所和全體人員 小心「護鯨」。

忙找專家護它出海

不過保護歸保護，問題還是沒有解決。這條鯨在港裡四處游蕩，難保不被船隻撞上，再不然，它也可能影響港區右側漁港的漁船作業。所以，除了保護之外，還得研究如何請它出海「回家」。
這一「研究」，問題就來了。鯨類在國內是「冷門」動物，專家難求，不必說港務局沒有這方面的專才，就是附近的臺灣省水產試驗所鹿港分所，也無力支援。最後還是水試所基隆總所李燦然所長，在十四日開了一次緊急會議後，才決定把「護鯨」的大任，交給高雄分所的研究員楊鴻嘉。
楊鴻嘉曾經在日本東京大學，隨世界聞名的鯨類專家西脇博士研究，是國內少數的內行人士。

鯨成了新聞焦點

就在港務局、水試所緊急動員保護的同時，傳播媒體也開始一場追逐賽。
確定它為鯨之後，聯合報與中國時報首先刊出新聞，而且附有圖片；接著，其他新聞媒體也陸續跟進，一時鯨成了報紙、廣播、電視新聞的要角。
「從那時起，課裡的電話就響個不停，都是打聽鯨的消息。記者們更是天天守著它，」臺中港公共關係課長韋年華說：「還有很多人只是想瞧瞧它的模樣，也紛紛湧到臺中港。臺中港是不對外開放的，但你又怎能拒絕遠從中興新村、苗栗來的人？」於是以本地居民為主的遊客，絡繹不絕的抵達港區，無論什麼時候，岸邊總會保持二、三百人。來的人，大多數也要看看個把小時，才「夠本」離去。
這些遊客多半是自備交通工具，就算是事先沒有得到消息，只要往人羣方向去找，也一定可以看到鯨。
前幾天，鯨都待在航道中，「觀衆」看得還不甚眞確。後來，它索性游到貯木池附近岸邊，讓人一睹眞面目。
它在水裡呈8字型迴游，每隔半分鐘就上浮呼吸，而且浮出水面的地點幾乎相同。
觀衆站在岸邊，專心的注意它那黝黑的軀體在水裡游來游去，每當它快要上浮呼吸時，都興奮得互相通告，不時發出陣陣驚嘆。
這時雖然岸上熱鬧得很，可是大家只知道它是鯨類，却還不知道它眞正的「大名」，也不知道要怎麼辦。

鯨「音感」好，循聲而來

元月十五日上午，剛拔掉一顆臼齒的楊鴻嘉和助理研究員陳守仁得到任務，由高雄北上，當晚趕到臺中。

擬虎鯨噴出晶瑩的水花。

The whale's crystalline spray.

"It's a porpoise!"

"No, it isn't. The papers say it's a whale!"

"It must be five yards long!"

"More like eight!"

"Shut up and take a picture — it's spouting!"

The crowd on the shore gesticulated animatedly, their eyes fixed on the creature in the water, a false killer whale. (False killer whales are true whales that resemble killer whales.)

Heedless of their remarks, the vast animal swam leisurely about in the deep, spraying a sparkling jet of water into the air whenever it surfaced.

Special guest: "I saw it first," claims Lin Hsien-tsung of the Taichung Harbor Bureau's Public Relations Office. Lin recalls that he was giving a tour of the harbor to some visitors from a machinery company on the afternoon of January 11 when he spotted a "big fish" in the channel by pier 12. It was too far away to make out exactly what kind.

Not until two days later, when it decided to swim in closer and reveal its true identity, was the "fish" — judging from its huge size, regular surfacing and frequent spouting — determined to be a whale.

The news made a big splash. Over 8,000 vessels had entered Taichung harbor since it was opened nine years ago, but this was the first whale.

Previously, whales that showed up around Taiwan often met with premature demises. A 7.5-meter blue whale was caught and killed by fishermen off the north coast last May.

As a result, when harbor bureau director Shen Lin learned of the arrival of the "special guest," he immediately ordered harbor personnel to guard it from harm.

In search of an expert: Protection was all well and good, but problems remained. The whale kept swimming around the harbor, where it might be struck by a ship or interfere with the fishing boats. So besides protecting it, they also had to study how to send it back home.

There lay the rub. Cetology isn't a popular subject in Taiwan, and whale experts aren't easy to find. Neither the harbor bureau itself nor the branch station of the Provincial Water Conservancy Bureau in Lukang could come up with one. Li Tsan-jan, director of the Water Conservancy Bureau's headquarters in Keelung, held a meeting on the 14th and decided that the right man for the job was Yang Hung-chia, a researcher at the Kaohsiung branch.

Yang had studied under a world-renowned cetologist at Tokyo University and was one of the few whale experts in the country.

Media attention: Just as the authorities were busy mobilizing personnel, the news media went into a feeding frenzy.

Once *United Daily News* and the *China Times* broke the story, complete with pictures, the rest of the media — newspapers, radio and television — soon followed suit. Suddenly, the whale had become a major news item.

"The phone never stopped ringing. They all wanted to know about the whale. Reporters were posted on it day and night," says Wei Nien-hua, the harbor's chief of public relations. "And people who just wanted to get a look swarmed in too. Now Taichung Harbor isn't supposed to be open to the public, but with people coming all the way from Miaoli and Chunghsing Hsintsun, how could we turn them away?"

By this time, visitors — mainly locals — were

十六號上午，楊鴻嘉在擁擠的羣衆和電視訪問中與「它」照面，一眼就瞧出是「擬虎鯨」。

「驗明正身」後，下一步是推測它從何、又爲何誤入臺中港？

「它可能是從臺灣地區最大的鯨類聚居處——彭佳嶼，被寒潮逼到臺中港外海，再聽到熟悉的螺旋槳打水聲而跟了進來，」楊鴻嘉分析。

根據港務局的記錄，首先發現擬虎鯨的日期爲元月十一日，當天只有一艘「桃園輪」入港，因此推斷它可能是跟著「桃園輪」進來的。

經過仔細商討之後，楊鴻嘉擬出了「護鯨方案」。

「用鯨類喜歡聽的聲音引誘它出海最理想，可是國內沒有這種錄音帶，向國外借，又遠水難救近火，而且就算借到，也沒有播放的設備。所以我們決定原始方法——敲打金屬器，」楊鴻嘉說。

他的計策是，用八艘竹筏圍成凵字型，缺口對準港口。然後由工作人員帶著鋼管，在竹筏上敲敲打打，製造鯨類不愛聽的噪音，迫使它游向缺口，回到大海。

「唯一要注意的是，不要亂敲或敲得太用力，以免傷害它的耳膜，」楊鴻嘉再三叮嚀參加「勤前教育」的工作人員。

萬事具備却缺擬虎鯨

十六日晚上，一切就緒，只待天明行動。而擬虎鯨却在這緊要關頭失蹤，使得工作人員陣脚大亂。

港務局港務長宮湘洲立刻出動六艘船，在港區內搜索，遍尋不著之際，臺中港北邊約十五公里處的頂庄海防部隊來電，說在沙灘上發現一條擱淺的鯨。

原來這條擬虎鯨是在當晚偷偷「溜」出臺中港的，沒想到方向沒有掌握好，一早就擱淺在沙灘上。幾位漁民看到這條鯨，以爲是上天帶給他們的意外之財，紛紛拿出魚網，想把它弄上岸來，不料緊要關頭，閃出一隊海防人員，大鯨才得免「上岸」。

看準浪潮，推鯨出海。
Watching the tide and pushing the whale out.

不久，楊鴻嘉、陳守仁和港務局的工作人員也趕到現場，只見擬虎鯨在淺灘上無助的搖尾、掙扎，爲了助它一「鰭」之力，許多人索性脫下皮鞋、捲起褲管，涉水走到擬虎鯨身旁，希望利用潮水，將它推送出岸。

但是天不助「鯨」，雖然大家費了九牛二虎的力氣將它推出淺灘。可是過不了幾分鐘，它又載浮載沈地被潮水沖回沙灘。如此三番兩次推送無效，只好宣布停止援救行動。

當晚七時半，它終於「離開」了一羣關心它的人。

由進入臺中港到最後的夜晚，在這七天當中，「它」帶給人們幾許興奮之情，也有無限的悵惘。

（原載光華七十五年三月號）

streaming into the harbor area. Two or three hundred were on hand at any one time. Many of them would watch for a good hour or so before calling it quits and heading home.

They came with their own means of transportation. No directions necessary — all they had to do was follow the crowd and they wouldn't miss it.

The first few days the whale stayed out in the channel, but then it swam in closer to shore, near the lumber storage pool, giving everyone a good look.

It swam about in a figure eight and surfaced twice a minute to breathe, almost always at nearly the same spot.

The spectators on shore gazed attentively at the black body swimming about in the water, calling to each other every time it was about to surface and then oohing and ahing.

It was a lively scene, all right, even though no one had any idea exactly what kind of whale it was or what should be done about it.

Lured by a sound: On the morning of the 15th, Yang Hung-chia, who had just had a molar pulled, left Kaohsiung with his assistant, Cheng Shou-jen. They arrived in Taichung that evening.

Early the next day, Yang, surrounded by television cameras and crowds of spectators, first met up with "it." One glance told him that "it" was in fact a "false killer whale."

Now that the whale had a name, the next questions were: Where had it come from? And why had it entered the harbor?

"It may have been carried to the coastal waters off Taichung by a cold current from Agincourt Island — that's where the most whales congregate in the Taiwan area. It may then have followed the familiar sound of a ship propeller into the harbor," Yang speculated.

According to harbor records, the only propeller ship to have entered the harbor on January 11th, when the whale was first spotted, was the Taoyuan, and so that is the ship it most likely followed in.

After intensive consultation, Yang came up with a "whale escort plan."

"The best thing would have been to lure it back out with a sound whales like to hear," Yang says. "But we don't have that kind of tape recording here, and even if we'd tried to borrow one from abroad, it would have taken too long to get here and we didn't have the right kind of equipment to play it with. So we opted for a more primitive approach — banging metal rods together."

The plan was to box the whale in on three sides with bamboo rafts and then, beating and pounding the rods together, to drive it in the direction of the open side toward the sea. "Just don't bang too hard and break its eardrums," Yang cautioned the recruits.

All set to go but the lead player is missing: By the evening of the 16th, everything was set and ready to go into action the next morning. But at this critical juncture, the whale disappeared, setting the team in a tizzy.

Kung Hsiang-chou, chief of harbor operations, immediately dispatched six boats to search the harbor. As they were fruitlessly searching, the Coast Guard at Ting-chuang, 15 miles north, reported that they had found a whale grounded in shallow waters on a nearby beach.

It turned out that the whale had slipped out of the harbor that night but, disoriented, had ended up stuck in the shallows in the morning. Some fishermen who had seen it, not ones to look a gift whale in the mouth, had cast out their nets and tried to drag it ashore, but the Coast Guard had come by just in time and stopped them.

Soon thereafter, Yang, his assistant and harbor personnel rushed to the scene, where they found the beached whale helplessly flopping its tail and struggling to swim free. Bystanders had kicked off their shoes, rolled up their trousers and waded out, trying to push it back out to sea on the tide.

But the fates were not kind. Although the volunteers managed by strenuous efforts to free it several times, the whale washed right back again a few minutes later. After three or four fruitless attempts, they had no choice but to call off the rescue.

That evening at 7:30, the whale finally "left" the crowd of people who were so concerned for it.

During the seven days since it entered Taichung harbor up until that last night, "it" had brought us a wealth of excitement and more than a passing touch of sadness at its parting. **S**

(text & photos by Cheng Yuan-ching/ tr. by Peter Eberly/ first published in March 1986)

近來大臺北興起一股飼養長臂猿、白鼻心等野生動物之風。(黃麗梨攝)
Metropolitan Taipei has recently succumbed to a fad for keeping long-armed gibbons, Formosan gem-faced civets and other wild animals as pets. (photo by Huang Li-li)

別把野生動物「寵」壞了

Don't "Pet" the Animals

文•張靜茹　圖•張良綱

爲了保護野生動物，去年六月臺灣已通過討論多時的「野生動物保育法」。然而在許多人爲動物爭取權利的同時，由進口的長臂猿、紅龍魚、變色龍，到本土的石虎、老鷹、白鼻心……，臺灣近來也吹起一股飼養「珍禽異獸」風。

若說宰殺各種野生動物進補等行爲，給臺灣帶來虐待動物的惡名；那把野生動物當「寵物」，是表示臺灣逐漸要轉型爲「動物天堂」？或者只是野生動物的另一個樊籠？

The long-discussed Wildlife Conservation Law was finally passed last June, Meantime, Taiwan has been swept by a fad for keeping exotic pets — from imported gibbons, chameleons and Asian bony tongue fish to indigenous eagles, ligers and gem-faced civets.

If the use of wild animals in food and medicine has brought the island ill repute over the years, then does keeping them as pets mean that Taiwan is becoming an animal paradise? Or simply another form of cage?

臺灣近來吹起的「珍禽異獸飼養風」到底有多強？

ㄅㄆㄇ猴園的負責人楊永賢，以其和動物商的熟識與了解大膽地估計，臺灣私人飼養的長臂猿，數量不會少於全世界動物園加起來的總合。

「每艘由印尼來的船上，至少都有十隻紅毛猩猩，」獸醫祈偉廉接觸過許多帶各種野生動物來醫治的人，他認爲大臺北市已成爲這股強風的「颱風眼」。

大臺北最熱門

建國花市的一角彷彿舉行動物大展。狗、貓、鳥、兎子之外，石虎、蛇、菓子狸紛紛出籠，一位圍觀者對著籠中看來一臉無辜的臺灣獼猴，又是同情、又是好奇，與老闆殺起價來。

× × ×

「進口蛇！不咬人，可當寵物又可防身！」人來人往的西門鬧市，一個手上纏了蛇、腳邊籠中還裝滿蛇的賣蛇人高喊著。大膽的過路人接過蛇在手上「把玩」，一邊則夾雜著同伴的嘻叫聲。

× × ×

常逛和平西路寵物店的人會發現，店中的「產品」多樣化起來，紅毛猩猩、長臂猿外，還有標上「臺灣特有種」的鳥類，號稱摩

Just how popular is the fad for keeping wild animals as pets in Taiwan?

Based on his knowledge and understanding of local animal dealers, Yang Yung-hsien, owner of the ABC Monkey Park, boldly asserts that as many long-armed gibbons are kept by private individuals in Taiwan as in zoos all over the world.

''Every ship that arrives from Indonesia has at least ten red haired apes on it,'' says veterinarian Chi Wei-lien, who has had people bring him all kinds of wild animals for treatment and believes that metropolitan Taipei has become the ''eye of the storm'' for this fad.

Hot spot — Taipei: One corner of the Chien Kuo flower market in Taipei is given over to what seems to be a menagerie show. In addition to the standard dogs, cats, birds and rabbits, there are also snakes, ligers and gem-faced civets on display. One onlooker, gazing with a mixture of pity and curiosity at an innocent-faced Taiwan macaque in a cage, starts bargaining with the vendors.

* * *

''Imported snakes! Won't bite! Perfect as pets or self-defense weapons!'' shouts a man in the bustling Hsimen district with a snake wrapped around one arm and a cageful of them next to him. A bold passerby takes the snake in his hands and plays with it, to the amusement and surprise of his friends.

* * *

People who regularly browse the pet shops of Hoping West Road have found that the merchandise has diversified recently: There are now red-haired apes, long-armed gibbons and birds labeled

國人認為銀帶與紅龍魚會帶來財運，因此在臺灣大受歡迎。（黃麗梨攝）

Asian bony tongue fish are considered by the Chinese to bring good fortune and are highly popular in Taiwan. (photo by Huang Li-li)

登寵物的「變色龍」也有了專賣店……。

這股風是怎麼吹起的？

可以由動物學者與保育人士哭笑不得的事情分曉。「好像特稀有動物保育風氣愈盛，想養特稀有動物的人愈多，」中華民國野鳥學會臺北分會理事長陳葉旺舉例為證。

過去鳥店都是進口鳥，或人工繁殖的鳥；現在則可以看到臺灣較稀有的鳥種，如臺灣藍鵲、角鴞、藪鳥、小彎嘴畫眉……公然被放在架上。

「說穿了，不過是許多人想要養隻與衆不同的動物來『炫耀』，」師大教授呂光洋說的露骨，因此業者為了滿足顧客需要，就會想法子找來貨源。

很多人又相信動物會帶來好運，像紅龍魚被認為是財運的象徵，因而使臺灣成為紅龍魚的最大進口地。雖然有人是因為見了籠中動物，興起「同情」之心而購買，但這都「造就」了動物市場。

痛苦比樂趣多

若依人類對動物的利用方式，把動物當寵物，和利用具經濟價值的家禽與家畜、科學實驗的動物一樣理所當然。實際上許多動物也都不只具有一種「功能」。人類對動物的依賴幾乎可以寫一部歷史。

現代心理醫師更證明，和動物做「第一類接觸」——養寵物——有治療心理疾病的作用。現在家庭人口少，老人有動物為伴可抒解寂寞，兒童還可由與寵物相處中，學習愛和關懷他人。

但並非所有動物都適合當寵物。狗、貓和作為經濟利用的家畜，是因為長久來被人類有計畫馴化，也已具備一套飼養、管理方法；但許多未經馴化的野生動物，即使過去被飼養過，因為體型較大，飼養經驗未傳遞、

''rare Taiwanese species'' for sale, and one store specializes in chameleons, touted as the perfect modern pet.

How did the fad get started?

Zoologists and environmentalists don't know whether to laugh or cry. ''It's as if the more steam the animal protection movement picks up, the more people want to keep them as pets,'' says Chen Yeh-wang, president of the Taipei chapter of the R.O.C. Wild Birds Society.

The birds in pet shops used to all be imported or artificially bred, but now there are many rare indigenous species as well.

''The truth of the matter is, a lot of people just want an unusual animal so they can show off,'' says Lu Kuang-yang, a biology professor at National Taiwan Normal University, adding that what consumers want, businessmen will supply.

Many people believe certain animals bring good luck: Asian bony tongue fish, for instance, are considered symbols of wealth and fortune, and Taiwan has become the number-one importer of them. And while some people may buy animals because they feel sorry to see them in cages, that only exacerbates the problem by stimulating the market.

More grief than pleasure: If you look at the question from the perspective of use, keeping animals as pets is as natural as raising domestic livestock for its economic value or employing animals for scientific experiments. In fact, many animals have more than one function. Mankind's dependence on animals forms an entire history of its own.

Psychologists have verified that ''close encounters of the first kind'' with animals — keeping them as pets — can have a therapeutic effect. In an age when families are getting smaller and smaller, an animal companion can relieve loneliness for the elderly and teach children empathy and concern for others.

Not all animals are suited for the part, however. Dogs, cats and other domesticated species have been systematically tamed and bred over the ages, and a regular set of methods has been developed for their care and feeding. But attempting to keep wild animals, which may be large in size or never have been kept in captivity, in modern apartments often causes both animals and masters more grief than pleasure.

''Foundlings'' frequently turn up at classroom doors or at the office entrance to the department of zoology at National Taiwan University in the form of pangolins, flying squirrels, parrots and other creatures, most of them in parlous condition.

''A parrot that sells for NT$10,000 and up kept defecating and tilting its head to one side, as though it had lost its sense of balance,'' says Wu Hai-yin, a graduate student in the department. ''It kicked the bucket the very next day!''

Chi Wei-lien knows the feeling. The veterinary clinic he works in, which is on Chung Hsiao East Road, has treated long-armed gibbons, gem-faced civets and other exotic creatures. A little flying squirrel that was brought in once didn't seem sick, except it refused to eat. He gave it first aid and offered it substitute foods, but it died a few days later anyway.

Trial and error: Chen Chao-jen, a wildlife protection specialist in the Council of Agriculture, says that many animals have a high mortality rate and are already latently ill before they reach the buyers' hands because of a lack of proper food, water or ventilation or due to the shock of being captured, transported and confined.

Chien Hui-jung, a reporter with the *China Evening News* who is knowledgeable about animal smuggling in Taiwan, relates that half a shipment of red-haired apes that an animal dealer imported from the tropics one winter caught cold and died, while many of the remainder failed to adapt to the urban environment and died before he could sell them.

Even if the animals are fine and healthy when sold, most owners don't know how to take care of them and dealers are interested only in getting them off their hands. The result is, the owners have to learn through trial and error. ''Animals are always sacrificed as guinea pigs in the process,'' Chi says bluntly. The most common problem owners have is they don't know what to feed them. Some animals are willing to eat substitute foods, but many others simply starve.

What's more, some animals that are normally very docile undergo massive behavioral changes as soon as mating season arrives, leaving their owners at wits' end. A gem-faced civet gobbled up the newspaper spread on the bottom of its cage. ''Fortunately, we were able to operate on it in time and save its life,'' says Chi Wei-lien, recall-

或根本無法馴化，養在現代公寓裏，往往給雙方帶來的是「痛苦比樂趣多」。

同樣是大臺北市，在一些較「不爲人知」的角落，場景就有了變化。

臺大動物系辦公室或教室門口，常有人送來「棄嬰」——穿山甲、飛鼠、鸚鵡……送來時大多已奄奄一息。「一隻市價上萬的鸚鵡一直下痢，平衡感已不佳，歪斜著腦袋，」臺大動物系研究生吳海音比了個手勢：「第二天就死了！」

祈偉廉的經驗更豐富，他位在忠孝東路的獸醫院，醫過白鼻心、長臂猿……。曾有一隻未成年的小飛鼠病患，看起來沒病，就是不吃東西，給予急救和代用食物，幾天後仍然休克而死。

錯誤中學習經驗

農委會保育科技士陳超仁表示，許多動物在到達買主手中之前，由於捕捉或運送過程，可能因爲饑餓、脫水、擠壓、驚嚇過久，或在船倉中悶了好幾天，都已潛伏病因，飼養的死亡率很高。

一位對國內走私進口動物頗爲了解的中時晚報記者也提到，某次有一動物商進口一批熱帶地區來的紅毛猩猩，由於進口時是冬天，結果有半數因爲感冒死掉，加上不適應人爲環境，在動物商手中又死掉一部分。

即使健健康康抵達買主手中，由於國人對野生動物的了解太少，動物商也只管把動物賣出去，買者通常都缺少飼養常識。因此國人多是「在錯誤中學習經驗」，祈偉廉說得直截了當：「總有些動物被當成實驗品折騰、犧牲掉。」

最常見的就是不知應該給牠們吃些什麼？有的動物也許願吃代用品，但仍有許多寧可餓到嗚呼哀哉。

一些平時很溫馴的動物，一到發情期却性情大變，主人沒有任何提防，就有一隻白鼻心因此生吞主人墊在籠底的報紙，「所幸及時送來開刀，撿回一條命，」祈偉廉回憶自己碰到的一件緊急情況。

野生動物不似馴化動物，很難配合主人的

生活作息，像小飛鼠是夜行性動物，晚上你要休息，牠想活動，結果只好日夜都被關在籠中。

小時可愛，大未必佳

不好過的不只是動物。未經馴化的野生動物，往往小時候還好應付，但一長大就便成「好動兒」，主人被整得精疲力竭。研究臺灣獼猴的吳海音以現在最多人養的猴子爲例，成年後的猴子活動量很大，破壞力強，還有所謂的「折枝行爲」，不但會把樹木花草全整死，到了發情期還會攻擊人，甚至對主人有不雅的動作，「小時可愛，長大變得可恨」。

祈偉廉就有位朋友花十二萬臺幣買了隻紅毛猩猩，最後因關不住，八萬元又賣回給原來的主人。

祈偉廉的例子不勝枚舉，有隻被人從小養大的臺灣黑熊和主人玩，結果把主人抓的皮開肉綻，主人一氣之下把牠賣給動物商殺掉了。另一個也是熊黑的例子，則是長大後請獸醫將之去爪，以防發生意外。許多養大型鳥的人，爲防止鳥類跑掉，就將鳥翅膀連尾骨一起剪掉，以防羽毛再生。

爲了「和平共存」，結果都是「做些違反動物意志的事，」陳超仁替動物抱不平。

獸醫遭受考驗

由於獸醫界的研究範圍很少涉及野生動物，一般獸醫對其所知有限。因此萬一動物生病，「往往連獸醫也束手無策，」家裡養有紅隼、長臂猿的陳連興表示，他曾把生病的鸚鵡送給鄰近的獸醫看，醫生說沒有毛病，結果鸚鵡當晚就「過去」了。

對野生動物醫療頗有興趣與研究的祈偉廉以爲，一般性的醫療都還算好，但他仍覺得自己「每回都在接受考驗，」他說，野生動物醫療在全世界都算是一門新領域，仍然在摸索中。

國內，臺大獸醫系開設有關野生動物醫療的課程不久；國外經驗雖然比我們豐富，但國外獸醫多半不願醫治被養在籠中的野生動

大陸畫眉入主臺灣天空，臺灣畫眉面臨難以維持純種的危機。(郭智勇攝)

Since the mainland *hwa-mei* entered the skies above Taiwan, the Taiwan *hwa-mei* has faced the crisis of losing its unique identity as a species.

ing an emergency he dealt with personally.

Nor do wild animals fit in well with human's living habits. Flying squirrels, for instance, are nocturnal by nature and active just when their owners want to sleep. The result is, they wind up locked in a cage 24 hours a day.

Amiable to odious: The animals aren't the only ones who have it tough. Undomesticated wild animals are often easy enough to deal with when young, but they become ''hyperactive'' when they're older. Wu Hai-yin, an expert on the Formosan rock-monkey, takes monkeys, a popular pet these days, as an example. Mature monkeys are extremely energetic and can be quite destructive. Not only is their so-called ''branch-breaking habit'' a constant threat to trees and plants — during mating season they may attack people or even make advances on a female owner. ''They're cute when they're young, but hateful when they're older.''

Chi Wei-lien has a friend who paid NT$120,000 for a red-haired ape and ended up selling it back to the original owner for NT$80,000 because he couldn't control it.

Similar examples are too numerous to detail. An owner who was mauled playing with a Formosan black bear that he had raised since it was a cub was so mad he sold it to an animal dealer to have it slaughtered. Another owner asked to have his black bear declawed. Many owners of larger birds have their wings and tails clipped off to prevent them from flying away.

In the name of peaceful coexistence, ''their rights and proclivities are sometimes violated,'' Cheng Chao-jen says, speaking up for the animals.

A test and a trial: Since veterinary training rarely covers wild animals, most veterinarians have a limited understanding of them. Should an animal happen to fall sick, ''even if you find a vet, they often don't know what to do,'' says Chen Lien-hsing, who has raised a falcon and a long-armed gibbon. He once sent a sick parrot to a nearby animal clinic. The doctor said nothing was wrong, but the parrot died that evening.

Chi Wei-lien, who has considerable interest and experience in treating wild animals, believes that standard veterinary practices work well enough on them but still feels that ''each time is a test and a trial.'' The treatment of wild animals is a new field of medicine that is still being explored in countries around the world.

In Taiwan, a course in wildlife medicine was instituted at the department of veterinary medicine at National Taiwan University only a short time ago. Veterinarians overseas have more experience with wild animals than we do in Taiwan, but most of them prefer not to treat wild animals held in captivity because they don't want to become accomplices in ''aiding and abetting'' the practice in keeping wild animals merely for show or display.

Transmitting diseases: Many vets, including Chi Wei-lien, worry about the spread of disease from wild animals to humans. A rumor two years ago that Tibetan mastiffs smuggled into the island had rabies created a panic at the time, but at least there are shots for rabies, which is not the case with many other infectious animal diseases.

Snakes are sometimes touted for sale to women as personal defense weapons, but there are no serums available in Taiwan for many of the venomous snakes imported here. If the snake hasn't been defanged and the owner is bitten, the intended bodyguard may turn into a lethal assassin.

In addition, there is the problem of diseases

貓熊已瀕臨絕種，如果大家再不考慮飼養條件，都想抓隻來養養，下一代眞的只能抱玩具貓熊長大了。圖攝於東京上野動物園。

Pandas are facing extinction, and if people keep trying to catch them and raise them regardless of how they should be fed and cared for, then future generations may really have only toy ones to grow up with. The picture was taken at Ueno zoo in Tokyo.

物，因爲不願「助紂爲虐」，成爲一些只爲炫耀、展覽而豢養野獸，卻不加照料者的幫凶。

成爲傳染病媒介

許多和祈偉廉一樣的獸醫，更擔心的則是疾病會波及人類。前年走私藏獒傳聞帶有「狂犬病」，造成國人的恐慌，但狂犬病還有「預防針」；野生動物的傳染病就防不勝防了。

蛇常被標榜可做爲女性的防身物，但許多進口毒蛇在國內還沒有血清，萬一未拔除毒牙，飼主被咬傷根本回天乏術，怎麼可能靠牠保護主人？

還有「人畜共通疾病」的問題，尤其是和人類同屬靈長目的衆多猴輩。去年十二月一批由菲律賓走私的馬來猴（食蟹猴），被懷疑帶有非洲出血熱病原體，轉口的國家荷蘭非常緊張，並通知我國農委會防範。非洲曾因這個病毒，造成感染者死亡率達百分之五十以上的慘劇。

後來衞生署指出，這種疾病是危險性及致死率極高的病原，操作此病毒實驗必須在所謂的「D-4」高度安全室，但國內根本無此級的設備。

目前所知靈長類的傳染病就多達八、九種以上。祈偉廉表示，國內海關對動物檢疫的能力都嫌不足，就曾有過因檢疫單位畜舍不良，造成動物相互傳染的事；更遑論未經任何檢疫手續的走私動物。萬一是由疫區來，等於一顆未被清除的炸彈上了飛機。即使未對人造成危害，但若傳染給畜產業，如口蹄疫、肺炎，進而使臺灣成爲疫區，畜產業必將遭受嚴重打擊。

溪流中發現食人魚

許多動物在主人發現有問題，或「不好養」「養不好」、新鮮感消失後，就想把動物「送走」，如此一來，動物園或大學動物系，就成了大家認爲「理所當然」的收容所。

木柵動物園被要求收容過雉雞、孔雀、猴子……，種類琳瑯滿目，數量難以計數。動

common to man and animals, especially from primates, which are close to humans biologically. Last December a group of Malay monkeys smuggled in from the Philippines were suspected of carrying the pathogens for hemorraghic fever. Authorities in the transit country of Holland were alarmed and notified the R.O.C. Council of Agriculture to take precautions. The virus once broke out in Africa with a fatality rate of more than 50 percent.

The Bureau of Public Health warned that the disease is extremely dangerous and highly fatal. Experiments on the virus must be performed in what are called D-4 high-safety labs -- which are not available in Taiwan at present.

There are currently more than eight or nine diseases in primates that are communicable to man. Local customs is unable to quarantine animals sufficiently, Chi Wei-lien says. Animals have infected each other because they were improperly quarantined together — not to mention the danger from smuggled animals, which don't go through any quarantine procedures at all. Animals from an epidemic area are like bombs waiting to go off. Even if they don't pose a threat to humans, they could deliver a severe blow to the livestock industry if the diseases they carry, such as pneumonia or hoof and mouth disease, spread to other animals.

Piranhas in the stream: When owners find that wild animals are hard to keep or when the feeling of newness wears off and they want to get rid of them, zoos and university zoology departments become the ''asylums of first resort.''

The Taipei municipal zoo in Mucha has been asked to take in pheasants, peacocks, monkeys and a host of other species. Chen Chien-chih, who works in the animal division, explains that the zoo can't accept all comers because it has to consider the origin of the animals and whether or not they have passed through quarantine. In addition, its capacity and manpower are limited.

Due to a lack of space, they once placed a Formosan rock-monkey that someone had sent them on monkey island, where it came under constant assault from the other monkeys. ''We often get yelled at when we turn people down,'' Chen says, feeling the zoo is misunderstood and ill treated.

Some animals have been freed by their owners to fend for themselves. Fishermen have reported

物組的陳建志解釋，動物園不可能照單全收，因爲必須考慮這些動物的來源，和是否經過檢疫；此外，動物園的容許量和人力也都有限。

由於地方不够，他們曾把別人送來的臺灣獼猴安置在猴島，結果老是被其他猴子痛打。「我們也常常因爲拒收而挨罵，」陳建志覺得動物園很冤枉。

有的動物則被主人「放生」，自求多福去了，近來釣魚客在溪裏發現食人魚和銀帶魚的事情就屢有所聞。專門拍攝臺灣野生鳥類的生態攝影者陳永福，在野外拍照，鏡頭中曾出現巴西雀、八哥、南美洲產的鸚鵡、大陸畫眉等外來鳥種的倩影。

不受歡迎的舶來品

早年常帶人打獵，如今仍不時往郊外跑的楊景星，在一次「爲野生動物保育法把脈」的研討會上表示：「宜蘭平原已被巴西龜佔領了」。

不管這些動物是被主人放生、自行「逃亡」，或是因佛教善男信女所爲，非本土動物已逐漸對臺灣環境造成影響。

「賞鳥時看到大陸畫眉與臺灣畫眉的交配種並不稀奇，」賞鳥經驗豐富的陳葉旺說。

許多保育人士擔心巴西龜會像過去的吳郭魚，被引進後，臺灣原來溪中的溪哥等本土魚種無法與之競爭，逐漸絕跡。

近來國內還興起一股養鷹、放鷹風潮，放鷹時，鷹的逃亡率很高。老鷹是食物鏈中的最高消費者，獵捕能力強。放鷹時，萬一被鷹成功逃亡，到野外獵食小型鳥類，也可能造成鳥類生態系的變化。

許多國家都以美國引進歐洲椋鳥的例子爲戒：因爲椋鳥大肆掠奪野外原有鳥種的巢穴，加上當地沒有天敵，牠們大量繁衍，最後聚集上萬隻，擾亂到人類生活安寧，形成鳥害。

要養就養「特稀有」？

在世界各國都希望能保存自己特有的人文、風俗、物產，並競相展開動植物基因庫的保存時，我們許多原生種類卻可能已被混種，或因競爭不過外來種而在食物鏈中消失。更何況很多野生動物都已瀕臨生存危機。

師大生物系教授呂光洋指出，爲何動物會成爲稀有？就是因爲牠們在野外的族羣已很少，生存很困難，若再因爲「稀少」或「特別」而被看中，族羣維持下去的機會就更加渺茫。

許多稀有或特有種動物，早已被國際保育組織或國內野生動物保育法列入保護，想進口或買賣都必須經過極嚴格的程序。因此市面上見到的動物，很多都是違法捕捉或走私進口的，買者事實上也觸法了。

求偶的基本權利被剝奪

如今國內仍很流行養畫眉，並有一年一度的「鳴唱大賽」，但畫眉鳥只有雄鳥會鳴叫，因此雄畫眉遭人大量捕捉，假使導致野外陰盛陽衰，無形中等於剝奪牠們配對、繁衍的權利。

鳥類的膽子小，被捕捉、上網時「嚇死」的機率很高，根據國外的估計，存活率只有百分之十，因此養一隻野生鳥的社會成本其實很高。

專門處理動植物業務的陳超仁說，一隻表面上看來很可愛的野生動物，背後可能都有個坎坷的「身世」。許多人也會發現，這些野生動物有不少因爲誤入陷阱，不是缺手斷脚，就是瞎眼、沒有尾巴。

三年前我國動物園進口小金剛，曾遭國際輿論指責，除了因爲牠是世界公認稀有保育類動物、我們未經規定程序購買外，看過電影「迷霧森林十八年」的人就會恍然大悟：要得到一隻小金剛，得殺死許多大金剛才換得來。

把野生動物都變成家畜？

不考慮飼養條件而硬要養，這是政府、民間所犯的共通毛病。

近來更有人在養出興趣後，專門以人工繁殖野生動物爲業。如白鼻心、黃喉貂，甚至臺灣獼猴都有人做繁殖，除了買賣，並自稱

catching piranhas in local streams recently, and nature photographer Chen Yung-fu has sighted mynah birds, Brazilian cockatoos and other exotic species on outings around the island.

Unwanted imports: Yang Ching-hsing, who used to take people hunting and still heads off for outings in the countryside frequently, told a seminar called ''Taking the Pulse of the Wildlife Protection Act'': ''Ilan Plain has been taken over by the Brazilian turtle.''

No matter whether they have been abandoned by their owners, have escaped on their own accord or have been freed by devout Buddhists (for whom ''releasing living creatures'' is a traditional virtuous practice), nonnative animals are having a deleterious effect on the local ecology.

''It's not unusual anymore to spot *hua-mei* thrushes from the mainland interbreeding with those from Taiwan,'' says experienced bird watcher Chen Yeh-wang.

Many environmentalists worry that Brazilian turtles will produce the same effect that Tilapia fish did when they were introduced to local streams several years ago: native species will be unable to compete and gradually die out.

A fad for falconry has arisen in Taiwan recently. Falcons are excellent hunters and stand high up in the food chain. If one escapes, which frequently happens, it hunts smaller birds and creates changes in the avian ecology.

The introduction of the European starling to the United States can serve as a warning. They robbed the nests of many indigenous birds and, lacking natural enemies, multiplied to such an extent that they disturbed the local populace and posed the threat of a ''bird scourge.''

Why pick on rare species?: Countries around the world are trying to protect their special cultures, customs and resources and are setting up gene banks to preserve rare animal and plant species, yet many of our own special indigenous breeds may already be dying out through interbreeding or an inability to compete with outside species, not to mention the fact that many types of wildlife are endangered anyway.

National Taiwan Normal University biology professor Lu Kuang-yang points out that the reason a species becomes rare in the first place is it has difficulty surviving in the wild. If it then becomes specially prized as a pet because it is rare or special, its chances of survival become even smaller.

Many rare species are protected by international organizations or subject to strict procedures for import or sale under the Wildlife Conservation Law. Many of the animals seen on the market have been illegally captured or imported, and buyers in fact may be violating the law.

Depriving them of their right to find a mate: Raising *hua-mei* thrushes is popular in Taiwan nowadays. There is even an annual singing competition, but since only the males can sing, they have been captured in greater numbers than the females. The resulting imbalance between the sexes is equivalent to depriving the females of their right to find a mate and reproduce.

Birds are easily frightened and many of them die of fright when they are netted or captured — some foreign estimates say the rate of survival is only 10 percent — which means that raising a wild species of bird entails an incalculable social cost.

Chen Chao-jen, who specializes in plant and animal affairs at the Council of Agriculture, says that a wild animal pet that looks cute and cuddly on the outside may have a rocky history behind it. Many people find that quite a few that lack paws or claws because of falling into traps or are blind or tailless.

Three years ago, Taiwan ran up against a storm of adverse public opinion when a zoo here tried to import a small gorilla. Besides the fact that the animal is recognized around the world as an endangered species and that authorities had failed to go through the proper procedures to buy one, people who have seen the movie *Gorillas in the Mist* will realize with a jolt: to get one little gorilla, many big ones have to be killed first.

Turning them all into domesticated animals?: Insisting on keeping wild animals in captivity without considering the proper conditions for caring for them is a common fault with the government and public alike.

With interest in exotic pets on the rise, some people have gone into the business of breeding wild animals in captivity, claiming not just to make money but to be protecting them from extinction as well. ''There's some thinking here that needs correcting,'' says zoology professor Li Ling-ling.

Animals bred in captivity are often infirm or

人們千萬不要看到動物就想抓來據爲己有，否則動物怎能「放心」與人親近？

If people keep trying to catch animals and keep them, how can they learn to be friendly with man?

是保育動物，使牠不致絕種。「這其中有許多觀念需要導正，」臺大動物系副教授李玲玲說。

野生動物由於數量少或不易捕捉，常發生近親交配，繁殖出一羣體弱多病、血統不良的後代。若有人好心爲使野外族羣不滅亡而將之放生，較好的情況是因無法回歸自然而死亡；但萬一和野外族羣交配，可能破壞野外族羣的品系，「影響野外基因的純度，」李玲玲憂心地說，也可能影響臺灣特有種動物基因的保存。

臺大畜牧系教授宋永義表示，不可把野生動物的人工復育和家畜的人工繁殖混爲一談。「家畜的人工繁殖已不像樣！」他說，原有的基因都已被改變。因此不能只看到可以多量繁殖，就沾沾自喜。

根據國外的估計，一個成功的人工復育需要十五年（請參考本書「復育大自然的少數民族」一文），所需時間、人力、金錢和養育知識，都不是隨便拿幾隻動物來交配那麼簡單的事。

「大家也不願所有的動物都變成家畜吧？」李玲玲說，若眞的愛護動物，最好的方法是保持自然現況，不要再去干擾破壞。爾後請我們的國家公園、自然保護區幫我們好好照顧著，我們再前往大自然一親芳澤，「爲何一定要把動物裝在自己口袋裏？」

讓孩子和大自然一起長大

許多養寵物的人會說，就是因爲現代工商社會離大自然太遠了，才想要養隻動物，不少父母也都希望孩子能有寵物陪伴長大。但一隻動物關在籠中或在侷促的空間，所能展現的只是單純的行爲，在自然界則有覓食、求偶、禦敵等活動，充滿無限生機。

只要盲目的開發不繼續，大自然不見得遠在天邊。大臺北市民可以就近到關渡平原走走，所得到的樂趣絕非在家養動物或到動物園看動物可以比的。「與其讓孩子和寵物長大，不如讓他們和大自然一起長大，」也是國中生物老師的陳葉旺衷心希望。

（原載光華七十九年五月號）

genetically inferior because of inbreeding or confinement. If well-intentioned but misguided people set them loose in the wild, the best thing that can happen is for them to die off before they can mate with their cousins in the wild and spoil the genetic pool.

Sung Yung-i, a professor of animal husbandry at National Taiwan University, says that the artificial propagation of wild animals can't be compared with that of domesticated animals. ''The artificial propagation of domesticated animals is already out of control!'' he exclaims, explaining that the original genes have already been changed. Not every species can be propagated in large numbers as we may wish.

According to overseas estimates, a successful artificial propagation program takes 15 years (see the chapter ''Restoring Nature's Dwindling 'Minorities'''). A lot of time, manpower, money and know-how are required — it's not simply a matter of taking a few animals and mating them together.

''Nobody wants all animals to become domesticated, do they?'' Li Ling-ling asks rhetorically. If we really love animals, she says, the best thing we can do for them is to preserve the natural environment instead of spoiling it and interfering with it. If our national parks and nature sanctuaries are well managed, we can go there and see the animals as they should be, in the wild: ''Why try to stuff them in our pockets?''

Many owners say it is precisely because modern industrial society is so far removed from nature that they want to keep pets — to give their children a companion while growing up. But an animal shut in a cage or confined to a room can only display a small part of the behavior it exhibits in the wild, where it seeks food, chooses a mate and defends itself against its enemies.

Nature is not really that far away. Residents of metropolitan Taipei who take a nature walk in nearby Kuantu will find that watching animals in the wild is much more thrilling than keeping them in an apartment or seeing them at the zoo. ''Instead of having your children grow up with an animal, it's better to have them grow up with nature'' is the fervent hope of Chen Yeh-wang, who is a junior high school biology teacher by profession.

(Chang Chin-ju/photos by Vincent Chang/ tr. by Peter Eberly/ first published in May 1990)

不僅家庭中養寵物，近年來人們對動物園的存在與否，也有強烈的辯論。因爲動物展示產生了市場需求，野外形成大肆捕捉野生動物的殺戮戰場。今天動物園也要由動物的「消費者」轉型爲物種保護的角色。

Not limiting themselves to household pets, in recent years people have been seriously debating whether there should even be zoos. This is because zoo displays create market demand and make the wilds into giant hunting and trapping fields. Today zoos want to go from being animal "consumers" to animal "preservers."

都市不是牠的家？

Surviving in the Urban Jungle

文・張靜茹　圖・邱瑞金

雖然飼養動物帶來許多問題，但是不是除了狗、貓和兔子這樣的「家畜」外，野生動物都養不得呢？國內動物學者與生態保育人士，對此有不同的看法。

師範大學生物系教授呂光洋以爲，卽使像狗這種被人類豢養歷史最久，最「好養」的動物，國人的責任感都令人質疑。大街小巷充斥被丟棄的癩痢狗——「我們還有什麼資格養可能造成更多問題的野生動物？」他因此極力反對。

野鳥學會臺北分會理事長陳葉旺則表示，養動物本身很難說好不好、對不對；而應看時代與環境的背景，決定該採取的「分寸」。今天自然環境被嚴重破壞，並危及到人類本身，我們只有儘量把自己對自然界造成的破壞行爲，減到最低程度，因此「應有限度地養野生動物」。

臺灣大學動物系副教授李玲玲和一些保育人士就都認爲，若眞的喜歡動物，就不應該一定要挑選所謂的特稀有品種。他們建議養狗和貓這些離開人類無法獨立生存的動物，如此也可避免飼養野生動物的麻煩。

中國人養珍禽異獸是有長久歷史的。古代帝王還有專用的苑囿，不僅爲賞玩，也爲打獵。

但此一時，彼一時也，過去野生動物可以供人類奢侈的利用，今天全世界大部分地區的野生動物已變成「弱勢團體」，人類對大自然的態度自然也應有所調整。

動物生命應受到尊重

如今卽使具有「教育意義」的動物園，也常因動物生活空間狹窄和設計不佳而受到批評。國外做爲科學實驗的動物，不管野獸、家畜或寵物，也都有非常嚴格的法令管制。

今天保育人士更開始推展高層次的生態保育觀；對動物，他們提倡動物也有生存權、動物的生命應受到尊重、任何利用動物的行爲都要考量「動物福祉」、把動物養在籠中是不人道的種種說法，來抵制人們不合理地

Keeping dogs, cats and rabbits as pets may seem like trouble enough — can wild animals be kept as pets, too? Zoologists and environmental experts hold various views on the subject.

Lu Kuang-yang, a professor of biology at National Taiwan Normal University, objects strongly. When the streets and alleyways are full of mangy, abandoned canines, which are relatively easy to keep, ''What right do we have to raise wild animals, which can create even more problems?'' he asks.

Chen Yeh-wang, president of the Taipei chapter of the Wild Bird Society of the R.O.C., believes it is hard to say whether keeping animals is right or wrong in and of itself. It all depends on the time and place and how it is done. At a time when the environment has been spoiled so seriously that even man himself is threatened, we must adjust our behavior to cause the least possible harm to the natural world, and so ''the raising of wild animals can be engaged in to a limited extent.''

Li Ling-ling, an associate professor of zoology at National Taiwan University, and others believe that if people truly like animals, they shouldn't insist on keeping rare varieties as pets. She recommends sticking with animals like cats and dogs that depend on humans to survive.

The Chinese have a long history of keeping rare animals. Emperors in ancient times had a special park for viewing them as well as for hunting them.

But times change — that was then and now is now. Wild animals may have served luxurious and extravagant purposes in the past, but now that they have become ''the powerless and the disenfranchised'' in most areas of the world, we humans have naturally had to adjust our attitude toward them.

Respect for animals' lives: Nowadays, even zoos with ''educational significance'' are often criticized for poor design or for keeping animals in cramped quarters. Overseas, the use of animals for scientific experiments — no matter they are wild or domesticated — is controlled by strict rules and regulations.

Wildlife activists today advocate a higher order of thinking. They maintain that animals have a right to live too, that animal life should be respected, that any use of animals must take into account their

你知道「野生動物保育法」對人們虐待、騷擾野生動物的行爲，就如將一隻正常、健康的獼猴關閉在鐵籠裡，將處有期徒刑？

Did you know that under the "Wildlife Conservation Law," mistreatment or harassment of wild animals, such as putting monkeys in iron cages, could land *you* behind bars?

對待動物。

但人類依賴與利用動物的歷史很長，和動物的關係密不可分，層次過高的生態保育觀念在許多地方都還不易被接受。加上目前的生態保育觀是「保育是爲了使人類能長久利用」，而把動物當寵物也是利用動物的方式之一。因此，只要有合理的法令規範，並不是絕對不准養野生動物，仍是目前被比較多人接受的想法。

立機關、淸戶口

現在國際上對野生動物的保護，訂有「華盛頓公約」。它考量世界上現存野生動物的數量和生存環境，把需要保育的動物一一列出，並分成幾類做規範，如第一類根本不准交易，第二類只有學術單位或動物園可購買……，列在公約中的動物幾乎都不准進行私人交易，但這份名單會隨著動物數量的變動加以調整。

不列在此名單上的動物，則由各國自行管理。一些比較重視生態保育的國家，也都有法令列出他們認爲不准捕捉、交易的動物，但通常會把經「人工繁殖、飼養」的野生動物排除在外。

在美國，許多動物雖未被列在保育類，但想要養某些動物的人，卻有像養狗一樣的一套管理制度須遵循。比方立機關，然後把每隻動物都列入登記，飼主想交易、繁殖都須報備。

如此政府可以知道這些動物的增減情形，

welfare, that keeping animals in cages is inhumane and other principles to prevent people from treating animals unreasonably.

But mankind has an intimate relationship with animals and a long history of using and relying on them. Too demanding a conception of wildlife protection isn't likely to win favor. In addition, the current view is that "ecological protection is aimed at facilitating long-term use" and treating animals as pets is one of the methods of using them. As a result, the type of thinking acceptable to most people at present is that the keeping of wild animals as pets should not be prohibited — as long as reasonable rules and regulations are in place.

Keeping track of numbers: Internationally, the protection of wild animals is governed by the Convention of Washington, which lists all the various protected species under several categories, such as those for which trade is absolutely prohibited in any form, those that can be purchased only by academic institutions or animal protection agencies, and so forth. Almost all of the animals listed are exempt from private trading, but the lists can be updated according to changes in species populations.

The management of animals not listed in the convention is up to the individual country. Most countries that stress environmental protection have regulations identifying species that should not be captured or traded in, but wild animals that are bred or raised by man are commonly excepted.

In the United States, many wild animals are not listed as protected species, but people who want to raise them must still adhere to a prescribed set of methods similar to those for keeping more conventional pets. Organizations and associations have been set up for owners to register their animals and to report sales and breeding.

In that way, authorities can keep track of increases or decreases in their numbers and provide owners with information on vaccines or other treatment in case of epidemic diseases.

Mistreating animals is a crime!: The United States, Britain and several other countries in Europe have strict limitations on people keeping animals. To take falcons as an example, besides stipulating that only varieties that are not endangered and have stable populations in the wild can be raised in captivity, U.S. law requires that falconers must be over 14 years of age and pass a test on ecology, falcon care and related legal knowledge.

Falconers are divided into three grades: a two-year apprenticeship, ordinary level and master level. Beginners must be guided by a falconer at the master level, and even masters can raise no more than three falcons.

They are not allowed to gamble on falcons. Special methods must be followed in breeding, and there are regulations relating to where they are kept and how they are raised, such as the size of the cage, the materials used for the cord to bind them with, how long they can be shut in a cage each day . . . all to try to ensure as safely as possible that the birds will not be mistreated.

Pet shops controlled too: People who keep exotic pets are also encouraged to join clubs and associations, which serve to supervise their members. "The rules in clubs and associations are usually stricter than the government's," says Chen Hui-sheng, a falcon fancier who relates that members of the U.S. Falconry Association who are caught violating its rules may not only be expelled but also may be subject to legal prosecution.

Animal shops in Europe and North America are also deserving of study and imitation, says Wu Hai-yin, a graduate student in zoology at National Taiwan University, pointing to a store in Michigan that specializes solely in lizards legally permitted for trade. Store owners there are subject to regulations on the size of the cages, the ventilation and so forth.

In advanced countries, authorities are headed in the direction of "coming up with a set of management methods for any animal that people want to keep." But in Taiwan, pet shops lie outside government control, and dealers seem to care only about selling the animals and take no responsibility for what happens afterward.

Reasonable use: "But there's a precondition that allows them to do it that way overseas," veterinarian Chi Wei-lien says. They spend a lot of time studying how the animals live in the wild.

For instance, there are specialized research agencies that study wild flocks and herds and estimate their numbers. If the animals are not in danger of extinction, then the restrictions on raising them don't have to be very strict. The agencies regularly perform follow-up studies, and if they find that the population of certain species is showing signs of

萬一有某些疾病流行，也可追踪和提供疫苗、醫療，便於管理。

虐待動物要坐牢！

美、英與歐洲一些國家對飼養者也有嚴格限制。以養鷹爲例，除了規定只能養野外族羣繁衍還很穩定、無瀕臨絕種之虞的種類外，美國法令還限定飼養者必須年滿十四歲以上，且先要通過有關動物生態、飼養、法律常識的考試，才能申請養鷹。

他們並將養鷹人分三等級。首先是兩年的實習生，然後是普通級與指導級。初級生必須有最高級的人指導；卽使成爲養鷹高手，最多也只能養三隻。

此外，不准拿放鷹做賭博；若要繁殖小鷹，也有處理方法，而養鷹的環境、方式也都有細則，譬如籠子的大小、綁鷹的繩具材質、每天關在籠中不能超過多久……，儘量把規則訂到使動物不致被虐待。

寵物店也要管理

政府並鼓勵養同一種動物的人組成協會，大家相互監督。「會內規定往往比會外規定嚴格，」對國外養鷹做過深入了解的陳輝勝指著美國養鷹協會出版的刊物說，若有人不遵守規定被發現，除個人名譽大壞、被剔除會員資格，可能還要受法律的懲罰。

臺大動物系研究生吳海音，也提到國外的情形可資借鏡。美國密西根州有蜥蜴專賣店，提供各種法令允許交易的蜥蜴。寵物店本身也受到規範，比方獸欄多大、店內空氣流通情況……。

先進國家已朝「有人養的動物，就想法要發展一套管理方法出來」的方向走。國內的寵物店則根本沒有管理，而動物商也只管把動物推銷出去，之後就不負任何責任了。

合理的利用

「但國外能這樣做，是有前提的，」獸醫祈偉廉表示，他們花許多時間了解自己的動物在野外生長的情形。

比方有專門的野生動物研究單位，對野外族羣調查，精確估算數量。那些目前不擔心絕種的動物，有人想養，則不需太嚴格限制。研究單位還隨時做追踪調查，萬一發現某種動物族羣量有減少跡象，就及時規定不能再捕捉。

「若能知道自己有些什麼東西，並定出完備法令，別說養，要打獵也沒問題，」祈偉廉說，英國某一地方在對自己區內每個山頭鹿的生長情形大致了解後，就規定在合理範圍內，一個冬天可以打幾隻；年齡太小、鹿角叉數不够的不能打；而且限制每個打獵者的獵獲量，並須繳費做爲鹿羣保育的費用。

獵物到手，還必須送警局登記，若研究單位索取鹿器官做研究，也不得拒絕。因爲他們可據此研究鹿羣野外數量的變化、是否有疾病傳染，並作爲下一年度可被獵捕數量的參考。

國內去年已通過「野生動物保育法」，詳細列出不得買賣的動物，並規定現有保育類動物都必須登記，和不得以藥品、器物或暴力騷擾、虐待野生動物。並將設立野生動物研究所，專門對國內野生動物的族羣、疾病、生態習性等等進行研究……。對我們的野生動物可說是喜事一樁。

需要貨眞價實的動物研究所！

但目前擁有保育類動物、向主管機關登記的人數「大概只佔百分之一，」祈偉廉估計後解釋，因爲許多人擁有的動物來源不明，大家都怕登記了會被處罰或沒收。「登記其實只爲便於管理，」農委會保育科技士陳超仁表示，法令既往不究，唯有大家確實登記，才能杜絕走私和違法獵捕，野生動物的管理才能眞正上軌道。

另外，由於許多人不諳野生動物研究所成立的意義，而把它當成「家畜研究所」，農委會畜牧處爭相把牛、豬、雞、鴨等家禽家畜列入其中做研究，使許多動物學者「頗爲無奈」……。

看來，國人想要有一套合理的野生動物飼養辦法，還有一段很長的路要走。 S

（原載光華七十九年五月號）

患了出血性腸炎的小狗接受治療。照顧馴化的寵物都問題不斷，更何況是養野生動物？(黃麗梨攝)
A puppy is treated for an intestinal haemorrhage. Taking care of domesticated animals is a lot of trouble — how much more so wild ones? (photo by Huang Li-li)

decline, then further capture can be prohibited in time.

"If you can find out just how many of each kind you've got and can set up appropriate rules and regulations, you can even allow hunting if you want," Chi says. After ascertaining the status of its deer, a certain area in Britain allows a reasonable amount of hunting, stipulating just how many can be culled each winter along with minimum age and antler size. Each hunter can only take so many, and they have to pay a fee, which goes toward maintaining the herd.

Any kills must be registered with the police, and hunters may not refuse any agency requests for deer organs for research. In this way, authorities can keep track of changes in the animals' population and discover whether or not they have any communicable diseases — information that is used in determining the next year's limit.

Last June the R.O.C. passed the Wildlife Conservation Law, which lists in detail the species of animals that cannot be bought or sold, calls for the registration of protected animals currently being raised in the country and prohibits wild animals from being abused or mistreated with drugs, instruments or forms of violence. A wild animal research institute will be set up to carry out specialist research on the island's wild animals, their habitats and related fields of study.

A real research institute needed: The proportion of people owning wild animals who have registered, however, is "probably only one percent," Chi Wei-lien estimates, explaining that most owners are afraid of fines or confiscation because the sources of their animals are unclear. "Registration really is only for ease in management," says Chen Chao-jen, a wildlife protection specialist in the Council of Agriculture, adding that the only way smuggling and illegal capturing can be stopped is if everyone registers.

Another problem is that many people mistake the wild animal research institute that is being set up as a research institute for domestic livestock. The animal husbandry department in the Council of Agriculture has nominated oxen, pigs, chickens, ducks and other domesticated animals for study, making many wildlife experts feel rather at a loss.

It seems that we still have a long way to go if we want to manage the raising of wild animals properly.

(Chang Chin-ju/tr. by Peter Eberly/ first published in May 1990)

人類一直在修正與動物的相處之道，進了六福村動物園，分不清是人看熊，或熊看人。

Mankind continues to "refine" its ways of getting along with animals. At Leofoo Village Safari Park it's hard to tell just who's looking at whom.

與自由貿易「唱反調」

談野生動物進出口交易

Against Free Trade — In Wild Animals

文・張靜茹　圖・張良綱

在世界各國交通愈來愈便利、貿易愈來愈頻繁的今天，有一種國際貿易却限制愈來愈嚴。

近兩年，澳洲野生袋鼠繁殖過量，不僅危害農業，袋鼠本身也缺乏糧食。澳洲政府最後決定射殺，而不願出售或送給別國——怕影響他國當地的生態環境。

日本多摩動物園為了使來自澳洲的幾隻無尾熊賓至如歸，花了三億日幣為牠們設計最適合的家；後來一隻無尾熊死了，多摩動物園受到輿論攻擊，園長因此自殺。

動物園的經營理念也在改變。今天，處心積慮弄到世界各種珍禽異獸，已不被先進國家的動物園認為是驕傲的事。即使萬「物」聚全，動物沒有和原居地相同的生存環境，也無法眞實展現原有的生活習性，達到動物園教育的功能。所以未來的趨勢是，動物園盡量本土化，以經營好和當地生活有直接關係的動物為先。

許多例子顯示，人對動物與環境的關係愈來愈重視。今年五月，世界野生動物保育基金會（WWF）正式委託我國自然生態保育協會擔任亞洲地區的「生態警察」——監察

In recent years, as the population of kangaroos in Australia has exploded, not only have people's crops been threatened but the kangaroos themselves have lacked sufficient food. Finally choosing to shoot some of them, the Australian government decided against selling or giving them away to foreign countries. The fear was that they might affect the ecology of their adopted country.

In order to make koala bears from Australia feel at home, Japan's Tama Zoo spent 300 million yen to design and build a most suitable home. When a koala later died, the zoo was roundly criticized and the zoo director committed suicide.

The concept of a zoo is radically changing. Today, the zoos of advanced countries no longer boast of their great variety of exotic wild beasts. Even with all the species in the world, without simulating the animal's native environment, zoos can't show the living behavior of the animals, which is the true aim of a zoo. Hence the trend is toward localization, and priority is shifting to properly handling the animals that have a direct relationship with the lifestyle of the local community.

There are numerous examples that point toward people placing ever greater emphasis on the relationship between animals and the environment. In May of 1987 the World Wildlife Fund (WWF) formally mandated that the R.O.C.'s Society for

、紀錄非法野生動植物，尤其是動物的進出口貿易，並設法阻止或通知該會採取對策。

話說最多的當風紀股長？

WWF已有廿六年歷史，成員包含世界各國的科學家、動物學家、生態學者與關心環境人士等，影響力遍及全球；他們先後拯救過三、四十種野生動物免於絕種，在世界促成過近三百個國家公園成立。

WWF之下設有專門緝查非法野生動植物交易的組織TRAFFIC，分支據點遍佈五大洲，亞洲總部設在日本。我們是繼日本之後，第二個受托爲生態警察的亞洲國家。

爲什麼選上我們？有人認爲有「話說最多的當風紀股長」的作用；因爲亞洲國家最先雀屏中選的日本，就是目前亞洲最大非法動物貿易國。我國則在新加坡去年雷厲風行、嚴格限制非法野生動物進出口後，不少生意見風轉舵而來。

犀牛角拒絕往來

但也有保育界人士認爲，我們近年來的努力，是更重要的原因。例如，我們不停進行動植物生態研究計劃、主動與國際保育團體接觸、參與合作「國際鳥類繫放計劃」等，正逐漸扭轉外人對我的看法。

前年，WWF會長英國菲利蒲親王就曾盛讚我國致力保育工作。國內保育團體也早已主動提供動植物轉口貿易資料給TRAFFIC，我們近年進行的保育工作也給他們留下好印象。

據行政院政務委員張豐緒表示，最大的原因是我們禁止了犀牛、犀牛角的進口。由於國人重食補、食療，長久來以犀牛角爲中藥、殺老虎秤斤論兩賣，受到不少國際保育人士指責，尤其犀牛數量在最近十年減少了百分之七十，早被國際列爲受保護的動物。

在國內保育人士大力奔走下，前年行政院核定嚴禁犀牛、虎、獅等五種瀕臨絕種動物進口；國貿局也將含有虎骨、虎骨膠、犀牛角等的中藥材列管進口。

嚴禁非法動植物及其附產品的國際交易，就是TRAFFIC主要的工作，處理原則係根據「華盛頓公約」而來。

懷璧其罪

國際間野生動物的貿易，已進行了好幾百年。根據現存的荷蘭東印度公司檔案記載，十七世紀荷蘭人據臺時代，平均每年出口十萬張以上梅花鹿皮，終至今天臺灣特有亞種野生梅花鹿絕跡。

廿世紀初，西方仕女流行戴綴有鳥羽的帽子，受羽毛美麗之害而被濫補的鳥類不計其數。

到了一九六〇年代，全世界交通更便利、貿易更頻繁後，野生動物交易的數量也劇增。根據統計，那時國際間每年鱷魚皮交易最高曾達一千萬張，一九六七年單單美國進口的各種虎、豹皮就將近十六萬件；一九七二年非洲肯亞出口一百五十噸象牙，創下新紀

Wildlife and Nature serve as an ''ecological policeman'' for the Asian area. It was charged with keeping track of illegal trafficking in wildlife (particularly the import and export of animals), stopping it when possible and otherwise notifying the WWF to adopt countermeasures.

Blowing the whistle: The WWF, whose membership includes scientists, zoologists, ecologists and environmentally concerned people from around the world, has a 26-year history and influence of global proportions. The fund has rescued some thirty or forty species of wild animals from extinction and has facilitated the establishment of nearly 300 national parks around the world.

Under the WWF is an organization called TRAFFIC (Trade Records Analysis of Flora and Fauna in Commerce). Specially charged with investigating illegal trade in wildlife, TRAFFIC has branches in strategic locations on every continent. Its Asian headquarters is in Japan, and the R.O.C. was the next Asian country to be asked to serve as an ecological policeman.

Why were we picked? Some people think it's like making the naughtiest pupils class monitors. Japan is the largest trafficker in illegal animal trade in Asia and was the first picked. Since Singapore clamped down on illegal wildlife trade last year, a lot of business has been funneled to Taiwan.

Refusing rhino horns: Many environmentalists, however, believe that the R.O.C.'s efforts in recent years — its continuing ecological research, its taking the initiative to make contacts with international conservation organizations, etc. — are a more important reason.

Several years ago, Great Britain's Prince Philip, head of the WWF, commended the R.O.C. for its efforts. Local conservation groups have long been supplying TRAFFIC with information on wild life transshipment, and the R.O.C.'s ongoing conservation work has also made a favorable impression.

Chang Feng-shu, chairman of the R.O.C. Society for Wildlife and Nature, says that the R.O.C.'s decision to ban the importation of rhinoceroses and rhino horns in 1985 was the major reason for the improved image. Conservationists had been pointing their fingers at Chinese for using rhino horn as a major ingredient of many traditional medicines and for buying tiger meat like ground beef — by the pound. The rhino in particular has declined in number by seventy percent in the past decade and has been internationally designated as a protected animal.

With the lobbying of local conservationists, the Executive Yuan prohibited the importation of rhinos, tigers, lions and other endangered animals and restricted the importation of Chinese medicine containing rhino horns, tiger bones, etc.

Shutting down the illegal trade in animals, plants and their by-products is the principal work of TRAFFIC, according to the Convention on International Trade in Endangered Species of Wild Fauna and Flora (CITES) signed in Washington.

A crime for beauty's sake: International trade in wildlife has been going on for centuries. During the Dutch occupation of Taiwan in the 17th century, over 100,000 pelts of Formosan sika, or "plum-blossom deer," were exported a year, according to records of the Dutch East Indies Company. Today the animals are already extinct in the wild.

At the beginning of the 20th century, Western women's penchant for feathers in hats decimated numerous bird species with beautiful feathers.

By the 1960's, when transportation around the world became much more convenient and trade boomed, trade in wild animals likewise grew tremendously. In that decade as many as 10 million crocodile hides were sold in a year; in 1967 the U.S. imported nearly 160,000 tiger and leopard skins; and Kenya recorded exports of 150 tons of elephant tusks in 1972.

The monetary figures are equally astounding. According to CITES statistics, U.S. trade in wild animals in 1981 reached nearly US$1 billion; one South American puma skin fetches as much as US$40,000 in West Germany; a single specimen of a certain species of Central American parrot can go for as much as US$5,000 in the international market; and ground rhino horn may be worth more than its weight in gold in many oriental medicine shops.The result has been the gradual disappearance of wildlife around the world. Experts have estimated that 20 percent of wild animal species will become extinct by the year 2,000, and flowering plants will decline in number by one-fourth.

CITES: Ecologists from around the world have repeatedly been pointing to warnings that the extinction of animal species poses a great threat to the existence of people. For example, by controlling the number of insects, birds prevent bugs from competing with people for food. With this line of reasoning, conservationists have been calling for the protection of wild animals and plants. Some have even taken it a step farther to say that animals and plants also have a "right to exist." For private gain, no group has a right to drive an animal extinct.

The import and export of wild animals and plants can also threaten human life and property. For example, snakes have been imported in high quantities in recent years. If an exotic species bites someone, the proper serum may not be available to save the bitten person's life. The *Ampullarius canaliculatus*, a snail which has caused great damage to agriculture and is virtually without natural enemy, is another example of exotica going awry.

If all of the world's countries can strictly control the export of wild animals, this would also help to reduce excessive catching of wild animals. Taiwan, "The Butterfly Superpower," had its image battered in the past because of the unrestricted import and export of live butterflies and butterfly specimens.

With this set of circumstances, 21 countries met in Washington D.C. to form the Convention

臺灣蝴蝶工業大興於日據時期。圖中仿米勒名畫「拾穗」的壁飾，完全以蝴蝶羽翼製成。
This reproduction of Millet's *The Gleaners* has been made with butterfly wings.

錄。

交易金額也十分驚人。據華盛頓公約記載，美國一九八一年的野生動物進出口貿易額將近十億美元；一件南美山貓皮在西德的價碼高達四萬美元，一隻中南美洲大金剛鸚鵡在國際上的價格爲五千美元，一盎司的犀牛角在東方藥舖可以比一盎司黃金還貴。

因爲貿易利益而濫加捕殺的結果，野生動植物逐漸在地球上消失。專家估計，到公元二千年將有兩成野生動物絕種，植物亦然，如開花植物就會少掉四分之一……。

華盛頓公約

各國生態保育人士於是紛紛提出，動植物滅絕將對人類生存產生重大威脅的警告，例如鳥的存在能控制蟲數，免得蟲與人爭食……，來呼籲保護野生動植物；許多人更進一步指出，動植物也有「生存權」，任何團體都不可爲了私利，將其趕盡殺絕。

野生動植物進出口也可能危害人的生命、財產。例如國內前幾個月有商人大量進口蛇，萬一其中有我們沒有的蛇類咬傷了人，根本無該種蛇的血淸可救治；福壽螺引起農作損失，却無天敵可剋的事件，也是一例。

世界各國若能嚴格管制野生動物出口，也可減少濫捕野生動物情形。使臺灣「蝴蝶王國」的美譽已大不如昔，就是因爲過去蝴蝶與其標本的進出口沒有管制。

就在這些背景下，一九七三年，廿一個國家在華盛頓簽署成立了「瀕臨絕種野生動植物貿易公約組織」（CITES），制定瀕臨絕種野生動植物之國際貿易規則（華盛頓公約），目前已有九十六個國家加入該組織。

華盛頓公約中把若進行國際交易，將造成絕種或可能絕種的動植物名册列出，標明禁止交易、管制交易（如限制數量）的項目。前者如大狒狒、藍鯨、貓熊、雪豹，臺灣的雲豹和特有種鳥類藍腹鷴也在其中，後者如穿山甲、浣熊。

其餘未列入的則爲第三類。第三類與各種生物標本，各國可自行決定是否交易，但需設管理機構管制與核發輸出、入許可證。我

on International Trade in Endangered Species of Wild Fauna and Flora (CITES) in 1973, establishing regulations governing the international trade in endangered animals and plants in what is known as the "Convention of Washington." Currently more than 96 nations have joined.

For the purposes of conducting international trade, CITES lists threatened species in a booklet, either prohibiting trading (category one) or restricting trading (category two) by establishing quotas, etc. The big baboon, blue whale, panda, snow leopard, Taiwan clouded leopard and Swinhoe's Blue Pheasant, a species unique to Taiwan, were among the animals listed under category one. The pangolin and raccoon were among those listed under category two.

The remaining animals, those not listed, are category-three species. It is left up to individual countries to decide whether or not to trade in them or the stuffed animals of any category, but any such trade requires proper export and import permits. CITES has authorized the Executive Yuan to appoint the Board of Foreign Trade of the Ministry of Economics to handle the granting of these permits.

Illegal trading: According to statistics of the Council of Agriculture, there is a startlingly large amount of trade in wild animals done in Taiwan — whether one is speaking of importing, exporting or turnaround trade. As many as 200,000 birds are traded a year. One animal trader, who wished to remain anonymous, estimated that several million more are traded on the black market, including trade in species prohibited by the Convention of Washington.

While the Conservation Office in the Forestry Department of the Council of Agriculture was examining information provided by the Bureau of Commodity Inspection and Quarantine about the quarantine of animals, it discovered listings for six Swinhoe's Blue Pheasants, which are protected as unique species under the "Cultural Resource Preservation Law."

Tsai Hsi-sung, the president of a company that trades in animals, says that there are four or five large companies engaged in the animal trade on Taiwan and numerous smaller firms which import or export tropical fish, birds, ivory and stuffed animals. But they are mostly operating in ignorance,

國就由華盛頓公約組織委託行政院指定經濟部國貿局擔任簽證的核發。

殺頭生意有人做

但根據行政院農業委員會統計，去年我國進、出和轉口的野生動物數量驚人；其中鳥類更高達廿幾萬隻。一位不願透露姓名的動物商估計，黑市數量可能更多達幾百萬隻，且其中不乏華盛頓公約不准進出口的種類。

農委會林業處保育科在整理商檢局動物檢疫資料時，就發現去年有六隻我國「文化資產保存法」中列爲保護的特有種鳥類「藍腹鷴」。

所以如此，經營動物進出口的銘得企業有限公司負責人蔡溪松認爲，國內目前具規模的活動物商有四、五家，其它鳥類、熱帶魚、象牙、動物標本商的數量可觀，但都悶著頭做，許多商人根本不清楚什麼是華盛頓公約。

也有人不以爲然。「只要從事過這個行業，多少會了解到野生動物買賣的限制；」農委會保育科一名技士說：「但是厚利當前，就只有猛鑽漏洞了。」

貿易商乘虛而入

例如國貿局進出口貨品分類表，對野生動物類別分得很籠統，像鳥類只分野鴨、野雞、雉雞、帝雉、藍雞五種，鹿只分長頸鹿與鹿兩種，又沒有圖片可供對照，申報的貿易商只要把國際上列名禁止進出口的種類，隨便塡上其中一種，就暢行無阻。

此外，許多動物種類只要銀行簽證即可放行、五百美元以下的餽贈物品亦無需申請簽證即可帶入境，也都是貿易商可鑽的漏洞。

國貿局過去也未嚴格要求貿易商要取得輸出國的輸出許可證，證明動物非走私管道出來，給了國外商人可乘之機。今年三月中，有一批華盛頓公約第一類的鱷魚皮從巴拉圭賣到日本，就利用這個漏洞由我國轉口；他國的交易，後來遭到華盛頓公約組織抗議的却是我國。

我們雖然不是華盛頓公約組織的正式會員，但瀕臨絕種動物的非法交易一經暴露，國際指責之聲四方交相而來，除嚴重影響國家形象，也可能因此帶來更壞的後果。

都是小金剛惹的禍？

最近臺北市立木柵動物園，就因不諳華盛頓公約規定，買了不少第一類的動物，且因所托非人——承包的外國動物商不按正常管道買小金剛，而引起國際保育人士不滿。

即將於八月三日在美國召開的ＴＲＡＦＦＩＣ年會，因此將提案要求大會剝奪我簽發野生動物貿易簽證的權利；萬一國貿局無法核發野生動物貿易簽證，以後國內動物商將因無法取得國際上公認的進出口許可證，斷了與國外交易的路。

針對這種狀況，國內有關單位已迅速採取幾項措失。國貿局在今年三月規定動物商不准從事野生動物三角貿易。以後凡野生動物均不得到銀行簽證，而免簽證的五百美金餽贈品，也將不包含野生動物。

農委會正進行翻譯華盛頓公約內容，開列我國野生動物禁止出口名單，並附上圖鑑，供國貿局、海關查對之用。六月廿七日起，凡華盛頓公約規定的第一類，一律禁止進出口，第二類野生動物及其產品之進、出口案，將嚴格檢驗出口國的簽證並依圖查核。

與自由貿易唱反調

但是，想大刀濶斧改革的主管單位也面臨國內動物園與貿易商的壓力。

由於華盛頓公約未限制學術機構與有教育功能的單位（如動物園）交換或相互買賣第一類動物；不少動物園希望農委會仍准許他們交易。

動物商則要求農委會給他們一段緩衝期，專門進口象牙的象國工業公司負責人梁奕星舉新加坡和香港爲例，他們在前年施行進口必須附對方輸出許可的規定時，曾給予動物商半年時間，否則有些已付訂金，對方却拿不出輸出許可證的生意會大賠老本。

也有動物商問：「臺灣特有的梅花鹿，野生種早已滅絕，可是現在有不少人養殖，爲

and many don't even know what the Washington treaty is.

Others maintain otherwise. "Anyone in this business knows something about the restrictions governing the buying and selling of wild animals," says one specialist in the Council of Agriculture's Conservation Office. "But when they see the profits to be made, they go for every loophole."

Finding the loophole: For example, the trading bureau's chart for the import and export of products puts animals into broad categories. Birds, for instance, are separated into such categories as wild ducks, wild chickens, pheasants and mikado pheasants. Ruminants, on the other hand, are only split into two: giraffes and deer. And since customs lack photographs for comparison, those importing restricted species need only to write a species traded legally for smooth sailing.

In addition, many animals can be imported through visas that are applied for at banks, and if goods are valued at less than US$500, they don't need visas at all. These loopholes allow traffickers to get around the restrictions.

In the past the Board of Foreign Trade did not strictly require trading companies to apply for export permits from the country of origin to prove that the animals had not been smuggled. In March of this year, a shipment of alligator (first-category) skins were shipped from Paraguay to Japan, making use of loopholes to go through Taiwan. The R.O.C. was the nation eventually rebuked by TRAFFIC.

Though we are not a formal member of CITES, as soon as cases of trade in endangered species surface, a chorus of international blame is heard from every direction, doing damage to the national image and perhaps leading to even more disastrous results.

All because of "Baby Kong": Unfamiliar with CITES regulations, recently the Taipei City Zoo in Mucha bought a number of category-one animals. And conservationists were unhappy that its foreign supplier didn't go through proper channels to purchase a young gorilla, dubbed "Baby Kong."

As a result, on August 3, 1987 at the annual meeting of TRAFFIC in the United States, it was proposed that the R.O.C. be stripped of its rights to issue visas for trade in wild animals. If the Bureau of Trade were not permitted to grant these visas, then R.O.C. animal businesses would be unable to obtain internationally recognized import and export permits and would not be able to trade in wild animals with other countries.

To prevent just such a situation from arising, R.O.C. government agencies took various steps. The Board of Foreign Trade had already in March of 1987 stipulated that animal traders were not permitted to engage in the turnaround trade of wild animals between other countries. Then it was declared that banks would not be allowed to grant visas for shipments of wild animals, and undeclared gifts worth less than US$500 could not include animals.

The Council of Agriculture translated the content of CITES, made a list of the animals prohibited from being exported and provided an illustrated handbook for the Board of Foreign Trade and customs. On June 27 of 1987 it banned entirely the trade of CITES category-one animals, and it called for strict inspection of visas with checks against reference photos for the importing or exporting of category-two animals and their products.

Against the grain of free trade: Yet the agencies pushing for reform are feeling great pressure from domestic zoos and traders.

Because CITES does not restrict the trade or sale of category-one animals among academic or educational institutions (such as zoos), many zoos hope that the Council of Agriculture will still permit them to trade.

Animal traders, on the other hand, requested that the Council of Agriculture give them a grace period. Liang Yi-hsing, the owner of Hsiangkuo International, which specializes in importing ivory, cited Singapore and Hongkong, which gave their traders a half year before requiring them to obtain export permits. Without such a grace period, traders would have to forfeit the deposits already made on shipments without the proper papers.

Animal traders also ask, "Sika deer, which are unique to Taiwan, long ago disappeared in the wild, but now that many are being bred and raised, so why can't they be exported?"

Whether for importing or exporting, there are many reasons why the authorities must be careful. Wu Kuan-tsung of the Council of Agriculture's Conservation office explains that currently all zoos go

打獵者喜歡將獵物製成標本，只是，還有多少動物可供狩獵？
Hunters like to hang up stuffed trophies, but how many animals are left to supply them?

什麼不准出口？」

不論進口或出口，有許多理由使主管單位不得不小心。農委會保育科技士吳冠聰解釋，目前動物園買動物都經過貿易商，許多動物並非來自國外的動物園，而是收了訂單再去捕捉。且華盛頓公約規定，賣方必須還有足够的動物可延續族羣，買方要附上該動物未來管理員與獸醫的履歷、動物園設備、建築材料……，供華盛頓公約組織查核，再決定可否買賣；並非雙方動物園想交易就可成交。

目前所以不開放梅花鹿出口，也是防範有人利用梅花鹿之名，出口國內還有的野生種水鹿。

「動物商必須了解，動植物進出口生意只有愈來愈難做，舉世皆然，」吳冠聰說，一、二十年前那種不需多少手續，就可進口野生動物的情況已不可能再發生。

你也可能是劊子手

在糧食不足、工業不發達的地區，最豐富的資源就是野生動物，不准他們出口很難確實執行。

像現在華盛頓公約規定產象牙國家，每年只准出口象牙廿噸，但在非洲私下進行的交易，數量遠超過規定數目。如此對改善他們的生活十分有限，例如一枝象牙在非洲才賣新臺幣一百元，一經轉手，價錢就高出上百倍，好處全進了貿易商的荷包。除了貿易商的推波助瀾，另一個「我不殺伯仁，伯仁卻因我而死」的肇事者，是消費者。

今年在臺北舉行的自然生態保育座談會中，WWF的代表帶來不少幻燈片當場放映。其中有一張是雙美麗的脚，穿著精心設計的蛇皮皮鞋，姿態優雅地站在光可鑑人的地板上，旁邊打出的字幕却是：她已涉入違法事件而不自知！

我們呢？你是否以踩著蛇皮皮鞋、揹著鱷魚皮皮包，在亞熱帶的臺灣披著皮草為榮？若是如此，你也可能成為動植物滅絕的劊子手！

（原載光華七十六年八月號）

through animal traders when buying animals. Many of these animals don't come from foreign zoos. Rather, they are caught after the orders for them are received. Among other conditions, CITES stipulates that the sellers must still have enough of the animals to ensure that their own breeding efforts will continue, and the buyers have to enclose for inspection the work histories of the animals' future caretakers and vets and information about the zoo's equipment and building materials. It's not as simple as just two zoos wanting to make a deal.

The reason sika deer are not permitted to be exported is to prevent people from capturing still wild Formosan sambar and exporting them as sikas.

"Animal dealers must understand that the trade in plants and animals is getting more and more difficult all over the world," says Wu Kuan-tsung. The lack of regulations that characterized the trade a decade or two ago is forever a thing of the past.

You may be the guilty party: In many regions where agriculture is poor and industry undeveloped, wild life is the greatest resource. Forbidding these areas from exporting wildlife is hard to do.

For example, although the Convention of Washington sets a quota of 20 tons a year for ivory producing nations, trafficking on the black market far exceeds that figure. Yet as a result, the trade doesn't do much to improve the living standards of those living near the elephants. An ivory tusk will cost only NT$100 in Africa, but after it changes hands, it will rise in price more than 100 times, and this is money that lines the pockets of the middlemen.

The driving force behind all the trade is, of course, the consumer. At an ecological protection seminar held in Taipei this year, a WWF speaker showed a number of slides, including one of a pair of feet in elegant snakeskin shoes standing on a polished marble floor. The subtitle was, "An unwitting participant in crime!"

And ourselves? Do you wear snakeskin shoes, carry an alligator purse or wallet, or have a fur coat hanging in your subtropical closet? If so, you too may be the murderer of an endangered species!

(Chang Ching-ju/photos by Vincent Chang/ tr. by Jonathan Barnard & Peter Eberly/ first published in August 1987)

沒有人是一座孤島
——地球日談生態關懷

An Earth Day Plea for Environmental Concern

文・林俊義口述／張靜茹整理　圖・張良綱

四月廿二日爲「世界地球日」，國內環保團體與世界地球日同步，廿二日這天在臺北中正紀念堂，舉行了盛大的地球日活動；許多人也響應地球日，穿著綠衣外出。而在這之前，環保團體也已在社會各階層展開一連串的生態教育。

但地球日要傳達的，不只是技術上的生態保護，或共襄活動而已；要長久的對生態關懷，必須人們在理念上有所調整。

本刊採訪了此次活動的總召集人、東海大學生物系教授林俊義，他以知性和感性兼具的情懷，娓娓道出做爲一個人對生態與生命應有的尊重，有助我們了解地球日的眞正意義。

To mark Earth Day on April 22, ROC environmental groups held a massive gathering at Taipei's CKS Memorial Hall, and many people responded by spending Earth Day outdoors dressed in green. Well before this event Taiwan's environmental groups had already begun saturating all sectors of society with ecological education drives.

What Earth Day seeks to achieve is more than simply technological protection of the environment or getting people to join in events. If people are to show an enduring concern for the environment, there must be a radical change in their thinking.

Here we are printing an interview given by Professor Lin Chun-i of the biology department at Tunghai University, the coordinator of this event in Taiwan. A man genuinely concerned for the environment, he speaks of the respect every person should have for the ecology and for life, and helps us to appreciate the true meaning of Earth Day.

今天大家已漸漸了解生態學橫跨很多領域，在農業、工程各方面也都有機會接觸，但通常都因爲實用需要而來接觸生態。

由實用科學導致出來的生態理念，往往有所妥協，但我認爲生態關懷是不可妥協的，爲什麼呢？讓我就由生態關懷說起。

完整的生態關懷，應該是人、自然與社會三者關係密不可分的。一個社會若缺乏對這

Today all of us have come to realize that ecology involves many different fields. It is encountered in agriculture and engineering alike. Normally it is practical necessity that brings us into contact with the environment.

Ecological concepts derived from practical science frequently involve compromises, but I believe we cannot afford to compromise when it comes to the environment. Why not? Let me try to explain.

全世界一百多個國家在四月廿二日連線舉行地球日活動。臺灣區地球日活動在中正紀念堂展開。
Over 100 countries worldwide held events marking Earth Day on April 22. Taiwan's Earth Day event took place at the CKS Memorial Hall, Taipei.

三者關係的反省和了解，將是個粗俗、不深入、不成熟的社會。

我們可以把人類對待自然的歷史簡單分為幾期：早期人和自然關係較單純，後來透過資本主義、宗教、社會運動、科技發達……，使人和自然關係發生變化；到今天，大家發現這樣對待自然造成了嚴重後果，導致地球危機，因此又回過頭來關心自然，所以有「地球日」的出現。

事實上，一九七〇年世界地球日在美國成立時，還沒有人想到廿年後地球會有這麼大的變化。那時從溫室效應、全球氣候改變、酸雨、臭氧洞、雨林破壞、野生動植物的保護，到沙漠化、海洋生物匱乏現象……都還很少人討論，並未受到重視。沒有人料到，短短廿年，地球面貌竟有了這麼大的改變。

沒有人是一座孤島

由人類演化史來看，人是無法脫離自然生存的。不僅實質的生存離不開自然，精神上亦然。

很多描述自然的書籍都說「自然是智慧」，莎士比亞說：「自然就是我的書。」只要到森林走走，摸摸樹葉、看看蝴蝶，就知道智慧的存在。

即使不由生態學，而由宗教來看，聖經中也說：「和大地說話，大地就會教導你；和鳥說話，鳥會教導你……」「透過自然，就可以了解上帝的存在。」

美國詩人惠特曼在七十歲時說過：「現在我明白了，創造完人的秘密，是讓人在自然中長大，與大地一同生活。」自然的功夫就在這裏，人的品質和自然的關係是非常密切的。

常常我們在探討人類生命與自然關係的過程中，許多文學、哲學家的作品可以提供解答。我在學校教「島嶼生態學」這門課程時，提到十七世紀英國詩人 John Donne（但恩）的一首詩說：

沒有人是一座孤島，
可以把自己隔離，
你我都只是大陸的一部分。
每個人的死亡都使我少掉一點，
因為我和人類整體是牢牢不可分的。
不要問半夜的喪鐘為誰而敲，
喪鐘是為你而敲的呀！

詩人深知自己不能脫離自然、人羣，因此每個生命的失去，對他而言，都造成損失。

史懷哲醫生上了年紀之後，也覺得這世界若要有希望，就要人類對生命有所敬畏。我們只有對生命尊重，才能談及其他。由我對自然生態的了解，覺得敬畏生命之外，還要讚美生命。之餘還要對生命感受到某種程度的奧秘，了解生命中很多事是超越你的控制的，我將之稱為「奧秘生命」。

只有認識生命中有很多你不了解的神秘性，而你也不能脫離自然、人羣而活，你才能更成熟、謙虛，成為一個「好人」。一個不曾感受生命中有不可控制的力量、自以為一切都在掌握中、不了解其他生命對自己重要性的人，會變得傲慢、無知。

當人與自然脫節

現在的臺灣正是一個人與自然完全脫節的社會，大家不了解自然、不敬畏生命。自然是幹什麼的？是讓人剝削、開發的；取之不盡、用之不竭的觀念，仍存在許多人腦中。

人們也沒有融入自然——像卡通片中，在森林手舞足蹈——的能耐。有詩人說，他在野外看到花開，感動地想哭都哭不出來，因為那種感動太深沉，反而讓人哭不出來。但我們在教育和成長的過程，都沒有被教過、啓發過這樣對自然的感受。如此我們怎會讚美自然？怎會覺得自然神秘？因而只會剝削，看到動物就是殺、吃，而不懂得欣賞、愛護。

在不正確的社會、人文、科學教育下，產生動植物是依賴人生存或相互依賴的錯誤概念。實際上是人依賴動植物生存，若所有動植物消失，你我也將隨著消失；但若人類消失，恐怕動植物還會額手稱慶呢！

今天臺灣的社會已變得缺乏人倫、生態倫理、未來倫理……而只剩錢的倫理。

To have a complete concern for the environment, you cannot separate man, nature and society. If a society does not engage in self-examination and lacks an understanding of the interconnectedness of these three, it is a coarse, superficial and immature society.

The history of man's treatment of the environment can be simply divided into several stages. In the early stage, man had a fairly uncomplicated relationship to the environment. Later, with the arrival of capitalism, religions, social movements and scientific advance, man's relationship to nature underwent a transformation. Then people discovered that this treatment of nature was creating serious consequences and leading to a crisis for the biosphere, and so they did an about-face and began to care for nature again, which is when Earth Day appeared on the scene.

In 1970, when the first Earth Day was held in America, no one imagined that in 20 years' time the world would have changed so much. From the greenhouse effect, global climatic change, acid rain, the hole in the ozone layer, destruction of the rain forests, conservation of wild animal and plant species, to desertification and the reduction of life in the sea, all of this was very little discussed and paid scant attention. No one could have expected that within two short decades the world would change so dramatically.

For spiritual needs as well as simple survival, man cannot exist in isolation from his environment — as is well shown by the history of human evolution.

Many writers equate nature with wisdom. Shakespeare said, "Nature is my book." All you need do is walk into a forest, touch the leaves and see the butterflies to become aware of this wisdom.

From the viewpoint not of ecology but of religion, the Bible also says, "Speak to the land, and the land will teach you; speak to the birds and the birds will instruct you." Nature helps you to realize the existence of God.

No man is an island: At the age of 70, the American poet Walt Whitman said: "Now I understand that the secret of creating a perfect man is to let this man grow amid nature, living together with the land." This is the power of nature, and the quality of a person is closely bound up with his relationship to nature.

When we examine the relationship between human life and nature, the works of many writers and philosophers can often provide us with answers. When I was teaching island ecology, I recalled the words of the seventeenth-century English poet John Donne, "No man is an island, entire of itself; every man is a piece of the continent, a part of the main. Any man's death diminishes me, because I am involved in mankind; and therefore never send to know for whom the bell tolls; it tolls for thee."

Poets know that they cannot separate themselves from nature and the human community, and thus they feel any loss of life is a loss of their own.

As an old man, Dr. Schweitzer also believed that if there was to be hope for the world, mankind must have respect for life. Only if we first respect life can we turn to other things. My understanding of nature and the ecology is that in addition to respecting life we should also praise and glorify life, and appreciate the mystery of life. We should understand that many things in life are beyond our control — this I call the profound mystery of life.

Only when a person begins to see that there are many incomprehensible mysteries in life and comes to realize that he cannot live in isolation from nature and human community, will he grow more mature, more humble and become a "good man." A man who cannot see that there are forces in life beyond his control, who thinks everything lies in his own hands and who doesn't understand the importance to himself of other living things, is bound to grow proud and remain blind to understanding.

When people are alienated from nature: Taiwan today is an example of a society where people are utterly alienated from nature, where nobody seems to understand nature or have respect for life. What is nature there for? It is there for man to exploit and develop. The inexhaustibleness of nature is an idea that still lurks in many people's minds. People are also not immersing themselves in nature — like by dancing in the forest in Disney cartoons. There was a poet who said that he saw a flower bloom in the wild and he was so moved that he wanted to cry but couldn't. He couldn't because being so touched has a profundity beyond tears. But in the process of our education and growing up, we are never taught to appreciate nature in this way. How are we to praise nature? How are we to feel the mystery of nature? And so we know only how

我「吃」，故我在

笛卡兒說：「我思故我在！」我思考，才知道我的存在。知道自己的存在是人的本質，是其他生物所沒有的。可是臺灣已變成是：我「吃」故我在！我吃大餐時，才知道我自己的存在；否則就是我逛街、購物才知道自己的存在。這種習慣性甚至已變成許多人心靈的一部分，有人眞的是不去買東西、花錢，就會懷疑自己的存在！逛街變成一個認識自己存在、別人存在的活動；和人擠來擠去，才知道「噢！原來我還在！」

一個社會扭曲到如此，就是和自然完全脫節的結果。很多人解釋臺灣的現象，由政治、經濟、教育、人文等層次來分析，我則認爲，這就是一個社會反自然的結果。

人與自然脫節的原因，在於我們是反自然的社會，透過各種經濟政策、教育政策、文化政策等等，使我們脫離了自然，認爲自然無用。

社會一旦反自然，人性就變了，我們的生活環境快要看不到一棵綠樹、看不到廣濶的綠地了，而綠對人很重要，人類是由綠地孕育出來的。在都市生活中，我們常說要到野外走走，排遣一下生活。道理很簡單，我們由綠地來，定要回歸到綠地，絕對不能脫離自然生存。

反自然導致反文化

但這一代的年輕人，成長過程中無法眞正接觸自然，感受自然的生命性！而脫離自然愈久，面貌愈醜惡，人心愈敗壞，如何創造最好的人？

說句笑話，現在的小孩連牛奶從那裏來都不了解，問他牛奶那裡來？他說「7-eleven

「人類孕育於綠地，終也要回歸綠地，」林俊義說，當生活環境脫離自然，人性也將岌岌可危。(鄭元慶攝)

"Mankind originated in the green land and ultimately needs to return to it again," Lin Chun-i says. If man's surroundings are divorced from "nature," human nature is in danger. (photo by Cheng Yuan-ching) to exploit. When we see animals, we kill them and eat them but don't know how to enjoy and care for them.

Faulty education in the social field, as well as in the humanities and sciences, has led to the misconception that animals and plants are dependent on man, or that their existence is mutually interdependent. In reality man is wholly dependent on animals and plants for his existence; if all animals and plants disappeared, you and I would soon disappear too. But if mankind were to vanish from the face of the earth, it might well come as a great relief to the animals and plants!

I "eat," therefore I am: Today's society of Taiwan has lost its humane moral principles, ecological moral principles, and future moral principles. The only principle left is ''money''.

Descartes said, ''Cogito, ergo sum'' (I think, therefore I am). Only when we think do we become aware of our existence. To be aware of his own existence is man's unique nature, something that other living creatures do not possess. But in Taiwan we lack this awareness. Our credo has become ''I consume, therefore I am!'' Only when we eat a big meal, go window shopping or go out buying things do we feel we are really alive. This habit has become ingrained in many people's thinking; there are people who really believe that unless they are buying something or spending money they are not living, and then they begin to doubt their own existence! Window shopping has become a way of affirming one's own existence and confirming that others also exist. Only when we're rubbing shoulders in the crowd do we reassure ourselves, ''Ah! So I'm still here!''

Such perverted distortion in a society is the product of man's total alienation from nature. Many people's explanations of these phenomena in Taiwan are based on political, economic, educational and cultural analyses, but I believe they are the result of our society having turned its back on nature.

The reason for man's alienation from nature is that we are a society that is anti-nature. All sorts of economic, educational and cultural policies alienate us from nature and lead us to believe that nature is useless.

Once a society pits itself against nature, people change. Many of us live in an environment in which it is virtually impossible to see a single green tree or an expanse of green grass. Yet greenery is extremely important to man, for mankind was originally nurtured by the green land. We city dwellers often say we want to go out into the countryside and take some relaxation. The reason is very simple: We belong to the great outdoors and must return to nature's embrace. We absolutely cannot exist apart from it.

Anti-nature turns into anti-culture: Yet the youth of today have had no contact with nature in the course of their development, no way to be moved by the life of nature! And the longer people are severed from nature, the meaner their looks and the more depraved their hearts. How can they become good people?

To put it another way, children today don't even know where milk comes from. If you ask them, they'll say ''7-Eleven.'' When it has gotten this bad, society has certainly pit itself against nature and is completely lacking in ecological principles.

But that is not all. Being at odds with nature leads to the formation of an anti-culture. Genuine culture is the result of patient interaction between man and his natural environment. If nature is destroyed, culture dies: Alternatively, a culture arises which is completely dislocated from the environment. As a case in point, just how much does the contemporary culture of Taipei city have in common with the actual land of Taiwan?

I remember years ago being in Hsintien for what was called ''the smelt season.'' The young and old from all over the city gathered on the shores of Green Lake to fish and chat. But with the destruction of the environment, the smelt disappeared. For this and other reasons, people have become strangers to each other.

Pollution is just the tip of the iceberg: Why are people no longer humane? Why are murders and thefts ever more common in society? As a result of being alienated from nature, people are becoming alienated from each other. Where men are alienated from themselves, they can act in strange and self-destructive ways. To put it another way, if there is not a proper relationship between man and nature, the relationships between people won't be right either.

Being anti-nature will certainly set society down a path of being anti-human and anti-culture. Why do I care about Taiwan's environmental problems?

」商店。搞到這樣，這個社會一定違反自然、缺乏生態倫理了。

不僅如此，反自然繼而還會導致反文化現象。文化，是人和自然環境慢慢相互運作結果產生的，一旦自然破壞，文化就無法產生，否則就是產生和環境完全脫節的文化，就如現有的臺北文化和臺灣這塊土地有什麼關係？

我記得從前在新店有過所謂的「香魚季節」。新店大街小巷的老老少少在那陣子羣聚碧潭抓香魚，大家談天說地。但環境破壞、香魚消失，加上其他原因，人與人已成陌路了。

汚染只是冰山一角

人性爲何不再良善？社會殺人、搶刼惡行爲何不斷？因爲反自然的結果，人和人也會產生脫節。人和自己也沒有了自我了解的過程，會做出怪異、自暴自棄的行爲，因爲他不認識自己。換句話說，人和自然的關係若不正確，人和人之間的關係也不會正確。

反自然的現象，定會將一個社會導向反人性、反文化的方向。我爲何關心臺灣的環境問題？空氣汚染、水汚染、噪音等等問題，其實都只是表徵；自然破壞所反應出的反自然、反人性、反文化才是危機所在，才眞正令人擔心。

大家都在討論臺灣民主化的問題，但不論政治結構改變多大，如果臺灣仍然繼續違反自然，臺灣社會仍是無法拯救的。

當然，空氣汚染令人懊惱、影響健康，可是假如你只關心這一點，仍是很膚淺的。

愛這片土地，先要了解它

臺灣的情況看不出生態倫理，生態倫理看起來也很抽象，我舉一個例子吧。美國某個地方，有個小孩在海邊發現一隻斷了脚的海龜，就將牠帶回家。鄰里街坊知道後，大家捐出一點錢，請外科醫生爲牠檢查，並爲牠裝義肢，然後又造個池塘，讓牠可以在自然環境中復健。等牠康復了，村人就讓兒童抱著海龜到海邊讓牠回歸大海，村民齊聲爲海龜歡呼、加油，鼓勵牠回到大海！這個景象就是最好、活生生的生態教育！

幾年前有一隻鯨魚誤闖聖地牙哥海灣，當牠被引導出海時，所有的人都在海岸邊、電視前歡呼！這件事情之後，也有一隻鯨魚闖進臺中港，結果却靠一個外國記者照會國際保育組織，通知我們的外交部要保護，外交部才下令臺中港務局必須保護鯨魚，防止牠被漁民捕捉。

爲了把鯨魚引導出海，必須知道牠的習性，但却發現國內沒有人研究鯨魚的習性。由北到南的八個水試所，加上海洋、中山大學和中研院動物所，都沒有人研究。

我們說立足臺灣，却對臺灣的事知道很少；我們是個島，却對四周的海洋生物一無所知。年輕朋友知道萊茵河、多瑙河、密西西比河，却不清楚濁水溪、大肚溪、高屛溪在那裡，它和我們的關係如何？如果我們不敎孩子了解這片土地，却要他們愛這片土地、尊重這片土地，那是不可能的。

簡樸的生活才是永恆

一個社會對自然不尊重，它也就對人、對自己都不尊重。對待自然沒有溫柔、敬畏的態度，不管它經濟多好、政治多清明，仍是不完整的社會。一個社會的文化、經濟、民主程度都反應出它對自然的態度，也就是說，沒有任何社會是經濟、政治很好，但却是反自然的。

今天臺灣經濟很好嗎？錢多並不是經濟好。一個關懷社會、人類的富商，和暴發戶是不一樣的！假如臺灣的經濟奇蹟導致今天大家不敢出門、不敢坐計程車，談情說愛要懷疑對方的態度，人的品質低落，四處聲色犬馬，那我們寧可吃粗飯、穿舊衣、過簡樸生活，人與人的關係却諧和，大家對自然、文化有期許。

物質生活的目的，是要讓人思考自己存在的意義，它只是工具，不是目的。我們却把物質生活當成人生目的，結果大家拚命追求物質生活，忘掉生存的本質，也導致生態環境更惡化，更惡性循環——我們每個人消耗

The problems of air pollution, water pollution and noise pollution are in fact all only scratching the surface. The anti-natural, anti-cultural and anti-human orientations, the result of the damage to nature, are really the crux of the crisis, really what people should be worrying about.

Everyone has been discussing the democratization of Taiwan, but no matter how much the political structure changes, if Taiwan continues to go against nature, there's no way to save Taiwan's society. Of course air pollution makes people upset and affects their health, but being concerned only about that is also superficial.

To love the land, understand it first: Taiwan lacks ecological principles. ''Ecological principles'' sounds very abstract, but let me give you an example. In the United States, a kid discovered a sea tortoise with a broken leg and brought him home. After the neighbors heard about it, they all made donations to have a veterinary surgeon examine the tortoise and put on a false leg. Then they made a pond for it so it could recuperate in natural surroundings. When it recovered, the people of the village let the kid take it back to the beach and return it to the ocean. The people of the village hooted and hollered, encouraging it to swim back out to sea. This is an example of a most desirable and lively ecological education!

Several years ago, a whale entered San Diego Bay. When it was enticed back out to sea, everyone was at the beach or in front of their televisions cheering! After this incident, another whale entered Taichung Harbor, and the result was that a foreign reporter informed international conservation organizations. Then these foreign organizations notified our Ministry of Foreign Affairs that the whale needed protecting and the Ministry of Foreign Affairs notified the Taichung Harbor Bureau to protect the whale and prevent it from being caught by fishermen.

To entice a whale out to sea, you've got to know its habits, but it was discovered that no one in the country understood the habits of whales. From the north to the south there are six water testing institutes, as well as the zoological institutes of National Taiwan Ocean University, National Sun Yat-sen University and the Academia Sinica, but there wasn't a single expert in this field.

We live in Taiwan, but we know very little about it. We're on an island, but we know nothing about the sea life around us. School kids know about the Rhine, the Danube and the Mississippi but aren't too clear about the location of the Choshui, the Tatu or the Kaoping or about the relationship these rivers have to us. If we don't teach children to understand this land, how can we expect them to love and respect it? It's impossible.

A simple life is for eternity: A society that doesn't respect nature, won't respect people or itself either. If nature is treated without tenderness or respect — no matter how thriving the economy or how good the political situation — the society is incomplete. A society's culture, economy and level of democracy all reflect its attitudes toward nature. This is to say that there is no country with an excellent economy and political situation that is anti-nature.

Does Taiwan have an excellent economy today? A lot of money doesn't necessarily mean a good economy. Being a rich merchant who cares about society and people is different from being *nouveau riche*! Suppose that Taiwan's economic miracle led to a place where no one dared to go outside or sit in a taxi, where people doubted each other when in love, where the quality of life was lowered, where a terrible clamor surrounded you on all sides. If that was the case, it would be better to eat simple food, wear old clothes and live a simple life. At least in that way, people would grow in harmony with each other, and everyone would have hopes for nature and culture.

The purpose of material life is to let people examine the meaning of their existences. It's only a means, not the end. We've got it backward, regarding material life as the purpose of living. The result is people exhaust themselves in the pursuit of material well-being and forget the essence of existence. This leads to a further degradation of the ecological environment, an ever more vicious cycle. The more we consume, the more garbage and pollution there is. Thus, a simple life is the way toward everlasting life.

Deep ecological concern: People's attitude toward nature has got to change from one of domination over it to an attitude that we are a part of nature and reliant on it. Through social movements, thought and introspection, this has got to slowly emerge. Hence, an even higher aim of Earth

愈多，垃圾、污水愈多。因此簡樸的生活是一條永遠正確的道路。

深沉的生態關懷

人對自然的態度，要由主宰者轉變爲：我是自然的一部分和我依賴自然生存，必須透過社會的運動、思考、反省，慢慢引導出來。因此地球日更高的目的是，一個理念的改變。社會行爲的改變，必須透過理念的改變才有成果。比方需要有一羣社會人士在理念上對環保有清晰認知，環保政策才能落實。

今天生態關懷也絕不是幾個人的力量可以做到的，因爲它牽涉了整個政治、社會、人文的問題，必須透過對整體的了解才能解決問題。

因此我期待大家能用更深沉的情感，關心臺灣的社會、環境，更深入地了解，人的問題、文化問題和關懷生態是不可分割的，這樣的關懷，也才更成熟、更完整。 S

（原載光華七十九年六月號）

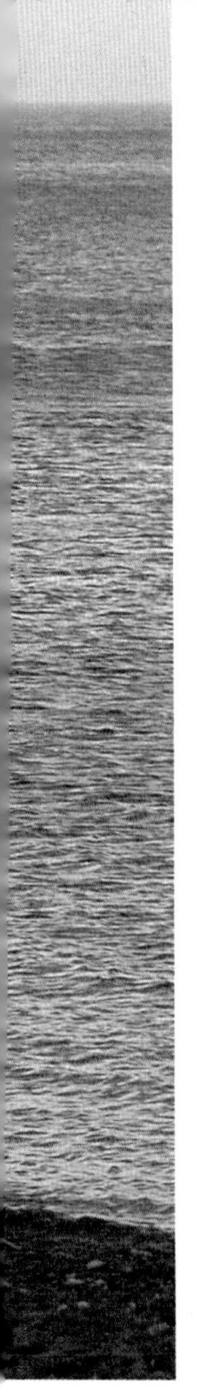

保護自然環境，也就是保護人類的生命與前途，左爲臺灣的蘭嶼東清灣，右爲遭到汚染的東北海岸一角。（本刊資料）

Protecting the natural environment also means preserving the life and future of mankind. At left is Iraralay Bay on Taiwan's Orchid Island, at right a polluted section of Taiwan's northeast coast. (photo *Sinorama files*)

Day is to change people's conceptions. A change of social behavior needs a change in principles before bearing fruit. For instance, only when there are a group of people in society who have a clear understanding of the principles of environmental protection, can environmental policy be effectively implemented.

Today, environmental protection cannot be successful with the energies of only a few because it involves all of the problems of politics, society and culture. An overall understanding is required before the problem can be solved. I myself understand that idealized academic concern is not only superficial but also has no way of solving the problem.

And so I am waiting for the day when everyone cares about Taiwan's environment, waiting for a society with deeper feeling and deeper understanding. People's problems, cultural problems and ecological concern cannot be separated. A concern of this kind is more mature and complete.

(As told to Chang Chin-ju by Lin Chun-i/ photos by Vincent Chang/ tr. by Andrew Morton & Jonathan Barnard/ first published in May 1990)

附錄 Appendix

近年來國內相關野生動物研究調查報告與出版物

論文名稱	計劃主持人
臺灣陸生哺乳動物學研究史	林俊義、林良恭
臺灣哺乳類動物地理初探	林俊義、林良恭
墾丁地區臺灣獼猴的行爲與生態研究(1)、(2)	吳海音、林曜松
臺灣獼猴的分布與現有族羣之初步調查	李玲玲、林曜松
臺灣獼猴野外供餌之研究	李玲玲
楠梓仙溪林道臺灣獼猴分布與棲地利用	林曜松、盧堅富、李玲玲
中橫公路（太魯閣至文山段）沿線臺灣獼猴資源之調查研究	林曜松、盧堅富
宜蘭仁澤臺灣獼猴猴羣生態之研究	
臺灣長鬃山羊之生態研究	呂光洋、滕春台、葉冠羣
臺灣長鬃山羊生態學上之初步探討	呂光洋、黃郁文
臺灣長鬃山羊之生態研究(3)、(4)、(5)	
臺灣梅花鹿的復育	王　穎
臺灣黑熊之生態調查及其經營管理策略	王　穎、陳添喜
臺灣山羌之生態及行爲研究	王　穎
臺灣山羌的經營與管理	裴家騏
臺灣野豬之生態與行爲研究	趙榮台、方國運

A List of Recent Reports and Papers Published in the R.O.C. about Wild Animals

Thesis Name	Sponsors
The History of Research on Terrestrial Mammals in Taiwan	Lin Chun-yi, Lin Liang-kung
Preliminary Research on Zoological Geography for Mammals in Taiwan	Lin Chun-yi, Lin Linag-kung
The Behavioral and Ecological Study of the Taiwan Macaque in Kenting Area (I, II)	Wu Hai-yin, Lin Yao-sung
A Preliminary Survey on the Distribution and Population Status of Taiwan Macaque	Lee Ling-ling, Lin Yao-sung
Study of Feeding the Taiwan Macaque in the Wild	Lee Ling-ling
Population Distribution and Habitat Utilization of Formosan Macaque at Nantsu Hsien Hsi Logging Road, Yushan National Park	Lin Yao-sung, Lu Chien-fu, Lee Ling-ling
Survey on the Resources of the Taiwan Macaque along the East-West Cross-Island Highway	Lin Yao-sung, Lu Chien-fu
Study of Population Ecology of Formosan Macaque in Jentze, Ilan	
The Ecological Study of the Formosan Serow	Lu Kuang-yang, Teng Chun-tai, Yeh Kuan-chun
Preliminary Ecological Study of the Formosan Serow	Lu Kuang-yang, Huang Yu-wen
Ecological Survey on the Formosan Serow (III, IV, V)	
Restoration of the Formosan Sika	Wang Ying
Ecological Survey of and Managerial Policies on the Formosan Black Bear	Wang Ying, Chen Tien-hsi
The Ecological and Behavioral Study of the Chinese Muntjac	Wang Ying
The Management and Control of the Chinese Muntjac	Pei Chia-chi
The Ecological and Behavioral Study of the Formosan Boar	Chao Jung-tai, Fang Kuo-yun

論文名稱	計劃主持人
臺灣穿山甲之繁殖保存研究(1)	趙榮台
櫻花鉤吻鮭生態之研究	林曜松、楊平世
櫻花鉤吻鮭魚道勘查規劃	臺灣漁業技術顧問社
櫻花鉤吻鮭族羣生態與復育研究(2)、(3)	林曜松、張崑雄、張瓊文
七家灣溪櫻花鉤吻鮭保育區水域農藥殘留監視	李國欽
臺灣特有飛蜥種類之調查研究與復育試驗	鄭先佑、張睿琦
臺灣山椒魚之研究——歷史、分布與生態學之初步研究	陳世煌
臺灣山椒魚之研究(2)阿里山地區山椒魚之族羣生態研究	陳世煌、呂光洋
翡翠樹蛙之研究	
臺北樹蛙生殖行爲之研究	楊懿如
蘭嶼角鴞之生態研究	劉小如
蘭嶼角鴞之生態研究與經營管理	周延鑫、劉小如
蘭嶼角鴞之社會行爲及棲地利用	劉小如
大肚溪口鳥類保護區之研究	陳炳煌
大肚溪口水鳥保育區整體發展構想研究	李瑞瓊
彰化伸港鄉海埔地鳥類保護區規劃報告	顏重威
澎湖貓嶼海鳥保護區之可行性研究	王　穎、陳翠蘭

Thesis Name	Sponsors
Research into the Propagation and Preservation of the Formosan Pangolin	Chao Jung-tai
Research into the Ecology of the Taiwan Trout	Lin Yao-sung, Yang Ping-shih
Examination of and Planning for the Aquatic Routes of the Taiwan Trout	The Taiwan Fishing Industry Consultation Society
Survey of the Population Ecology and Restoration of the Taiwan Trout	Lin Yao-sung, Chang Kun-hsiung, Chang Chiung-wen
Supervision of the Pesticide Residue in the Waters Reserved for the Taiwan Trout in Chichiawan Brook	Lee Kuo-chin
Survey and Restoration Experiments on the Agamidae Unique to Taiwan	Cheng Hsien-you, Chang Jui-chi
Research into the Formosan Salamander: Preliminary Survey of its History, Distribution and Ecology	Chen Shih-huang
Research into the Formosan Salamander (II): Survey of the Population Ecology of the Formosan Salamander in the All Mountain Area	Chen Shih-huang, Lu Kuang-yang
Research of the Feitsui Tree Frogs	
Research of Taipei Tree Frog Reproductive Behavior	Yang Yi-ju
An Ecological Research into the Langu Scops Owl	Liu Hsiao-ju
The Biology and Management of Lanyu Scops Owls (*Orus elegans hotelensis*)	Chou Yen-hsin, Liu Hsiao-ju
The Social Behavior and Habitat Utilization of the Langu Scops Owl	Liu Hsiao-ju
A Study of the Tatu River Estuary Bird Sanctuary	Chen Ping-huang
Survey on the Overall Development Planning for the Water Fowl Preserve at the Tatu River Estuary	Lee Jui-chiung
Report on the Planning for the Bird Preserve on Tidal Land in Shenkang Rural Township, Changhua County	Yen Chung-wei
Survey of the Feasibility of the Marine Birds' Preserve in Penghu's Cat Island	Wang Ying, Chen Tsui-lan

論文名稱	計劃主持人
東亞國際候鳥繫放先驅計劃	莊永泓
臺灣地區猛禽類之調查	林文宏
臺中縣鷺鷥鳥調查保護計劃	臺中縣政府
臺灣特有亞種鉛色水鶇的生態研究	王　穎
紅尾伯勞之遷徙、過冬領域及覓食行為	劉小如
太魯閣國家公園地區鳥相及其變化	王　穎、孫元勳
臺灣野生動物疾病防治計劃	邱慧英
蘭嶼珠光黃裳鳳蝶復育可行性之研究(1)	陳維壽
臺灣產青斑蝶之幼蟲食草及生物學研究	楊平世
臺灣越冬型蝴蝶谷之生態研究	
哈盆地區昆蟲相研究(1)	楊平世、吳文哲、許洞慶
大武山自然資源之初步調查	臺灣大學動物系、林試所
大武山自然保留區經營管理與保育計劃	李玲玲
玉山國家公園新康山動物資源調查	周蓮香
臺灣地區山產店對野生動物資源利用的調查(1)	王　穎、陳瑾瑛、林文昌
大甲溪石鰟之族羣分布研究	林曜松、張明雄
自然文化景觀保育論文集(2)：鮭鱒魚保育專輯	農委會林業特刊

Thesis Name	**Sponsors**
Pilot Banding Project of East Asian International Migratory Birds	Chuang Yung-hung
Survey on Predatory Fowl in Taiwan	Lin Wen-hung
The Taichung County Egret Conservation Plan	The Taichung County Government
The Ecological Survey of Endemic Subspecies on Plumbeous Water Redstart in Taiwan	Wang Ying
On the Migration, Wintering Territory, and Foraging Behavior of Brown Shrikes	Liu Hsiao-ju
On the Bird Population and Its Variations in Taroko Gorge National Park	Wang Ying, Sun Yuan-hsun
The Disease Prevention and Treatment Plan for Wild Animals in Taiwan	Chiu Hui-yin
Study of the Feasibility of Restoring the Orchid Island Monarch Butterfly	Chen Wei-shou
Taiwan Milky Weed Butterfly Larval's Consumption of Grass and Biological Research	Yang Ping-shih
Research into the Ecology of the Wintering Butterfly Valley in Taiwan	
Survey of the Insect Population in the Hapen Area	Yang Ping-shih, Wu Wen-che, Hsu Tung-ching
Preliminary Investigation into the National Resources on Tawu Mountain	The Department of Zoology of the National Taiwan University, Taiwan Forestry Research Institute
A Management and Conservation Strategy for the Tawu Mountain Nature Preserve	Lee Ling-ling
Survey on the Animal Resources in Hsinkang Mountain in the Yushan National Park	Chou Lien-hsiang
The Study on the Consumption of Wildlife Resources on Taiwan	Wang Ying, Chen Chin-ying, Lin Wen-chang
Study of the Distribution of the Population of *Acrossocheilus Formosanus* in Tachia River	Lin Yao-sung, Chang Ming-wei
Collection of Theses on the Preservation of Natural and Cultural Landscapes (II): On the Conservation of Salmon and Trout	Special publication of the Forestry Section of the COA

論文名稱	計劃主持人
自然文化景觀保育論文集(3)：野生動物保育專輯	農委會林業特刊
野生動物保育研討會專輯： 國家公園與自然保留區之野生動物	農委會林業特刊
臺灣野生動物保育史	農委會林業特刊
澎湖縣湖西鄉海豚資源之保育利用規劃	農委會林業特刊
臺灣野生動物資料庫：蜥蜴類、兩棲類	呂光洋、賴俊祥
臺灣東亞家蝠活動模式之研究	盧道杰
墾丁國家公園海域珊瑚礁及海洋生物生態研究	墾丁國家公園
南仁山生態保護區水域動物生態研究： 1.南仁山區之湖沼和兩棲爬蟲動物相研究 2.南仁山淡水魚類及水生無脊椎動物研究 3.南仁山水域動物性寄生蟲生態研究 4.南仁山區水棲昆蟲之初步調查報告	墾丁國家公園
墾丁國家公園南仁山生態保護區水域鳥類生態研究	墾丁國家公園
墾丁國家公園南仁山雁鴨保護區水生植物生態研究	墾丁國家公園
墾丁國家公園海珊瑚礁分類學及生態學之研究	墾丁國家公園
滿州地區狩獵灰面鷲之調查	墾丁國家公園
墾丁國家公園區昆蟲相及蜘蛛相調查研究	墾丁國家公園

Thesis Name	Sponsors
Collection of Theses on the Preservation of Natural and Cultural Landscapes (III): On the Conservation of Wild Animals	Special publication of the Forestry Section of the COA
Collection of Conferences on Wildlife Conservation: On National Parks and Wild Animals in Nature Reserves	Special publication of the Forestry Section of the COA
The History of Wildlife Conservation in Taiwan	Special publication of the Forestry Section of the COA
The Planning on the Conservation and Use of Dolphin Resources at Huhsi, Penghu County	Special publication of the Forestry Section of the COA
Taiwan Wildlife Database: Saurians and Amphibians	Lu Kong-yang, Lai Chun-Hsiang
A Study of the Behavioral Patterns of *Pipistrellus abramus* in Chutung	Lu Tao-chieh
Research on the Ecology of Marine Creatures and Coral Reefs in the Waters in Kenting National Park	Kenting National Park
Research on the Ecology of Animals in the Waters of the Nanjen Nature Preserve: (1) Research into the Amphibian and Reptilian Population on the Nanjen Mountain (2) Research into the Aquatic Invertebrates and Fresh Water Fish An the Nanjen Mountain (3) An Ecological Report on the Parasites in the Waters on the Nanjen Mountain (4) A Preliminary Investigation Report on the Aquatic Insects on the Nanjen Mountain	Kenting National Park
An Ecological Study on the Water Fowl in Nanjen Mountain Nature Preserve, Kenting National Park	Kenting National Park
An Ecological Study on the Water Born Plants in the Nanjen Mountain Family Aratidae Preserve, Kenting National Park	Kenting National Park
Research on the Classification and Ecology of Coral Reefs in Kenting National Park	Kenting National Park
Survey on the Hunting of Gray-faced Buzzard Eagles in the Manchou Area	Kenting National Park
Survey of the Insect and Spider Populations of Kenting National Park	Kenting National Park

論文名稱	計劃主持人
墾丁國家公園海域軟體動物生態研究	墾丁國家公園
墾丁國家公園日行性猛禽調查研究	墾丁國家公園
太魯閣國家公園沙卡礑溪哺乳動物資源之研究	太魯閣國家公園
太魯閣國家公園野生哺乳動物資源與經營研究	太魯閣國家公園
太魯閣國家公園華南鼬鼠之生態調查	太魯閣國家公園
太魯閣國家公園大合歡山地區山椒魚資源調查研究	太魯閣國家公園
太魯閣國家公園蛾類相研究	太魯閣國家公園
太魯閣國家公園陶塞溪、蓮花池和神秘谷鳥類生態研究	太魯閣國家公園
太魯閣國家公園烏頭翁及白頭翁分布調查	太魯閣國家公園
太魯閣國家公園烏頭翁及白頭翁生態及行爲研究	太魯閣國家公園
太魯閣國家公園文山、 天祥地區臺灣長鬃山羊棲息環境之調查	太魯閣國家公園
太魯閣國家公園之昆蟲相研究	太魯閣國家公園
太魯閣國家公園胡蜂之調查	太魯閣國家公園
太魯閣國家公園毒蜂調查研究	太魯閣國家公園

Thesis Name	Sponsors
An Ecological Study of the Mollusk in the Waters of Kenting National Park	Kenting National Park
Survey on the Predatory Fowl in Kenting National Park	Kenting National Park
Survey on Mammal Resources in Shakatang Brook in Taroko Gorge National Park	Taroko Gorge National Park
Survey on Wild Mammal Resources and Management in Taroko Gorge National Park	Taroko Gorge National Park
An Ecological Survey of the Chinese Mink in Taroko Gorge National Park	Taroko Gorge National Park
A Survey of Salamander Resources on Great Hohuan Mountain in Toroko Gorge National Park	Taroko Gorge National Park
A Survey of the Moth Population in Toroko Gorge National Park	Taroko Gorge National Park
An Ecological Research into the Birds in the Taosai Brook, Lotus Pond and Mystery Valley Areas	Taroko Gorge National Park
Survey on the Distribution of the Harter's Bulbul and the Styan's Bulbul in Taroko Gorge National Park	Taroko Gorge National Park
An Ecological and Behavioral Study ont he Harter's Bulbul and the Styan's Bulbul in Taroko Gorge National Park	Taroko Gorge National Park
An Investigation into the Formosan Serow's Habitat in the Wenshan and Tienhsiang Areas in Taroko Gorge National Park	Taroko Gorge National Park
A Survey of the Insect Population in Taroko Gorge National Park	Taroko Gorge National Park
A Study on the Wasp in Taroko Gorge National Park	Taroko Gorge National Park
Survey of Poisonous Bees in Taroko Gorge National Park	Taroko Gorge National Park

古者以文會友
今人因書結緣
擁有光華叢書，豐富一生

Good Books Make Good Friends
and
Enrich Your Life

光華叢書全部中英對照
以雪銅紙彩色精印
尺寸/7.5″×8.5″
The books contain both
Chinese and English
and printed in color
on fine paper
Size/7.5″×8.5″

臺灣風光新探

240頁，中英文／定價：平裝NT$230。
你登過玉山國家公園，走過八通關古道嗎？
近幾年臺灣開發了許多休閒新去處，本書帶您盡興一遊！

Taiwan's Favorite New Scenic Spots

(240pp. Chinese-English edition, softcover price US$ 8)
Have you climbed Jade Mountain in Yushan National Park? Or walked the ancient Pat'ungkung Trail? In the past few years many new recreation areas have been opened up on Taiwan. This book takes you there and shows you what to see and do.

畫說成語

160頁，中英文／定價：平裝NT$200。
小時候，課本會告訴你中國成語的典故，現在，讓"畫說成語"告訴新一代的小朋友，許多充滿中國色彩的成語故事。
漫畫式的成語故事，適合每一個小朋友閱讀。

Chinese Sayings Told in Pictures

(160pp. Chinese-English edition, softcover price US$ 8)
This book tells the stories behind common Chinese sayings through colorful cartoons. Since each story is told in both Chinese and English, adult language learners will find the book fascinating as well as children.

寶島的中國美食

246頁，中英文／定價：平裝NT$200，精裝NT$250。
各式各樣的中國吃食
糯米食的世界　中國麵食　台北市的名館名菜　中國豆腐　台南小吃

Culinary Treasures of China

(246pp. Chinese-English edition, softcover price US$ 8 , hardcover price US$10)
Chinese Festive Rice Sweets/The Many Ways of Wheat/A Guide to Restaurants in Taipei/Bean Curd The Chinese Taste Sensation/The Flavors of Tainan.

影像的追尋(上)(下)

120頁，中英文／定價：每册平裝NT$300，精裝NT$350。
透過十餘位民國30～50年代攝影家的眞實記錄，及張照堂的採訪整理，五十年前樸實無華的臺灣再度呈現。

In Search of Photos Past (Vol. I & II)

(120pp. Chinese-English edition, softcover price US$12, hardcover US$14 each)
Through the realistic works of fourteen photographers active during the 40s, 50s, and 60s, together with Chang Chao-tang's text and organization, the pure and simple Taiwan of fifty years ago comes to life once again.

臺灣特稀有生物

168頁．中英文／定價：平裝NT$200。

內容包括臺灣獼猴、翡翠樹蛙、蘭嶼角鴞、水筆仔等臺灣特稀有動、植物。作者以深入淺出方式寫來，兼具學術價值及趣味性。

Taiwan's Extraordinary Plants and Animals

(168pp. Chinese-English edition, softcover price US$ 8)

Contents include the Formosan Rock-Monkey, the Feitsui Tree Frog, the Orchid Island Homed Owl, the Kandelia, and others of Taiwan's fascinating rare plants and animals. The authors have written about their subjects in a clear, vivid style that brings sound scientific information to life. Full color pictures throughout.

中國造型

160頁．中英文／定價：平裝NT$200。

你知道民間咬錢金蟾的典故嗎？爲何在建築彩繪中處處可見獅子的造型？本書一一爲你說明。

Chinese Forms

(160pp. Chinese-English edition, softcover price US$ 8)

Do you know the classic popular tale of the Moon Toad biting money? And why are lions so prevalent in Chinese architecture and carving? *Chinese Forms* tells you the answers one by one.

結在異鄉的果實——焦點華人

192頁．中英文／定價：平裝NT$200。

他們的故事告訴你：
在異鄉如何出人頭地！
冰點之下找高溫——朱經武／他們說我勇敢——鄭念／黃建勳特寫／黃雙安與白嘉莉／這樣的中國人——李遠哲／園丁參議員——鄺友良

Blossoming in Foreign Lands — Chinese in the Spotlight

(192pp. Chinese-English edition, softcover price US$ 8)

Their stories show you how to make your mark in a foreign land.

當西方遇見東方——國際漢學與漢學家(一)

260頁．中英文／定價：平裝NT$240。

西方眼中的中國是何形貌？在講求世界觀的現在，你不可不知！漢學「西遊記」／美國的「中國研究」／輸出中國古文明／荷蘭漢學家許理和／美國的「中國通」龍頭—費正清

When West Meets East — International Sinology and Sinologists

(260pp. Chinese-English edition, softcover price US$ 9)

What does China look like in the eyes of Western scholars? What do they see as their role? Chinese and anyone interested in China will want to know.

世界著名大學巡禮(一)

244頁．中英文／定價：平裝NT$240。

「世界著名大學巡禮」預計出版三册，深入介紹世界(美國除外)排名前十五名，美國排名前十五名的大學，以及日本四大名校，是國內第一本翔實介紹各校特色與留學生在當地求學、生活情形的書。第一册包括：牛津、劍橋、海德堡、慕尼黑、愛丁堡、巴黎、哈佛、柏克萊加大、史丹福、普林斯頓、芝加哥、麻省理工學院等。

A Grand Tour of Famous Universities of the World (I)

(244pp. Chinese-English edition, softcover price US$ 9)

A Grand Tour of Famous Universities of the World, planned for three volumes, introduces the top 15 universities in the United States, the top four in Japan and 15 other leading universities from around the world. It is the first book to describe in detail the living conditions of Chinese students along with the special characteristics of each school.

世界著名大學巡禮(二)

232頁．中英文／定價：平裝NT$260，精裝NT$320。

延續世界著名大學巡禮(一)的編採風格，第二册介紹美國加州理工學院、康乃爾、哥倫比亞及耶魯大學；日本東京、慶應、早稻田、京都大學；泰國亞洲理工學院以及澳洲墨爾本和莫斯科大學。全部第一手資料，讓您了解各學術重鎮的精神面與生活貌。

A Grand Tour of Famous Universities of the World (II)

(232pp. Chinese-English, US$10 softcover, US$13 hardback)

Produced in the same format as that of the first volume, *A Grand Tour of Famous Universities of the World* (II) features in-depth presentations of Caltech, Columbia, Cornell, Yale, Tokyo, Keio, Kyoto, Melbourne and Moscow universities, all with completely firsthand information, to give you an in-depth understanding of the intellectual atmosphere and day-to-day way of life at each of these citadels of learning.

中國人的註册商標(上)

180頁．中英文／定價：平裝NT$220，精裝NT$260。

你能將中國人的文化特質做歸納嗎？什麼是代表中國人的價值觀、特殊發明及生活習慣？進補、麻將、划酒拳；印章、算盤、祖先牌位，這些「註册商標」，或許你我早已身在其中而不察，本書，將一一爲您細數它們的來龍去脈。

"Trademarks" of the Chinese (I)

(180pp. Chinese-English, US$9 softcover, US$11 hardback)

Can you sum up the special cultural characteristics of the Chinese? What are some of the Chinese people's most representative value concepts, inventions and living habits? This book examines the ins and outs of each of them in close detail.

中國人與動物

304頁．中英文／定價：平裝NT$280，精裝NT$320。

幾千年來，由於動物始終與人類共生，隨著文明的進步，卻造成無數的動物即將瀕臨絕種，因而「生態保育」的觀念，逐漸受到世人重視。

中國人傳統與現代的動物觀及目前我國生態保育的作法等，均在書中有深入的介紹，配合數百幀珍貴圖片，可使您對此領域有更深一層的認識。

Animals and the Chinese

(304 pages, in Chinese and English/Paperback US$11, Hardcover US$13)

For thousands of years, as man has made use of animals in every imaginable way, countless numbers of species have gone extinct. As the concept of ecological conservation has taken hold, people have been giving these issues greater attention.

The book makes a thorough introduction to traditional and modern views of the Chinese toward animals and current R.O.C. efforts in ecological preservation. The text is matched with hundreds of precious photos, which will give you an even deeper understanding of this realm.

劃撥帳號：0128106-5光華畫報雜誌社
地址：台北市許昌街17號14樓
訂購專線：(02)361-0577
P. O. BOX 8-398, Taipei, Taiwan, R.O.C.

《光華雜誌社叢書⑱》

中國人與動物

作者／張靜茹等 著
發行人／胡志強
總編輯／王家鳳
副總編輯／蕭容慧・鄭元慶
叢書編輯／陳麗珠・趙意玟
美術設計／倪淑雲
英文編輯／ Peter Eberly, Phil Newell, Jonathan Barnard, 劉瑞芬
出版者／光華畫報雜誌社
地址／中華民國臺灣省臺北市城中區100許昌街17號
　　　壽德大樓14樓
電話／(02)3123369.3888355
傳眞電話／(02)3615734
郵撥帳號／0128106-5
印刷／裕台公司中華印刷廠
　　　台北縣新店市寶強路6號
總經銷／臺灣英文雜誌社
初版／中華民國82年3月
定價／平裝　新臺幣280元・美金11元
　　　精裝　新臺幣320元・美金13元

登記證／局版臺誌字第4700號

Published by Sinorama Magazine 1993

14 F. No. 17, Hsuchang St.,
Taipei 100, Taiwan, Republic of China

國立中央圖書館出版品預行編目資料

中國人與動物＝Animals and the Chinese/張
靜茹等著.--初版.--臺北市：光華雜誌出
版；［臺北縣新店市］：臺英雜誌總經銷，民
82
面；　　公分.--（光華雜誌社叢書；18）（
生態保育系列）
ISBN 957-9188-18-1(精裝).--ISBN 957-
9188-19-X(平裝)

1.自然保育

855　　　　82001599